*f***P**

The
ELECTRIC MEME

A New Theory of How We Think

ROBERT AUNGER

The Free Press
NEW YORK • LONDON • TORONTO • SYDNEY • SINGAPORE

THE FREE PRESS
A Division of Simon & Schuster, Inc.
1230 Avenue of the Americas
New York, NY 10020

THE FREE PRESS and colophon are trademarks
of Simon & Schuster, Inc.

For information regarding special discounts for bulk purchases,
please contact Simon & Schuster Special Sales:
1-800-456-6798 or business@simonandschuster.com

Designed by Paul Dippolito

Manufactured in the United States of America

10 9 8 7 6 5 4 3 2 1

Library of Congress Cataloging-in-Publication Data

Aunger, Robert.
    The electric meme : a new theory of how we think / Robert Aunger.
        p.   cm.
    Includes bibliographical references and index.
    1. Social perception.   2. Memetics.   3. Thought and
        thinking.   I. Title.
HM1041 .A96 2002
302'.12—dc21   2002019392

ISBN 0-7432-0150-7 (alk. paper)

CONTENTS

INTRODUCTION

When I was doing anthropological fieldwork in central Africa, I encountered people who believe that witches can attack you in your sleep and eat your brain, turning you into a witch like them, with upside-down ideas like walking abroad at night, living homeless in the forest, and having sex with animals. In many cultures around the globe, similar stories are told: People can be haunted by supernatural agents that do them damage or make them into something new and strange. I hasten to add that these people are not "weird" in any other way; the individuals I knew were smart, caring, thoughtful. I grew very fond of them. And certainly they knew how to survive in their environment much better than I could. When they intended to kill an animal on the hunt, they understood the rules of physics well enough to fire arrows so that the animals died and they got to eat. And we were able to converse about many everyday things, despite my lack of belief in witchcraft, suggesting that many of our thoughts traveled common pathways. We shared the bond of being definitely and resonantly *human*.

Do these central African people feel any kind of cognitive dissonance between their metaphysical and physical worlds? Between the cultural beliefs they learn from others and what they experience through their own contact and experience with the world? Maybe these "crazy" witchcraft beliefs are some kind of parasite on their minds, able to perpetuate themselves somehow, serving their own needs. They certainly don't seem to make the life of anyone who holds such beliefs any better, since belief in witchcraft can make social relationships, even with your closest kith and kin, rather tense. You're always wondering whether some cross word or unintended slight will make someone angry enough to visit you in the night as an impossible animal that sinks its teeth into your skull.

1

Of course, you don't have to believe in witchcraft to get a vague sense that competing streams of thought are simultaneously burrowing their way through your head. Perhaps this feeling arises because some of our thoughts really are "alien" to us. Maybe what psychologists blandly call "cognitive dissonance" derives from the fact that at least some of our thoughts have their source outside us and come together somewhat unhappily inside our heads. Psychotic delusions—in which a person *consciously* hears unfamiliar voices echoing through his mind—might then begin when these alien thoughts become too numerous and too rancorous. It's not a wholly new idea; recall that stock cartoon image of an angel whispering, "Don't do it!" into some character's ear while a devil is shouting, "Aw, go ahead!" into the other.

So perhaps we are literally possessed by thoughts imported from those around us. To use a more medical analogy, maybe ideas are acquired as a kind of mental "infection" through social contact. We know that we can acquire terrible diseases in this way, from germs sneezed at us by someone else. What if we need to fear that something caught *culturally* from our compatriots can be dangerously infectious as well? We might become contaminated with treacherous brain pathogens just by talking with one another! In effect, through conversation, ideas might be able to move from brain to brain, replicating themselves inside our heads.

Why *do* we think the things we think? Do we have thoughts, or do they have us? This startling idea—that thoughts can think themselves—is the brainstorm behind a new theory called memetics. This theory is based on an important insight relevant to social species like humans. It begins by recognizing that many of our thoughts are not generated from within our own brains but are acquired as ideas from others. What memetics argues is that, once inside us, these thoughts then go to work for themselves, pursuing goals that may be in conflict with our best interests. These ideas have their *own* interests by virtue of having qualities that make them like biological viruses.

Social scientists have long remarked that the pool of beliefs and values held in common by the members of social groups—their culture, in short—appears to evolve over time. New varieties of belief—mutants—pop up with fair regularity and then are selected by individuals based on a wide range of criteria, such as their psychological appeal. This resemblance between cultural and biological processes led the eminent zoologist Richard Dawkins (now the Charles Simonyi Professor of the Public Understanding of Science at Oxford University) to suggest that cultural evolution might be described using the same principles as biological evolution. More particularly, he identified a unit of information that plays a role analogous to that of genes, the biological replicator. He coined the term "meme" as the name for these cultural particles, which he presumed could replicate themselves as people exchanged information. The upshot of this view is that memes are ideas that collect people like trophies, infecting their brains as "mind viruses." Maybe what we think hasn't so much to do with our own free will as with the ongoing activity of something like "thought genes" operating inside our heads.

Many have found the idea of memes attractively logical and have run with it. However, much of this speculation has been irresponsible, since the existence of memes remains to be established. Nevertheless if it could be shown that social intercourse regularly involves the replication of information, such a discovery would have important implications for the nature of human psychology and society. A concerted attempt to sort out what memes must be like is therefore warranted. In this book, I take seriously the notion that such cultural replicators exist. By identifying what memes must be like and where they can be found, I hope to hasten an end to the continuing rounds of conjecture about memes. If the possibility of memes is confirmed, an era of "hard" findings in the new science of memetics could then be initiated.

To help attain this goal, *The Electric Meme* begins with a chapter clarifying the core idea of memetics: that memes are replicators. Any evolutionary process, including the cultural kind, needs only to exhibit features that correlate from one generation to the next. This quality is what biologists call heredity. Replication is a more precise claim about how evolution works—it suggests that a special kind of agent causes the recurrence of cultural features: a replicator. Some evolutionary approaches—competitors to memetics, such as sociobiology and evolutionary psychology—invoke only *genetic* heredity in their explanation of culture. I disagree. Socially transmitted information is central to the nature of culture. But when it is transmitted, is it replicated? That's the crucial question. To answer it, we have to find some new sources of information that anchor our thoughts and keep our speculations from flying away with us.

What might be the proper grounds for a science of memes? How can we, in fact, determine whether replication occurs when we inherit cultural traits? First of all, we require a clear idea of how we can generalize Darwinian theory to cover the case of cultural evolution. In particular, we need a better idea of what we mean by replication in the first place. In this book, my first job is to firm up just what we mean by cultural evolution and to determine how it happens. For assistance in this task, it is reasonable to look to the other replicators we know something about—prions and computer viruses—for insight into how a cultural replicator might work. It turns out they work quite differently from genes, which considerably expands the possibilities for memes.

Replicators transmit information. But information has often been seen as a magical, protean kind of thing, capable of taking on any form a meme requires—in effect, enabling memes to flit through your mind and out into the world, and then to live long-term in books or monumental architecture, before zooming back into your brain. I suggest this jet-setting lifestyle is not one any form of information can sustain. We must stalk the wild meme and determine in exactly what kind of place it might be found. After considering alternative proposals, I conclude memes will be found only in the brain.

With such investigations completed, we move forward to a triumvirate of chapters at the heart of this book. These chapters tell a story that follows the evolution of memes since their beginning, possibly some hundreds of millions of years ago. Memes must have "started small," beginning their careers by replicating exclusively within individual brains. Following those early days, memes learned a trick that enabled them to move from one organism to another. Somewhat controversially, I argue they didn't do this by themselves hopping between brains. Instead they used signals like spoken phrases as agents to help them spread. These signals, once they penetrated the new host brain, initiated the reconstruction of the relevant meme from materials located there. Through this indirect process, memes effectively hurdled the gap of space between brains. More recently, memes learned to use artifacts such as books, CDs, billboards, and T-shirts as storehouses for their messages. This provided them with advantages in terms of longevity and the fidelity with which they could be transmitted as they journeyed from brain to brain.

This is a book that sets out a new way of thinking about how we think and communicate. Obviously, if we are zombies controlled by memes rather than free agents capable of independent thought, this fact has considerable bearing on our conception of ourselves, on what we say and do, and on the nature of the societies we construct. We need to find out about memes to answer these fundamental questions. Although it is unlikely to be the final word on the subject, this book aims to bring us a few steps closer to determining whether mind viruses are secretly and silently replicating inside our heads at this very minute, unknown to us—at least until now.

Chapter One

IN THE MIDDLE
OF A MUDDLE

*In 1953, a young girl of the Fore tribe, participating in funerary rites,
consumed pieces of her deceased grandmother's brain. The elderly woman
had died from an illness that progressively caused an uncontrollable
quaking of the limbs, loss of coordination, paralysis, and dementia. Four
years later, just as a brash young American doctor reached their village in
the Eastern Highlands of Papua New Guinea, the girl began to exhibit
symptoms of the trembling disease herself. A year later, she was dead too.
Most of the women in her village were soon suffering from what they called*
kuru, *the shaking. But then the young virologist, Carleton Gajdusek,
established a connection between participating in funerals and becoming
the subject of a funeral yourself. The cultural practice of eating brains soon
stopped when the news of this link spread, and* kuru's *devastating
consequences on Fore society gradually dwindled away.*

*In 1838, a young Charles Darwin, fresh from his trip circling the globe
aboard the* Beagle, *hungrily devoured the ideas of a dead man: Reading a
book by Thomas Malthus about the competition among individuals for
scarce resources sent a shiver of delight up his spine. A connection was made
in Darwin's mind to his own problem of explaining how the composition of
populations changed systematically over time. When another link was made
to the idea of inheritance, the theory of evolution by natural selection
among alternative traits was born. Darwin soon began scribbling away on
his own book* The Origin of Species.

7

In 1992, on March 6, a young boy in Cincinnati woke up and turned on his home computer. But suddenly a shiver of fear gripped him as he noticed the usual boot-up procedure wasn't executing. He quickly determined that most of the information on his hard disk had been mysteriously scrambled. Code composed by a hacker perhaps half a world away had overwritten sectors of the boy's hard drive with data taken at random from the computer's memory. He would later read in the local newspaper that the so-called Michelangelo virus (because March 6, its trigger date, was the famous painter's birthday) had probably been accidentally shipped with some software he had recently purchased and uploaded onto his PC.

These apparently unrelated vignettes, recounting events from varied times and places, all involve the *transmission of information*—in biological, cultural, or electronic form. *Kuru,* for example, is caused by a pathogen containing biological information, which spreads from person to person through infection. Information also spread in the second case. As Darwin's eyes passed over a piece of paper covered with patterned ink, his mind acquired Malthus's idea of ecological competition for survival. And the Michelangelo virus, a packet of digital information, was conveyed electronically to the Cincinnati boy's computer through a potentially long chain of physical links, involving a variety of storage media.

But there is more to the story than this. These events also involve the *replication* of information. In each case, the original copy of the biological, cultural, or electronic message remained with its source after transmission: *kuru* in the dead person's body, Malthus's idea on the pages of *An Essay on the Principle of Population,* and the Michelangelo virus in the hacker's computer (suitably anesthetized, no doubt). Thus these events seem to involve the duplication of a message in some other location besides its source. Such events are acts of communication.

But do they involve—or perhaps are they even caused by—*replicating* information? There's a big difference between something being duplicated through its own efforts rather than as the consequence of some other agency's activity. Why does this distinction matter so much? If a unit of information is replicating, it can be called a replicator. And if communication involves replica*tors,* and not just replica*tion,* then such events share a special kind of dynamic, aptly called the replicator dynamic. This dynamic underlies all evolutionary processes and can be described mathematically as a generalized catalytic reaction. Basically the replicator formula considers the particular means by which some object is able to produce a copy of itself. Typically this requires the help of a catalyst—a commodity that speeds up the production process without itself being consumed by it. In particular, the formula is concerned with the details of how many copies of the replicator can be made per unit of time. The speed of this process then depends on such things as whether the replicator is itself the catalyst or leaves that role to some other participant in the reaction. Such energy-absorbing events are momentary slices out of a longer history of duplications that are linked together to form an evolutionary lineage. Such lineages define a chain in which the same information gets passed from place to place and thereby persists through time. On the other hand, mere duplication may be an isolated, independent event, and so not part of such an evolutionary history. Whether a replicator is involved when a given bit of information is duplicated therefore makes a big difference indeed.

So are replicators required to explain communication events like reading a book or catching a disease, biological or otherwise? Diseases, most biologists agree, are indeed caused by a replicator. The question is which one?

In 1976, the doctor who ventured into New Guinea, Carleton Gajdusek, was awarded the Nobel Prize for medicine. Gajdusek had argued that the definitive feature of *kuru*—a gradual buildup of lesions and peculiar plaques in the brain—is produced by a growing colony of parasitic replicators. What was unusual about *kuru* from his point of view as a virologist was the length of time between the postulated date of infection and the first appearance of symptoms. His explanation, which won him the most prestigious award in science, was the suggestion of a new strategy, long-term dormancy, in an existing type of replicator, a virus. However, the Western medical community has not found a virus correlated with the appearance of the illustrative pattern of cortical plaques, despite more than 20 years of active research.

In 1997, Stanley Prusiner, another doctor, was also awarded the Nobel Prize for medicine, this time for a second explanation of *kuru* and related degenerative neurological diseases such as scrapie, bovine spongiform encephalopathy (BSE), and Creutzfeldt-Jakob disease (CJD). He believes a novel class of biological replicator causes these similar diseases:

prions (short for "proteinaceous infectious particles"). This is a more radical explanation than Gajdusek's because it violates the biological dogma that all pathological agents must include, and be fabricated by, nucleic acids (DNA or RNA). Instead these proteins are produced directly by other proteins (although they probably require the help of an unknown catalyst called "Protein X"). However, the existence of prions has not yet been conclusively proven either. Because they invoke alternative causes, either Gajdusek or Prusiner must be wrong—famously wrong.

What difference does it make whether a virus or a prion causes *kuru*? Both explanations seem to fit the basic facts. There is a link between consuming brain tissue and the later development of symptoms diagnostic of the disease. The causal pathway to disease is clear. But, in fact, a lot rides on the answer to this question. This is because prions don't achieve their evolutionary goals in the same way as viruses. For one thing, prions don't depend on getting translated back into DNA each generation: They don't pass through a genetic bottleneck in order to reproduce themselves. This allows them considerable independence and means they can be particularly malicious to the products of genes, like people. (All known prion diseases are fatal.) Their independence from DNA also allows prions to transfer rather easily to new kinds of hosts— for example, from cows to people. A harmless form of the prion protein is present in a very wide range of species. But a new species can become infected if enough of these protein molecules get into it, as when cows were fed ground-up sheep tissues (a farming practice in parts of Europe). The result: Sheep's scrapie was transformed into "mad cow" disease. (That is why we can get CJD, the human form of prion disease, from eating beef.) Further, if you look at population statistics for the course of these diseases, you begin to find differences in the values of parameters with epidemiological importance: the likelihood of infection from an exposure, the time delay expected between infection and the onset of affliction, etc. And how can it be cured? Anti-viral agents wouldn't have any effect on a disease caused by prions because prions aren't viruses. The moral here is that even when you invoke a replicator, you must make sure to get the *right* replicator to successfully explain a disease.

Let's move on to the second case of information replication described above, that of Darwin's infection by a "mind virus." Analogous options exist for explaining this event. Let's begin with the first, more conservative choice. This is the "Gajdusekian" assumption that an existing class of replicator, genes, is responsible for the phenomenon in question. It suggests that Darwin simply responded to Malthus's desire to persuade others of some novel ideas he had written down before he died. The assumption is that genes are simply using a novel strategy to make the organisms they produce more clever and hence better able to cope with whatever difficulties they might come across. This strategy is social communication, or the acquisition of information from other organisms, in this case mediated through the printed page.

On the other hand, perhaps Malthus's idea succeeded in lodging itself in Darwin's mind because the notion of ecological competition for survival *itself* had a variety of appealing features that made Darwin's brain—or any other—highly susceptible to it. That is, maybe ideas are replicators that have evolved abilities to get themselves planted in new host-minds and thereby gain a foothold for future replication.

This second type of explanation is more radical (and hence "Prusinerian") because it postulates the existence of a novel class of replicator. In this case we're not talking about genes, or prions, but rather what Oxford biologist Richard Dawkins called "memes." Memes are generally thought to be replicators residing in people's brains that are able to reproduce themselves during transmission between individuals. Thus memes in Malthus's head were somehow recoded as spots of ink that survived their originator's demise on the pages of a book, there to be visually picked up and copied into Darwin's active, living brain. Once inside, they became linked to other ideas already present in Darwin's mind to form a unique complex: the theory of evolution. (Exactly the same insight flashed into another brain—that of Alfred Russel Wallace—after a reading of Malthus some years later.) The success of this meme complex, from the point of view of the ideas themselves, has been truly extraordinary. Malthus's original idea, now recast in this larger intellectual framework, has survived through a number of generations of host individuals and become ever more prevalent with the passage of time—even to the point where it has been called a universal theory. Malthus's idea about competition was itself successful in a new kind of competition for survival: a battle among a new kind of replicator for places in a population of brains.

What about the third episode, concerning the boy in Cincinnati?

From the "genes-only" point of view, a malicious hacker in a dark corner of the world is able to interact indirectly with a large number of other computer users through the instrument of a program. Often, unwitting accomplices further the hacker's criminal goals by sending the damaging software along its way to other nodes in the network, where further havoc follows. But the sorry consequences of the hacker's action can be conveniently summarized by the malevolent intent of this one individual.

From the novel replicator perspective, once the hacker's brainchild has been let loose on the world, it develops a life of its own, infiltrating distant hard disks through its inborn abilities to manipulate networked computer systems. It creates a long chain of interactions in this way, and many replications of the program ensue. But many of the pathways the computer virus follows will be unintended by the hacker, causing unforeseen consequences at various destinations. If these become sufficiently serious to other users of the network—if the virus is "too successful" at replicating itself—its liberation into Webworld may even result in the hacker being tracked down and prosecuted. This is probably not what the hacker had in mind when she initially released it into this population of interlinked intelligences. It is rather a consequence of the virus's qualities as a replicator.

We have now seen that each of these vignettes can be explained from two different perspectives. The traditional view of communication—even when mediated through artifacts such as books or computers—is that the organization of such a social interaction, as well as the information content of what is exchanged, are determined by the wills of communicating parties (like Malthus and Darwin). This standard approach can readily admit that information is duplicated during social transmission, as long as people are the exclusive agents behind the process. Information was duplicated, yes. But because it was itself a replicator? No! Responsibility for duplicating information lies squarely and solely in the hands (or mouths or brains) of the communicators.

It is only when information *replicates* that an additional causal force becomes involved in the explanation of communication. This is the very essence of the meme hypothesis. The memetic suggestion is that there is an information-bearing replicator underlying communication that goes unnoticed by the traditional approach: a hidden homunculus acting as a second kind of agent, a puppeteer pulling invisible strings that direct aspects of the communication process. This puppeteer is the information packet itself, evolved to manipulate its carriers for its own ends.

Only one of these theories of information contagion can be correct: Either memes exist or they don't. Therefore the central question in this book is simply this: *What causal forces underlie the communication of information?* Which of these two ways of "reading" communicative events is correct? Does responsibility sit squarely with gene-based organisms, or do non-genetic replicators, with their own evolutionary interests, also play a role? Just as in the case of *kuru,* finding an effective explanation of social communication depends on picking the right replicator.

Quite a lot depends on this—no less than our whole conception of ourselves. Who's talking when I speak: the memes or me? Are my very thoughts something I was able to decide on, or are they just parasites attempting to get out of me and thus infect others? It's the importance of the prospect that memetics throws up—of people turned into zombies, with only the illusion of control over their own behavior—that should amply repay a serious investigation into the claim that memes exist.

GENES AND GERMS

Venture capitalist Steve Jurvetson coined the phrase "viral marketing" in 1997 to describe the strategy, first implemented by Hotmail, of tagging the end of every e-mail message with a promotion for its new service. This self-replicating advertisement helped the company itself become "hot." Hotmail experienced an epidemic rate of growth—12 million new users in only 18 months. Since then, every dot-com-and-Harry business flocking to the Internet has attempted to ride the wave of viral marketing by making use of similar gambits, like putting advertising banners on the "free" Web pages they offer customers. Demonstration downloads of software and reciprocal linking agreements with Web pages advertising related products are other strategies to boost one's customer base. The basic idea is to create a "buzz effect." Instead of marketing an idea or product to largely unknown and amorphous mass audiences through expensive ad campaigns, clever companies focus on key potential customers—early adopters—and let *them* market the product to everybody else. Information, it is felt, can be infectious. It can get spread through word-of-mouth (or its updated equivalent in this digital age, word-of-mouse) from customer to customer, rather than always being funneled through the business or the products it controls, such as advertising. The byword is: Just focus on those capable of setting the trend, the socially promiscuous and those with the power to influence others. These are folks itching to "tell a friend," to make use of the power of social networks.

Recently several popular books have attempted to take advantage of this idea, by attempting to sell us the idea of viral marketing. What are the trade secrets these books give away? How can we achieve this enviably cheap and effective promotion of ourselves or our ideas?

Malcolm Gladwell, in *The Tipping Point*, argues that effective word-of-mouth is something created by three rare and special psychological types, whom he calls Connectors, Mavens, and Salesmen. These are the people we must use to spread our message. Connectors are the gregarious gossipers who know everyone else and so speed information on its way through the population. But Connectors don't just know lots of people; they know lots of people in different walks of life. They provide links *between* social networks, or bridges to what would otherwise be isolated groups of people. Their activity greatly reduces the likelihood that an idea will circulate only in some media backwater; instead it will reach far and wide. Connectors also know those crucial people in each of these networks who will then be able to get the idea disseminated through their own groups. Gladwell calls this second type of people Mavens (from the Yiddish word for those who accumulate knowledge). They are helpful people with expert knowledge of those who are important in some area, and so they can introduce the idea to that strategic "someone" who can be vital to the further success of the idea or product in that area. Mavens are people who see trends in their earliest phases. The third important category of individuals who add "oomph" to the speed of spread is Salesmen: intuitive "people-people" who sweep you up with their passion and commitment. They convince you of the value of the new idea or product, converting mere contact into *effective* transmission. They add what Gladwell calls "stickiness" to an idea. An idea can be presented to someone, but unless they become convinced of its value—are truly infected by it—they won't then spread it on its way. Stickiness gives a message impact; it "sticks" to your memory-bank, so you can't get it out of your head. Salesmen are persuasive people who can take a basic vision and translate it into digestible form for the masses. This can require complex transformations to make ideas or products more appealing. Of course, increasing a product's "stick" is the basic role of advertising, so Salesmen are active, real-life advertisements.

If you can get these kinds of folks to look seriously at your idea or product, the epidemic nature of social networks will ensure your message sweeps through society, doing your work for you. In this way, a small group of these specialists can leverage widespread popular interest in what you have to sell.

Of course, those seeking to take advantage of the fact that new products and services can infiltrate the consuming public through interpersonal communication networks must live or die by the often mysterious dynamics of epidemic spread. Essentially, the buzz has to be good or you'll get burned. What a company wants to establish is a brand-name. Although gossip can make or break a reputation, it has to be backed up by solid business practices because networks are themselves out of a company's control: Good service and excellent product quality are a must for those daring to tap the power of viral marketing.

At bottom, viral marketing is simply the application of a more general idea to business practice. This idea is—to be only somewhat cute— the idea of infectious ideas. Since Richard Dawkins named such ideas "memes," this appellation has itself proven highly infectious, so that infectious ideas are often called "memes" nowadays. The *Oxford English Dictionary* defines a meme as: "An element of a culture that may be considered to be passed on by non-genetic means, especially imitation." This definition reflects a specialized biological expression derived from a Greek word meaning "that which is imitated." Dawkins provided some initial examples: "tunes, ideas, catch-phrases, clothes fashions, ways of making pots or building arches."

Dawkins was neither the first nor the last theorist to speculate that there might be something akin to a gene operating behind social communication. However, his coinage—a neologism that combines hints of "memory," "mimetic," and "gene" in one pithy package—has proved popular. And the analogy to genes embodied in the term is not only memorable but also ideologically appealing. As one commentator put it, "On the one hand, it holds out the tantalizing prospect of an elegant, universal theory of cultural evolution; on the other, it evades genetic deter-

minism by offering a parallel cultural process with interests of its own."
Memes are a second form of replicator that, although as "selfish" as any
replicator, are at least somewhat independent of the interests of our
genes. Memes are generally thought to be replicators residing in people's
brains that are able to reproduce themselves during transmission
between individuals. Memes arise as a consequence of social learning, as
in the vignettes recounted at the beginning of this chapter. As memes are
supposed to be acquired through imitation, Dawkins had to take imitation "in a broad sense" to encompass the possibility that memes could be
acquired from reading books or watching television, rather than through
direct, face-to-face interaction with other people. Since culture is widely
believed (these days) to be socially learned knowledge, memes appear to
be an account of how inheritance of this corpus of knowledge occurs. In
effect, memes become an explanation of cultural evolution.

Adherents of memes encompass philosophers, psychologists, sociol-
ogists, computer scientists, and, more generally, interested passers-by
from all walks of life. Most memeticists are *not* biologists or anthropologists by training, and so neither evolutionary nor cultural theory is their
professional expertise. This has given memetics a distinctively "populist"
flavor. It is the "people's choice" for an explanation of culture. There is
even a counterrevolutionary feel to it, against the Ivory Tower nature of
the other, academic approaches to explaining culture. However, this also
means that meme *aficionados* may have little awareness of alternative
evolutionary approaches to culture. Another downside of the vibrancy of
memetics is a certain lack of rigor, so that the general level of discourse
in memetics is somewhat low by "hard science" standards—a fact that
has been recognized by some of the more prominent exponents of the
approach. There are a few hints of formal theory, but no general system
for analyzing the evolution of memes has been adopted as legitimate by
the majority of memeticists.

This populist stance among memeticists has been rewarded with
antipathy from academics, who often greet the meme idea as a church
congregation would the arrival of an apostate. Outside academia, however, memes—these snippets of information learned from the cultural
surround—are all the rage. Like viral marketing, they have been the
subject of several recent popular books. And the viral marketeers even
acknowledge, in their humbler moments, that their ideas originate in

these books on memes. Gladwell "borrowed" his ideas from Aaron Lynch and Richard Brodie, prominent meme enthusiasts with earlier books on the popular market. By the time these ideas reached Seth Godin, the most youthful and brazen of the marketeers, Gladwell's trilogy of Connectors, Mavens, and Salesmen had been reduced to possessing a single common trait: They are adept at "sneezing" ideas. The complex process of social transmission is reduced in Godin's apocalyptically titled book *Unleashing the Ideavirus* to the activity of merely loquacious and gregarious people who influence product reputations. Similarly Gladwell's "stickiness factor" is reduced (somewhat ironically) to "smoothness": the ease with which new infections can be effected. In each case, the subtlety of Gladwell's presentation, taken more directly from the academic literature on social networks and idea contagion, becomes diluted. So already we have a chain of infection appearing in this corner of memetic theory—from memeticists to Gladwell to Godin. Basically the same ideas get marketed as a "new and improved" version with each iteration of the borrowing process. But unlike the promotional advertisements, the informational goods inside the books have in fact become increasingly weakened and noisy. Rather like the message in the game of Chinese whispers, by the third time the ideas have been reproduced, the rather elaborate analyses in the memetics books have been reduced to a small number of concepts, and these wind up being somewhat vulgarly expressed. This is itself an example of the kind of process that memetics purports to explain: The dissemination of cultural knowledge—in this case from author to author to author, and thence into the general public in various forms—depends on which version of the book-borne infection readers "catch."

The literature on memes uses two analogies to come to grips with the nature of memes. The more popular interpretation of memetics—as the titles of the books on memes (*Virus of the Mind, Thought Contagion*) and our discussion thus far both suggest—sees memes as microbes. In effect, the authors of the popular books on memetics adopt an epidemiological approach to the study of communication events, like the ones described earlier. Memes are the equivalent of a cold virus that, by causing sufferers to sneeze (à la Godin), succeeds in infecting everyone in the vicinity. So memetics is the cultural analogue to the study of how disease-causing pathogens diffuse through populations. The striking metaphor of memes as "mind viruses" (again originally due to Richard Dawkins) takes memes as particles of culture that parasitize human hosts, causing them to behave in ways conducive to getting copies of their information into the heads of other people. Memes, like viruses, are parasites because they make use of another organism's physical, chemical, and mental processes for their own transmission. Furthermore, both memes and viruses undergo vigorous competition for survival. Viruses must overcome the immune system and induce the host to transmit new virus particles to uninfected hosts; memes must overcome those memes previously existing in a host's mind and then induce her to transmit the meme to new potential hosts. Both of these processes have a great deal in common. In short, we don't have ideas; ideas have us! We are hosts to parasites feeding on our brains that cause us to behave in ways beneficial to them, not us.

As a result of using this epidemiological analogy, memeticists concentrate on how memes get transmitted from person to person. For example, they might tell a story about how a meme for suicidal imagery (as the analogue of a lethal microbe) spreads through a cultural group, perhaps becoming less virulent as the first wave of more susceptible hosts are killed off. The meme then achieves some kind of equilibrium presence in the population once all of those who can be, have been infected by it. (Some might say that, thanks to the widespread sales of books on memes, this stage has already been reached for the meme idea itself, which has become endemic among the literate population. Indeed these authors are making major reputations, and minor fortunes, for themselves as "sneezers" of the meme-meme.)

While it may be a fine basis for the art of advertising, a number of real weaknesses debilitate the epidemiological analogy as the foundation for a *science* of memes. Mostly it allows you to be intellectually lazy. Since epidemiology looks at the level of a population, you don't need to be concerned with the mechanics of how the pathogen actually gets from here to there. Can Vector X—say, a person—really transmit the pathogen from point A to point B in the time frame allowed? Do they move on foot, by train, or by plane? Do they have to climb high mountains or suffer cold temperatures? The epidemiologist cares only about the rate of spread, not the means of spread. Most epidemiological models also don't allow for changes in the nature of the pathogen, being concerned merely with describing variation in its distribution over time; so mutation can be completely ignored. The moral: Epidemiology need not even be evolutionary. But culture (the population of ideas and beliefs that become shared in a group as a result of communication among group members) is nothing if not evolutionary.

Relying on the epidemiological analogy has thus left the notion of the meme itself quite sketchy, and there has been little incentive for people to be precise. Memes have been variously suggested to exist as

— an idea in someone's head
— a repeatable piece of behavior like a spoken word, or
— embodied in the form of artifacts, like wheels

Reasonable scenarios for the duplication of any of these can be suggested, and the means by which they can be passed from person to person can be argued as well. You can simply assume that memes are one or more of these categories of "thing" and go forward from there. But this does not make for a very convincing story. Can a meme really be both a mental representation and a physical object? Or, if it can only be one of these kinds of things, why not the other? This vagueness about the physical nature of memes can lead to empirical confusion as well. People argue that memes are ideas and then count up postings to an Internet chat group to test the relative success of memes at spreading themselves. But a meme in the mind and its manifestation in computers are not necessarily the same thing.

Given these problems with representing memes accurately, the second major approach has been to suggest that memes are like genes. This viewpoint makes a fundamental point: Memes are cultural replicators. All of the interesting arguments made by memeticists—that there is some form of agency in these bits of information (a "meme's-eye view"), and that evolution occurs for the good of memes rather than genes— endow memes with their own evolutionary interests. What is a meme, after all, if it isn't a replicator like a gene? Just another name for a message bandied about by people in their social games.

In essence, this brand of memetics argues that cultural evolution cannot be explained without reference to a new replicator. Genes, with their well-known rules of transmission, cannot account for the ways in which information is passed around in human social groups, with their use of language and books. Even the minion of genes—brains—can't account for the rapid increase in knowledge in modern Western societies. Memetics argues that some other force is also at work—that there is a replicator underlying social communication.

Does Dawkins himself define a meme as a replicator? Although it's not perfectly clear from his writings on the topic, it appears that he does. At least in one place, he compares memes to DNA, the quintessential replicator that

> *makes copies of itself, making use of the cellular apparatus of replicases, etc. . . . [This] corresponds to the meme's use of the apparatus of inter-individual communication and imitation to make copies of itself. If individuals live in a social climate in which imitation is common, this corresponds to a cellular climate rich in enzymes for copying DNA.*

Daniel Dennett, Professor of Philosophy at Tufts University and Dawkins's close intellectual heir on such points, confirms that memes are replicators, saying that memes are "elements [that] have the capacity to create copies or replicas of themselves." He even uses the active voice. Susan Blackmore, Reader in Psychology at the University of the West of England and the admitted disciple of both of these figures, also says, "Imitation is a kind of replication, or copying, and that is what makes a meme a replicator" (because for her, memes are "units of imitation"). So it seems that the most prominent contemporary memeticists *do* identify memes as replicators.

Just how the two replicators—genes and memes—differ, however, remains somewhat obscure. Little work has been done in memetics using the gene analogy. In fact, there has been no extensive intellectual campaign to deduce the special qualities of memes as cultural replicators. No mention has been made of specific mechanisms by which memes replicate, are selected, vary, or get transmitted. Instead it has simply been argued that memes are passed among people through imitation, while evolving in the process.

Genes are, of course, the quintessential replicator. If memes are replicators, they must share many essential features with genes. But genes were first on the evolutionary scene and so are special; any replicators coming along afterward must live in Gene World and are (probably) going to be dependent on it. Further, genes are ancient, having had time to become complex, whereas subsequent replicators are much newer and hence probably simpler. So memes, even though they are also replicators, need not be the same as genes in every respect. Dawkins himself has suggested that the meme-gene analogy "can be taken too far if we are not careful." Fundamentally, since memes are parasites on the genetic process, they have no need to produce their own organisms to serve as hosts.

Thus there are those who advocate the contagion-like or viral metaphor and those who prefer the gene metaphor, with each group claiming the other is retarding progress in memetics. Basically, a war is underway between the "meme-as-germ" and "meme-as-gene" factions. Neither side can yet proclaim victory because there is no clear evidence being presented that favors one side over the other. There's the obligatory *Journal of Memetics,* which allows us to check on the state of play among the professionals. Indeed all of the claptrap surrounding a growing academic industry is in place. But something new will have to be done if memetics is going to advance and become a viable alternative to standard theories of cultural change.

No one knows what a meme is. Certainly the existence of one has yet to be demonstrated. That no one has sounded an alarm about this is astounding considering the controversy that has greeted the similar, but less radical, suggestion that prions explain *kuru* and related diseases. It simply has not been generally recognized that, at least from the perspective of evolutionary biology, claiming that cultural replicators inhabit brains (along with prions!) should be controversial. Meanwhile many people blithely debate possible features of memes, ignoring the fact that their existence first needs to be proven. Susan Blackmore, for example, argues in *The Meme Machine* that we can explain phenomena as diverse as the expansion of the human brain and tipping in restaurants as the direct result of memes working in our daily lives, while offering no evidence for how these hypothetical entities accomplish these things.

The ease with which the notion of memes has been accepted in many quarters might be thought to arise from differences in the standards of proof in the social as opposed to the biological or physical sciences—that is to say, such standards are lower in social sciences and higher in biological sciences—except that a prominent evolutionary biologist coined the word "meme." Moreover, much of the interest in the meme hypothesis has also remained among those relatively far from social scientific discourse, such as philosophers of biology and computer scientists. Maybe memes are just concepts useful in philosophical debates. What real use can we expect to gain from understanding them?

MEMES IN THE MUDDLE

The meme-based approach faces an immediate challenge: There doesn't appear to be any particular virtue in invoking memes to explain the cases of information transmission described at the beginning of this chapter. The social sciences have gotten along just fine for more than 100 years without invoking memes, and in 25 years of consideration, no major conceptual or empirical advances in memetics have appeared. This implies that memetics is not a very progressive program of research. There must be some underlying problem with the present conception of memes that accounts for this scientific stagnation and for their apparent lack of appeal to conventional social scientists. Surely, if memes exist, they leave traces in the world that could be found if only we knew where to look. It's possible that if we had a better image of memes we would begin to find them.

A more philosophical argument against memes is that it is just too complicated an explanation of how we think and communicate. The memetic perspective suggests that two lines of heritable information—genes *plus* memes—are required to explain information transmission in cultured species. This memetic line of argument—which is "Prusinerian" because it invokes a new causal force, the meme—violates the principle of parsimony, or Occam's Razor. This principle suggests that preferred explanations should invoke fewer causal factors to account for some domain, or involve a given number of factors but elucidate a wider range of phenomena than alternative accounts. A conservative "Gadjusekian" option would account for the same phenomenon, culture, by relying on a single line of inheritance: gene-built minds. Just as a new variant on an existing theme was first put forward to explain *kuru*, adding a dimension of complexity to explain communication and culture by positing a new class of replicators is only to be preferred if all the available alternatives fail.

The original evolutionary process to have arisen on Earth, that of biolife, *is* based on a replicator: genes. So the natural tendency has been to use genes as an intuition-pump for thinking about any kind of evolutionary process. The genetic case has been taken, at least implicitly, to define the necessary qualities of *any* evolutionary process, regardless of its physical substrate. But just because biological evolution happens to be based on a replicator does not mean *all* evolutionary processes have to be. In fact, it is possible that cultural evolution can be explained simply as the consequence of human beings going about the business of their genes, with big brains and all the other paraphernalia of culture, including artifacts, as *means* to get that all-important competitive edge over their fellows. Cultural information need not represent an end in itself; it is produced, by people, solely as an instrument to manipulate others for genetic ends. People and other clever creatures, in this view, have just evolved the ability to send and receive messages as a strategy for improving their biological fitness. People—and even people's own products for copying information, such as fax machines—may only be acting at the ultimate beck and call of genes. In fact, once genes are on the scene, evolutionary processes in cultural information become secondary and derivative. The ultimate responsibility for recent evolutionary jumps, such as the appearance of human culture, may remain with genes that simply become more and more remote from the scene as additional layers of organization get laid on top of them—proteins, organisms, behavior, and now culture. Genes can, in effect, single-handedly set the whole chain of events off and then sit back to reap the rewards. The new kinds of evolutionary process in gene products like proteins and brains could not arise were genes not in place, there, underneath it all. But they are. Although the spread of cultural beliefs appears to be a life-like process, the liveliness may all come from the humans' behavior, not the memes'. Cultural evolution certainly depends on genetic evolution, even if it is not, at least efficiently, reducible to it.

So with genes already present in the world, what need is there for an additional complication to the explanatory picture like memes? The answer, some would argue, is *none whatsoever*. Once you unpack the claims about memes, the role memes are meant to fill disappears.

In sum, the meme hypothesis finds itself in a considerable bind. This bind is centered on the issue of whether the recurrence of cultural traits depends on forms of inheritance that are, or are not, reliant on replicators independent of genes.

Chapter Two

A SPECIAL KIND OF INHERITANCE

The theory of cultural evolution [is] to my mind the most inane, sterile, and pernicious theory in the whole theory of science.

—Berthold Laufer

Evolutionism is the central, inclusive, organizing outlook of anthropology, comparable in its theoretical power to evolutionism in biology.

—Marshall Sahlins

During the course of human evolution, there has been a general increase in both the complexity and the diversity of cultural forms. While *Homo erectus* produced only rather uniform-looking stone tools, today we have a proliferation of cultural things, ranging from music genres (everything from classical to bebop, metal, grunge, and so on) to seemingly infinite brands of toothpaste and the plethora of sites on the Internet. How can this cultural diversification be explained?

Some kind of evolution was responsible for this process. Memes are often invoked nowadays to explain such cultural evolution. The basic problem is that no one is sure whether the *replication* of information required by a memetic explanation actually underlies cultural reproduction. The claim of memeticists that cultural phenomena depend on replication casts the validity of a meme-based approach to culture into doubt.

It is possible that culture evolves, with similarity between generations in cultural traits, but without information transmission at the social level—that is, directly from individual to individual. This may seem, at first sight, a peculiar stance for someone with an admitted sympathy to evolutionary accounts of culture. But it is a perfectly legitimate position. This is because evolution, as a general principle, can occur without replication (although *biological* evolution does involve the replication of genes). Why this is so needs to be understood clearly, since it means memetics is not necessarily correct about the nature of *cultural* evolution.

In general, evolution need only exhibit a certain number of properties. Briefly, these properties are

— *Heredity:* Entities of kind A usually give rise to another A, of kind B to more Bs, etc.
— *Variation:* Heredity is not exact—A sometimes gives rise to B, and
— *Fitness:* Variation must be associated with differences in the probability of survival to reproduction, on which selection may act

From this perspective, evolution requires a population in which there is variation among entities, differential reproduction among entities on the basis of their traits, and heredity of the traits associated with that differential reproduction.

Heredity, as the eminent evolutionary thinker John Maynard Smith defines it above, is the capacity of like to beget like. In biological terms, the mechanism generating heredity generally involves nucleic acid molecules—genes. In such a case heredity is due to descent with modification in a lineage of replicators. But the existence of these underlying entities that pass on their structure largely intact—replicators, in short—is not strictly necessary.

So heredity doesn't require replication. It simply measures a purely statistical correlation between the characteristics of parents and their offspring. Say Judy Smith has hairstyle A, while Amy Jones prefers hairstyle B. Then we find that the two girls exhibit the same hairstyle as each of their moms. This obviously has nothing to do with a different "hairstyle" gene being replicated in the two families. Still, the necessary resemblance can be found: One hairstyle A in the Smith family has given rise to another A hairstyle in that family, while Mrs. Jones, the B mom, has produced another B. This correlation can result from the operation of any mechanism, in principle. It may arise because both parents and offspring sample possible hairstyles from their cultural environment and by happenstance select the same way of wearing their hair. Or it could be that each mom has imposed her preferred style on her daughter. It doesn't matter how the correlation is achieved, so long as it exists. The important point for us is that the information need not be inherited through the genetic channel—not through DNA replication. If parents and their offspring are more similar to each other than randomly selected pairs of individuals in the population, then evolution by natural selection can occur: Those people with the favored kinds of traits will tend to increase as a relative proportion of the population over time.

Heredity is about *phenotypes,* the physical manifestation of a trait. More particularly, it is the degree to which the phenotypic traits in the parents' generation predict those of their offspring. This gives heredity the virtue of being about readily observable things, which is convenient.

Replication, on the other hand, is about *replicators,* which are typically so small you can't see them, except with specialized equipment like microscopes. If heredity is underpinned by replication, then the phenotypic traits being measured in a population are parts of the organisms produced by those replicators, their manifestation or expression in an observable trait.

Duplication, for example, is a feature of replication, but not necessarily of evolution. With heredity, you can just have sequential replacement instead. Imagine that a DNA molecule, in the effort to produce a copy of itself, was degraded in the process but left another DNA molecule behind in its stead. There is no point at which two DNA molecules exist and so no duplication has occurred, but information has been passed from a DNA "parent" to its "offspring." Heredity even allows cases in which there is no temporal overlap in the material. This would be as if the DNA molecule produced a protein that degraded and then was responsible for conducting a process resulting in the construction of a new DNA molecule: a dance of death and reconfiguration.

Replication is a special kind of heredity, one in which additional features apply. The requirement of heredity in traits affecting biological success is weaker than the requirement that replicators exist, and heredity is all that is needed for evolution.

In ideal circumstances, populations can theoretically grow exponentially. But it is almost always the case that, in reality, resources are limited. For this reason, population sizes tend to eventually stop growing. This was the main point of the theory formulated by Thomas Malthus in his famous *Essay on the Principle of Population*. What Darwin added to Malthus's demographic story was a recognition that individuals within populations vary; they have unique characteristics. He also supposed that an individual's characteristics can be passed on to its offspring—the inheritance principle. But change in environmental circumstances can occur, and sometimes this change results in offspring having slightly different characteristics from their parents. This is the variation principle.

For us, the point of variation is that heredity is not always precise. That is, A must be able to generate A, *and* if some variant of A arises, the new variant form must be able to generate copies of itself. This second point is crucial: The mutated, or variant, copy must retain—at least some of the time—its ability to inspire copies of itself being made. This is variation with a difference, a difference important to evolution because if this ability were lost, heritability would be lost.

An implication of this ability to vary is that replication, as one mechanism of heredity, sometimes doesn't work; an exact copy of the original is not the result of the replication process. But this occasional failure is almost more interesting than the normal outcome: Replication can produce something slightly different than expected, but which nevertheless has the ability to copy itself. So there is a sense of disjuncture: The old form is now dead. But there is also continuity: The new form will now go on to replicate itself over and over. This combination of old and new represents a branch-point in a lineage. You can think of this branching in two ways: Either the old lineage is dead and a new one has begun, or there has been a change of state in the lineage, which now goes under a new name, in effect. Otherwise the chain of replication ends, and with it the possibility of accumulating useful functionality, or what biologists call adaptations.

Some of the variants that arise will give the individuals with those traits a competitive advantage over their neighbors—that is, greater success in acquiring resources and reproducing. In biological terms, such individuals have higher reproductive success; they are likely to have more offspring in their subsequent family trees than these neighbors. If these favorable traits can be inherited by later generations and continue to be favored by selection pressures because the relevant environmental conditions remain the same, then evolution occurs: The traits of these relatively successful families increase in frequency and will eventually come to dominate the population as a whole.

If we add the further point that these selected traits are the expression of particular replicators, which thereby are favored by the evolutionary process, then we can say that such replicators are relatively "fit." Fitness denotes a replicator's ability to spread through a population. (The term "fitness" in this book concerns a replicator's success in out-reproducing competitors for places in succeeding generations. It is not about physiological well-being. Fitness is typically measured in somewhat rough-and-ready fashion as the number of offspring that an individual has, assuming that next-generation success is a good proxy for longer-term success.) This quality is a function of environmental conditions: The phenotypic traits that lead to success in the race for future representation in the population in one place and time will not necessarily be "good" for an organism somewhere else, or even the next time around if the situation has changed. There is no universal standard of value in biology; what "works" in evolution is always contingent on local conditions, and replicators may come and go.

According to this perspective, any entities in nature exhibiting the triad of fitness, variation, and heredity may evolve. They are evolutionary individuals. In principle, any level of biological organization can exhibit these three properties; thus any collection of units that can be grouped into a population has the potential to evolve by natural selection. So entities from any level of the biological hierarchy—genes, cells, or organisms—might play the role of an individual in a population. Even populations of organisms might serve as individuals in comparison to a "metapopulation"—that is, a population composed of multiple populations. An example of such a metapopulation exists in the form of ant colonies linked together via underground tunnels and the exchange of members.

WHAT IS CULTURE?

The major question for those seeking to understand culture from an evolutionary point of view is therefore this: Does culture replicate, or does cultural evolution just exhibit heredity? In other words, what we need to know is whether each generation resembles its predecessor because they are linked through the replication of cultural entities with the capacity to beget something like themselves through social transmission. And if cultural similarity *is* a social phenomenon, is it one involving the replication of particulate units of information between individuals, as memetics presupposes? Or is the march of history a process that is not itself caused by social forces, but which nevertheless results in a correlation of beliefs and values with what was seen before? Perhaps the acknowledged resemblances between cultural incarnations over time is due instead to genes acting through a phenomenon it reliably produces at the social level—for example, by constructing brains with universal features that consistently reproduce the same behavior.

Because this question is so central to the whole enterprise of this book, let me explain again the alternative possibilities. Cultural traits may consistently appear in each generation without being recreated each time by a replication process working at the cultural level—that is, without the relevant information being transmitted from one individual to another in a social group. The features called "cultural" may reliably reappear together without being *causally* related. After all, many kinds of phenomena coincide with high statistical regularity—for example, a barometer reading "rain" and environmental conditions in which water falls from the sky. However, the barometer does not cause rain, nor does rain directly cause the barometer to read "rain." Instead both are caused by a drop in atmospheric pressure.

The same might be true of a "cultural" phenomenon like the resemblance of young people's food preferences to those of their parents. Perhaps children learn what to eat by observing their parents' behavior and mimicking it. Such a scenario suggests that tastes for particular foods are memes being duplicated through social learning. But alternatively the correlation in consumption habits between generations could be due to a universal, inborn set of taste buds that bias the learning of each person as they acquire a knowledge of what foods taste good through individual trial and error. Such an explanation does not invoke a social channel of information transfer between individuals but rather a genetic one, which presumably underlies the predisposition to find certain kinds of foods tasty.

We can't really hope to make progress on answering the heredity question without first determining that the phenomenon we are trying to explain with memes—culture—in fact evolves in a way consistent with a replication process. So what is culture anyway?

This turns out to be rather difficult to determine because the notion of culture has been notoriously difficult to define, even though it is arguably one of the central concepts in modern social science. Indeed the entire discipline of anthropology is based on the idea of culture. Efforts by anthropologists to define culture began over a hundred years ago and have shown certain trends. It used to be fashionable to throw everything—including the kitchen sink—into the definition. For example, the first academic anthropologist, Edward Tylor, allowed that artifacts (such as kitchen sinks), kinship and marriage systems, and religious beliefs and rituals were all parts of culture. But this omnibus definition piles ideas, material objects, and behavioral practices into the same category. This is bad practice because it becomes difficult to distinguish what exactly is *not* part of culture.

Nowadays, thanks to the widespread success of the so-called "cognitive revolution" of the 1960s, there is considerable agreement that culture consists solely of things "in the head"—that is, of beliefs, values, ideas. Attention has focused particularly on those "mentifacts" that have been communicated and hence become common in a group. However, once one attempts to move beyond this starting point in defining culture, the degree of consensus falls off precipitously. Nevertheless we can say that, in the academy these days, culture is predominantly seen as a cohesive and coherent set of mental representations that is reproduced relatively intact through the enculturation of subsequent generations. A somewhat more operational form of this definition sees the "culture of the moment" as a collection of ideas, beliefs, and values that can be abstracted from individuals and considered as a pool of information at the population level. Each snapshot of a culture is a simple function of what was circulating the last time anyone looked. Tracing the history of a particular trait (or a set of linked traits) through these snapshots then defines a cultural tradition or lineage.

The cognitive notion of culture, then, is quite pervasive in contemporary social science. It has the definite advantage of making it clear that culture is the result of social learning, of being what we know thanks to transmissions from others. But there is an obvious tension present in this definition. On the one hand, the attempt to keep culture separate from genetics is pointed. The emphasis on enculturation is clearly meant to distinguish cultural inheritance from genetic inheritance—that is, "nurture" from "nature." What makes humans, the cultural species, unique from this perspective is extreme altriciality: Human babies are born full of plasticity and potential rather than having a recognizable nature or a fixed set of behavioral routines that are then repeated throughout life. The long period of childhood and adolescence seen in humans allows a prolonged cultural apprenticeship to parents and others. What we do in that long period is to imbibe norms and other situated knowledge through social learning, which we are particularly adapted to do by our evolutionary psychology. Thanks to this period of dependence, we are a cultural species.

At the same time, there is an evident desire implicit in the definition to take note of the resemblance of cultural and genetic transmission. This occurs in the reference to the iterative quality of cultural reproduction, which fits directly into the general framework of evolutionary theory. This tension is captured in the simple phrase "cultural evolution."

Unfortunately the general impression that culture evolves doesn't amount to real understanding. Our knowledge of culture is not so precise as to pinpoint a particular mechanism producing change in cultural traits. A number of candidates for a theory of cultural evolution have been put forward to fill this vacuum. Memetics is definitely not the only voice crying out for attention on this front.

We need to run through these theoretical alternatives to see if we can eliminate any of them on other grounds besides adherence to the Darwinian program. The challenge faced by memetics, as one of the applicants for the job of explaining cultural evolution, will come into much better focus once we have distinguished these alternatives.

SOCIOBIOLOGY

The first "modern" evolutionary theory of society and culture to arise was sociobiology. Considerable fanfare attended its major declaration in 1975, with the very public appearance of E.O. Wilson's authoritative book, whose title, *Sociobiology,* gave the discipline its name. Wilson devoted the last chapter of his book to human behavior, to show that the scope of his theory could encompass the cultural animal the same as any other. Wilson's lead was quickly followed, with many studies of sociobiological theory having now been applied to the case of humans.

The basic argument underlying such studies is that human behavior can be understood in terms of genetically acquired learning biases and rational calculation. Behavior, as a kind of phenotype (whether cultural or not), is a way to make responses more flexible. And the learning of new behaviors is a form of phenotypic plasticity, no more. The decision-making of an organism can therefore be expected to be adaptive, except for the odd, randomized mistake. The only principle one need invoke to explain the behavior of any organism, including *Homo sapiens,* is fitness maximization. Culture as an individual trait is unimportant. The more radical human sociobiologists don't even consider culture to be a word in their vocabulary: "I, personally, find 'culture' unnecessary," says one.

How then does the human sociobiologist explain the fact that people in Peoria don't act like people in Pongo-Pongo? Isn't that due to culture? Not according to sociobiologists. Behavior differs because there is variation in the kinds of stimuli that folks in these disparate regions receive. Why is there variation in environmental stimuli? Because people live in different kinds of places. So folks in Peoria, where it's cold, wear warmer clothing and have a word for snow, while people in Pongo-Pongo wear next to nothing and have never heard of a frightfully cold substance that lies about on the ground.

Sociobiologists claim that so-called "cultural" variation is the result of an interaction between genes and environment. The way in which a genetic trait manifests itself is not heritable; instead it is the product of a specific interaction. Uproot a Chinese man and fly him to Los Angeles. His kids, thanks to a change in diet and lifestyle, may have a higher risk of cancer than their father. Alternatively, the kids may imbibe the jogging culture in southern California and thus reduce their chances of this late-onset illness. The correlation between parent and offspring phenotypes will then be low, but only because the proclivity toward cancer, while inherited, reacts to local conditions. Removing the grandchild back to China would see a return of the original phenotype (again with a low correlation between generations). The phenotypic response is not genetically encoded itself, but the ability to respond correctly to different circumstances must be heritable and evolve by natural selection. Genetic and environmental factors alone are sufficient to explain the variation we see in the behavior around us.

At best, human sociobiologists admit that the social transmission of information occurs, but they suggest it can be ignored because it doesn't produce any novel evolutionary dynamics. Thus, the "grandfather" of human sociobiology, the eminent zoologist Richard Alexander, says:

> *To whatever extent the use of culture by individuals is learned . . . regularity of learning situations or environmental consistency is the link between genetic instructions and cultural instruction which makes the latter not a replicator at all, but, in historical terms the vehicle of the genetic replicators.*

Why do sociobiologists think there is no independent cultural replicator? Their argument is that there used to be a world without culture, prior to the evolution of the genus *Homo,* from which we derive. Culture must have arisen through a biological process because the ancestors of modern humans, like the rest of the animal kingdom, didn't have it. Acquiring information and passing it along is just a biological capacity. There is no reason to think, then, that having culture is different from exhibiting any other evolved trait, and evolutionary biology is sufficient to explain such traits in other animals. So culture and social learning can be reduced to the normal activity of natural selection on genes. Culture is simply another strategy to enhance the fitness of the behaving organism through learning. The net result of adopting a sociobiological position is that humans are placed squarely in the same conceptual box as other animals: No special dispensations are allowed. People are just like polecats, and evolutionary biology is all you need to explain them.

Sociobiologists, in effect, assume that any behavior which evolved as a response to particular circumstances is governed by evolved psychological or physiological mechanisms that produce the relevant conditional strategy—even for situations like playing the violin or making a soufflé. They suggest that inside anybody's head there must be a rule of the general form "In context X, do A; in context Y, switch to B." This rule is placed in the mind by natural selection favoring generations of those animals that have historically responded appropriately to both contexts X and Y. People tend to do the right thing, given their circumstances, because they were designed to make the right choice by everyday, run-of-the-mill evolutionary processes.

What about that other cause of behavior: our exercise of "free will" through decision-making? To the extent that human sociobiologists think at all about the mind, it is conceived as a general-purpose, all-weather, all-the-time information processor. It can rapidly learn about local conditions and respond adaptively to them. The brain is expert at inducing the locally optimal rule for behavior. The mind is not seen as the locus of adaptation; rather the behavior is, for it is behavior that is selected, and the gene for the behavior that is favored or not. Basically the brain is an invisible intermediary and can be ignored because it does its job perfectly, translating the needs of the organism into the optimally correct behavioral response. The presumption is simply that whatever is required, selection will have produced it.

A considerable number of studies show that various human behaviors *do* tend to maximize some measure of fitness, as expected by the tenets of sociobiology. Typically the empirical test is whether some characteristic correlates well with the number of offspring an individual has, which is taken as a good approximation of fitness. Richer people tend to have more children, and better hunters attract more mates. Sociobiological research has tended to concentrate on subsistence and resource exchange, on parental investment strategies such as birth-spacing and gender differences in parenting, and on reproductive strategies such as polygamy versus monogamy. These are, of course, the kinds of behavior most closely tied to biological fitness—the number of matings achieved, the number of babies produced—and so might be considered to evolve under more stringent constraints than craft traditions or religious beliefs. Tests have also typically occurred in foraging societies, where the problems of survival might be considered to keep culture on a rather tight leash. When sociobiologists turn their attention to "developed" Western countries, they tend to be faced with conundrums like richer people having *fewer* children, which reverses the pattern seen in "anthropological" populations. Their lack of a true theory of culture means sociobiologists simply can't address the availability and use of contraceptives, for instance, and must begin to mumble under their breath. As we will see, "modern" societies put considerable strain on other evolutionary social theories as well.

EVOLUTIONARY PSYCHOLOGY

Evolutionary psychology, a more recent but highly successful approach to explaining human behavior, adds a few wrinkles to the sociobiological picture. As the application of evolutionary theory to the domain of psychology, it is only natural that this school would argue you need not just biology, but psychology too, to explain human behavior.

From the perspective of evolutionary psychologists, their "home" discipline, psychology, was in trouble in the 1980s: No general theory of how the mind works was on the horizon. This made psychologists feel their discipline wasn't really scientific. After all, physics and biology had their overarching paradigms. Evolutionary psychologists see themselves as coming to the rescue of psychology by supplying such a paradigm. Their solution: Make psychology consistent with the other sciences by founding it on evolution. Evolution is, after all, the only scientifically valid theory that can explain complexity arising from simplicity, and the brain is the most complex object in the known universe.

To understand the evolutionary psychologists' position, it's worth setting out briefly the intellectual context from which evolutionary psychology arose. In the first half of the twentieth century, the default assumption in psychology was a largely implicit model of the brain as a "general information processor." The brain was seen as the organic equivalent of a computer and, like a computer, a machine capable of learning anything and everything with equal facility. The only problem was that this view turned out to equate poorly with some of the results coming out of behavior labs. It turns out the supposedly equipotent brain cannot learn some things, while other tasks come to it relatively easily.

John Garcia performed the classic experiment along this line in the early 1960s. He was trying to induce taste aversions in rats by various means. His lab animals readily learned to associate illness (created by radiation with X rays) with the consumption of particular foods, but buzzers or lights presented prior to giving the foods never meant "Watch out for this one!" to them, no matter how many times the experiments were run. On the other hand, if Garcia electrically shocked the poor rats after presenting them with a meal accompanied by lights and sounds, the rats learned to associate the shocks with the disco effects, but not the foods. Further, these learned aversions were quickly induced, and they persisted, even if long intervals lapsed between trials. This led to the concept of "prepared fears"—fears the mind has been made ready for by the species' evolutionary history. Blinking lights and buzzers simply don't happen in the rat's natural environment, so rat brains can't learn to associate them with something "natural" like nausea, even when they regularly recur in a lab environment. If such things weren't part of the creature's evolutionary history, they couldn't be relevant and hence aren't learnable. The results of such experiments could not be explained simply in terms of stimulus-response or conditioned memory; rather "intervening variables" (psychobabble for a mind) had to be invoked.

Such research eventually broke the back of behaviorism, until then the dominant school of thought in psychology. (An indication of this dominance is the fact that Garcia couldn't get his paper published for years.) Behaviorism asserted that there was no need to invoke internal structure in the mind to explain behavior: Brains are simply trained by experience, and all behavior is a learned response to environmental conditions. After the death of behaviorism, psychology needed to take the biology of organisms into account—and more particularly, to recognize the evolutionary constraints under which brains worked.

Simultaneously during the 1960s, the "cognitive revolution" occurred. This movement sought to import concepts like algorithm, representation, search, and solution-space from computer science into psychology. In light of this history, evolutionary psychology can be seen as a concerted effort to join two strains of thought current in the psychology of some years earlier: the information processing view at the foundation of cognitive science (which now dominates thinking in psychology) and the incipient movement to recognize biological constraints on thinking (which got shunted aside after some initial gains in the mid-twentieth century). So to evolutionary psychologists, the brain computes; it just doesn't compute everything, nor with equal alacrity. At the same time, the brain is, at bottom, an evolved organ for processing information. Where a sociobiologist sees the organism as a general-purpose learner capable of performing any behavior necessary to maximize biological fitness, an evolutionary psychologist sees a more restricted "adaptation-executer," limited in what it can learn by the innate structures in its brain.

In effect, evolutionary psychology flips the sociobiologist's perspective on its head: It's not behavior, but the decision-making *prior* to behavior that matters more to the course of evolution. Evolutionary psychologists emphasize that the incidentals of a particular situation—the conditions that produce a behavioral response—cannot be the focus of selective forces. Such circumstances never materialize in quite the same form again. But the brain, which reliably reappears each generation, *can* serve as a repeated target of selection. So the attributes of the mind that are responsible for producing successful behaviors can be rewarded. Over time, it is the brain that is expected to exhibit adaptations, not behavior. We must look for evidence of the (past) workings of natural selection, then, not in patterns of behavior, but in the brain.

A basic tenet of evolutionary psychology is that the human brain consists of a set of evolved psychological mechanisms designed by natural selection to solve adaptive problems that our ancestors faced recurrently during our species' evolutionary history. This point is Garcia's legacy. Since almost all of this history for humans was spent as hunter-gatherers during the Plio-Pleistocene—an epoch spanning several million years and ending only 10,000 years ago—the evolved structure of the human mind is basically adapted to the way of life of prehistorical hunter-gatherers. An important task for evolutionary psychologists is therefore to reconstruct the ancestral environment in which human beings evolved by finding the consistent selective pressures that influenced the evolution of our unique mental adaptations. For humans, this niche is small, isolated groups of kin, organized into nuclear families; a savannah habitat in which women gather and men hunt and sometimes raid other groups for additional women and territory; and the universal, monogamous marriage of young women to somewhat older men. At least that's what contemporary foraging societies suggest it must have been like. But perhaps instead mother-offspring groups formed around valuable ecological resources, while bands of males roved together through overlapping home ranges, as seen in chimpanzee societies today. The problem is that it is difficult to know what the relevant features of Plio-Pleistocene living arrangements were because social groups don't fossilize.

In any case, our Plio-Pleistocene ancestors would have faced an enormous number of adaptive problems—from basic survival skills, such as acquiring appropriate foods and avoiding predators, to gaining reproductive opportunities through the choice of appropriate mates, to protecting reproductive investments by proper parenting, to living with others. Because these problem areas are diverse, it is unlikely that a successful solution to one kind of problem could be transferred wholesale to another domain. Rather each type of problem would have selected for the evolution of its own dedicated problem-solving mechanism.

This makes the mind "modularized." Hundreds or even thousands of mental modules may have evolved to solve the myriad adaptive problems our ancestors must have faced. These modular reasoning or learning circuits are considered to have complex structures, so different neural circuits are specialized for solving particular adaptive problems. These evolved psychological mechanisms are designed to accept only certain kinds of input—say, stimuli looking like animals. These inputs are then processed using specialized algorithms. In the case of animals, figuring out the category in which to place the animal might run through a branching decision structure mirroring the nested hierarchy into which scientists have organized the "tree" of life-forms. The individual would then check off, say, "long," "no legs," and "head-with-hood-shape." The conclusion might be, say, "It's a cobra!" Finally these modules are supplied with substantial innate knowledge about their proprietary domain, so that prior experience is not necessary to output a response that has historically proven adaptive for the organism in such situations. This means you can be a competent problem-solver soon after emerging from the birth canal. This "rapid-response facility" comes in handy when you need to quickly solve a puzzle, as when your processor concludes that a cobra is present in the visual field and says, "Run!"

Many of these modules would have evolved to become what Steven Pinker, the prominent linguist at MIT and chief band-leader for evolutionary psychology, calls a mental "instinct." Such modules develop reliably in all normal human beings without effort or formal instruction, and they can be applied without conscious awareness of their underlying logic. The processing of information by such instincts is effortless, automatic, reliable, and fast precisely because we have all this complicated machinery dedicated to the task. These instincts are distinct from whatever more general abilities we may have for processing information.

For example, recognizing the face of someone you know well seems to be the easiest thing in the world. However, this accomplishment is achieved thanks to a multistage sequence of unconscious mental events that occupy nearly one-third of the entire cortex. Seeing *appears* easy only because all the highly efficient machinery is hidden from view—that is, from consciousness—because this makes it work even better. Having to "think" about seeing would not be adaptive because it would add time to that flight reflex after seeing a cobra. It might also give you the opportunity to wrongly second-guess what the world was trying to tell you and therefore allow you to make a silly behavioral response—like trying to make friends with the snake. Specialized kinds of reasoning suggestive of isolated mental mechanisms at work have been demonstrated for interpreting the behavior of objects (physical causality), living things (objects moving on their own), animals (mobile living things), human kinds (races), the beliefs and motivations of others ("theory of mind"), and the detection of cheaters in social contracts. In sum, the mind is a "Swiss Army knife," with many specific modules providing that quintessential human characteristic, behavioral flexibility.

Since natural selection takes a long time to design complex adaptations such as sophisticated cognitive mechanisms, the construction of mental modules must be a very slow process of cumulative selection, typically requiring hundreds of thousands of years. Any spontaneous mutations are likely to harm the well-oiled functioning of the brain. Thus their genetic basis must be universal and species-typical. This implies that everyone is endowed with the same general mental structure and that any psychological differences between people must be primarily induced by the idiosyncratic conditions in which an individual may have lived. Since selection typically makes complex adaptations universal in a species, evolved human psychology consists in a single, universal pan-human design, which is the root of our unique human nature.

Thus evolutionary psychologists argue that just as evolution by natural selection has created morphological adaptations that are universal among humans—walking on two legs, opposable thumbs, a range of tooth types to handle our omnivorous diet—so too has it created universal *psychological* adaptations, ways of thinking that will typically sort out the best answer for genes in a recurring kind of situation. The goal of evolutionary psychology, then, is to discover and describe the functioning of our psychological adaptations, which are the proximate mechanisms that cause our behavior. By providing an explanation of how the species-typical mechanisms of the human mind function, evolutionary psychology will provide the discipline with its own set of universal laws, comparable to those in the other high-status sciences.

This notion of psychological universals—a main tenet of evolutionary psychology—allows psychologists to ignore variation in performance, including that associated with cultural diversity. At the same time, the idea of "the psychic unity of humankind" appears easily refuted by cultural diversity and individual differences. But evolutionary psychologists do not claim that all human beings will behave in similar fashion all the time everywhere or that all humans share the same manifest attitudes and preferences. Rather they claim that all humans share common psychological mechanisms that generate different behaviors and preferences in response to the unique developmental and historical circumstances in which people find themselves. So the variation we observe between individuals and across cultures—from Peoria to Pongo-Pongo—is not the result of different psychological adaptations, but a universal set of rules—the same "Darwinian algorithms"—found in bodies at different stages of life and genders, responding in a contingent fashion to novel inputs.

Recent technological improvements have produced significant changes in the social and biological context of modern societies from those characteristic of our ancestors. An obvious example is medical contraception. The availability of the Pill and its "sister" technologies allow women to artificially limit their fertility and so violate the biological primary directive to go forth and multiply. However, due to the complexity of gene coding, we should expect inertia in the ability of the brain to respond to such changing circumstances. Evolutionary psychologists suggest, then, that maladaptive behavior, like using the Pill, is caused by ancestral responses to modern conditions. We are all trapped in "Stone Age" minds time-transported into modern conditions. The result is a mismatch between the kinds of inputs the mind expects and the actual inputs the techno-environment gives it, and hence maladaptive behavioral responses.

It should be clear by now that what separates evolutionary psychology from its colleague, sociobiology, is the claim that not just structure, but mental content too—the knowledge needed to function in society—can be implanted by a history of natural selection into an infant's mind: We are born with much of what we need to know to function as competent members of cultural groups. We just need the time, and occasion, to show it.

PUNCHING THE BUTTONS ON THE JUKEBOX OF LIFE

What social scientists call culture is just an illusion to sociobiologists. It is like the shimmering you see at a distance on a hot road through the desert. When you get closer to inspect the situation, the mirage disappears, and all you are left with is the hard asphalt under your feet. Sociobiologists see culture as just evolved behavioral responses, dressed up to look fancy by social scientists.

Evolutionary psychologists, too, largely dismiss the importance of culture—or at least what "the man in the street" calls culture: *transmitted* information. They distinguish between what they call "evoked" and "epidemiological" culture. Epidemiological culture is what textbook definitions of culture refer to; it is the product of learning from others— information spread through a population like a virus. Evoked culture, on the other hand, is innate information that resides in human heads and is expressed contingently in different environments. For evolutionary psychologists, what *appears* to be acquired through the exchange of information is mostly knowledge evoked by environmental stimuli from a commonly inherited storehouse of wisdom. This storehouse can be found in the head of each individual and is a gift of the genes, acquired by picking only the best products off the shelf as hominids have made their way through the shopping mall of prehistory.

Some of the most prominent evolutionary psychologists go further, arguing that basically *all* of the knowledge you need has already been placed in memory by evolution. In effect, these evolutionary psychologists not only reject the importance of social transmission but also belittle the importance of learning and experience more generally. They argue that gene-based structures in the brain, with which individuals are born, and which therefore precede individual experience, constitute a filter by which all experience is evaluated, even in the womb. At its most extreme, the idea is that babies are born with most of the information they need to get by as adults, except perhaps for the vocabulary they need to plug into their language module. Humans don't *learn* things from the environment but instead simply *recall* them from memory.

An image popular among evolutionary psychologists compares evoked culture to a jukebox. The mental "records" of every musical style are in every jukebox, but jukeboxes in Rio play samba while those in

Havana play salsa and those in London play pop. This is because what jukeboxes play is a function of which buttons are pushed by the people hanging around the machine; it's just that the buttons pushed in Rio are different from the ones pushed in London.

Thus evolutionary psychologists believe that the importance of information transmission is greatly exaggerated by most social scientists. Is this a reasonable position for anyone to take in explaining culture? It seems we have to answer the more fundamental question of whether culture is transmitted before we can move on to the question we have set ourselves: Does culture replicate?

The question before us, then, is whether evolutionary psychologists are right in their belief that the evolution of cultural traits can be accounted for by heredity without replication. For them, both variation and commonalities in culture are explained by recourse to the same causal machinery: the action of genes in producing big-brained creatures. Cultural similarities arise not through the diffusion of good memes across the vast distances that separate similar cultures; rather they are evoked in response to shared environmental conditions. It is convergent evolution based on shared experience, rather than the transportation of memes from one locale to another. This is a denial of direct information transfer from brain to brain. Instead the causal arrows point back to genes and their ability to produce clever organisms: us. On the other hand, cultural variation—although it results in different traits being expressed—is produced by the same causal mechanism: responses to the environment. In this case, it is just that ecological factors differ, and so therefore do responses.

This does seem to be a powerful view of the human mind, which may account for the growing popularity of evolutionary psychology and the spate of recent books on the topic. But does it jibe with the facts? What about the evidence? There are, in fact, considerable problems with ignoring the role of information transmission from individual to individual in social life when you get down to cases.

First, the position taken by evolutionary psychologists requires a very big jukebox to hold all the "records" that might be requested anywhere in the world. But the brain is indeed big—big enough, say the evolutionary psychologists, to hold everything you need. However, even a rough calculation would seem to rule out a strong version of the evoked culture argument that everything is pre-stored in the brain.

The human genome consists of about 1.5 billion nucleotide pairs, or about 150 megabytes of information. Of this, we share about 97.5 percent with chimpanzees, leaving 2.5 percent of our genome to code for the innate differences between our closest living relative and ourselves. Chimpanzees have some culture but much less than humans, so any recently evolved culture that is innately stored would have to be packed into this 2.5 percent of the genetic material. This equals about 3.75×10^7 base pairs. At 1 bit of information per base pair and 8 bits per byte, this translates into about 4 megabytes of information available for the storage of evoked culture in the human brain. But this is much less space than the database of cultural information any competent individual must command to function appropriately in any contemporary human society—much less any society in which humans have ever lived (remember, evolutionary psychologists believe *all* the possible human lifestyles are stored in the brain, not just the one popular in the society you happen to grow up in, because you could be born anywhere).

To flesh out this argument, let's assume that the assembled corpus of descriptive work in ethnography is a good measure of the total amount of cultural information the human species is capable of learning. There are written records of this information—the results of more than a century of work by professional cultural anthropologists as recorded in their ethnographies, or anthropological studies of the ways of life in different cultural groups. The Human Relations Area Files (HRAF) is the main repository for this ethnographic information. The primary computer-based database at HRAF consists of 1 million pages of information, with text entries on 365 cultures coded for many features of life in each of these groups. Just how much information does this represent? It takes around 5 bits to represent each letter, and if words average around 5 characters each, we get 3 bytes per word. With about 700 words per page, this comes to a grand total of around 2,000 megabytes of information in the HRAF database. These are summaries of the relevant ethnographies, and not every society that has been studied by anthropologists appears in this database. In effect, this is a considerable *underestimate* of the true amount of information humans might have to carry around in their heads. Of course, not every culture varies on every dimension from every other one; considerable similarity between cultures is also present in the database, which means that not every variable for every culture needs to be stored separately. Nevertheless the amount of information in this database is *three orders of magnitude greater* than we have judged a single human being has space to devote to this knowledge. It may be that genes somehow compress the information that will actually be stored in the brain, so there will always be some slop in this argument. Yet it does suggest that the evolutionary psychologists are overly hopeful in their estimates of how much culture can be evoked by circumstances. In fact, only a rather modest fraction of what we know can be present inside us at birth. This conclusion seems commonsensical enough. After all, where are the mental slots in the developing brain for recently coined slang, Game Boys, or body piercing?

There are other reasons to think the jukebox model just won't work. The crucial claim of the jukebox model is that environmental variation determines the pattern of cultural behaviors: Individuals are simply reacting to the stimuli thrown at them by the ecological circumstances in which they live. A number of studies have now been undertaken on the vexed question of whether environmental variation explains the differences in human behavior between cultural groups, as presumed by both sociobiology and evolutionary psychology. We can even use evolutionary psychologists' own studies to test this proposition, because a number of the best empirical projects in evolutionary psychology have been cross-cultural and so seem good candidates. These studies have concentrated on the features that each gender looks for in a potential mate, such as what shapes men prefer in the bodies of their women (in particular, the ratio between the circumference of their waist and hips) and whether one gender prefers faces in which the two sides are more or less symmetrical. Each of these analyses finds a consistent, expected pattern of variation within groups. But often the between-group variation is at least as significant as the trend studied by these researchers. This variation remains unexplained because evolutionary psychologists concentrate on universal patterns.

This tendency to ignore the problem of unexplained variation is shown clearly by one of the best examples of a cross-cultural study in evolutionary psychology, that conducted on homicide by the prominent evolutionary psychologists Margo Wilson and Martin Daly. This couple has worked intensively for years to show that the pattern of who kills whom in a variety of countries can be explained—as a kind of extreme response to interpersonal conflict—by the principles of evolutionary psychology. In particular, people should tend not to kill those individuals who might have copies of their genes. For example, they show that children living with stepparents (with whom they are not genetically related) are at a much higher risk of being killed than genetic offspring. This is in accord with expectation, since biological evolution should have favored those parental psyches that do not squander care on non-relatives.

It has also become clear through Daly and Wilson's work that the cross-cultural variation in homicide rates is truly staggering. Rates in the United States, for example, are vastly higher than those in neighboring Canada (the researchers' home country). Environmental and socioeconomic variables—even many cultural variables such as language, fashions, and media—overlap greatly in these two countries. So what's causing the big difference? Laws permitting gun ownership in the United States don't "legislate" their frequent use, and most killings are not accidental. The variation in homicidal behavior between the two North American countries requires another kind of explanation. The United States, it has often been said, is truly a "gun culture," where organizations like the National Rifle Association can acquire large memberships and influence law-making. Wilson and Daly don't acknowledge this kind of explanation, but then their research goals lie in other quarters.

Nevertheless this example and many others strongly suggest that culture is not just a straightforward response to stimuli presented by the environment, as evolutionary psychologists suggest. If this proposition were true, the research program called "cultural ecology," which was founded on the idea of ecological determinism, would have been much more successful than it proved to be. Human groups living side by side exhibit marked differences in beliefs and practices, despite facing common ecological problems.

Cultural variation is, in fact, both significant *and* independent of environmental variation. And this is true not just of humans. The existence of both ape and cetacean (dolphin and whale) cultures have been deduced by a number of field biologists, based on patterns of behavioral variation in time and space that cannot be explained by environmental or genetic factors in these animal populations. It therefore appears that evolutionary psychology cannot explain such variation as the result of different buttons on the universal jukebox being pushed by ecological variables, even in its own best studies. Instead cultural variation must be at least partly due to historical inertia or purely random factors such as "national character," neither of which is reducible to ecological responses. So, as was the case for sociobiology, evolutionary psychology falls short when it comes to explaining intercultural variation, much less intracultural variation.

From this argument we can conclude that cultural heredity is, in fact, due to the transmission of information between individuals in populations. This suggests that cultural replication is still an option. It is thus apparent that we need a legitimate theory of how cultural evolution occurs through the social transmission of information. Our search for a good model of cultural change must continue.

Transmission Happens

Sociobiologists and evolutionary psychologists have one final riposte to the contention that information transfer must be included in any reasonable picture of human evolution. They don't deny that information is exchanged between people. Humans are a highly social species. We spend a *lot* of our time interacting with one another. Transmission happens, of course. But these schools argue that culture is mere veneer, accounting only for relatively trivial traits such as styles of dress or taste in music—things that have essentially no impact on biological evolution. For this reason, evolutionists studying humans can ignore transmission and its effects.

On the other hand, cultural anthropologists contend that humanity cannot be understood *except* as a species embroiled in a unique way of life totally dependent on the exchange of social information. Culture, they say, is something that surrounds us like the air we breathe, determining the ways in which we interpret our every experience. We are born knowing nothing important; all is imparted to us by the social group in which we find ourselves. We must imbibe from this group the special habits, rituals, and traditions that make us fully human, as members of these special communities. What we pay attention to from among the welter of sensory inputs is determined by what is marked as important by our social group. We cannot even *see* the world around us except through the lens of culture. And without the group, we die.

Even from an evolutionary perspective, the question arises: Why should we spend so much time in apparently fruitless activities like gossiping if they aren't significant? For example, doesn't gossip function to maintain social reputations that are important for getting or keeping a mate, a business partner, or a new social role in the group—any of which can have a significant impact on one's biological reproduction? And aren't "superficial" cultural traits like the kind of clothes you wear markers of membership in some social clique, which determines access to resources available only to the members of such cliques? Perhaps no feature of cultural life is trivial in a species with significant social structuring.

So transmission happens, but does it matter? Which of these views about the importance of transmission is correct? There are strong grounds for believing that cultural transmission is not just an occasional oddity but an important component of everyday learning.

When some practice, technology, or idea is novel, it nearly always spreads through a population in the same way. If you graph the proportion of the population exhibiting the new trait in terms of time, this proportion always begins at very low frequency (because it is an innovation, after all). The rate at which people adopt the trait starts to increase slowly, then faster, until a point of relative satiation is observed, after which time the number of new adopters dwindles, causing the adoption curve to bend over. At some point, no new cases occur (perhaps because everyone is already doing it), and the population curve flattens out or even goes back down a bit (because sometimes a number of people go back to the "old way" of doing things). This "S-shaped" cumulative adoption curve is a very robust finding from several thousand studies on the diffusion of innovations. The types of traits following this pattern can be quite different and include the adoption of hybrid corn among Iowa farmers, bottle-feeding practices among impoverished Third Worlders, new governance practices among Fortune 500 companies, chemical fertilizers among small-scale farmers, novel approaches to mathematics training (the "new math") in secondary schools, and the practice of not smoking among Americans.

Recent formal models of the psychology underlying this diffusion process have proven an important point: Reproducing the S-shaped pattern requires that people not base their adoption of novel traits strictly on trial-and-error experience of their own. Direct learning by individuals from the environment, based on cost-benefit analysis, does not generally produce the characteristic S-shaped curves that dominate the empirical literature. Just observing one's neighbor engaging in the relevant cultural practice and copying it if it makes sense (a purely rational form of adoption) cannot explain how new traits spread over time through a population. Instead an S-shaped curve is created only if individuals *prefer* to copy their neighbors ("keep up with the Joneses"). Only if people are not just making judgments for themselves about the trait will the adoption process have a *social* dynamic—speeding up once enough adopters are around to serve as models for others. Only then will changes in behavior or belief have the correct temporal dynamics. This means individuals must adopt the trait regardless of its payoff to them—in effect, there must be a psychological *bias* toward the adoption of that trait. Typically this bias could favor traits exhibited by prestigious individuals or by the majority of those observed from the individual's social group (a process called "conformist transmission"). But such a bias must be in place, and be used by many people, for this period of very rapid adoption to take place. The fact that novelties are a ubiquitous component of any culture—just think of the constant turnover of fads and technology-inspired behaviors and ideas—suggests that cultural transmission is an important influence on how people in a group come to share their beliefs and values.

The implication of these models is that the social transmission of information is important indeed and should not be cast aside as a trivial footnote to evolution. It is to the description of approaches that begin with this admission that we now turn.

CULTURAL SELECTIONISM

The acknowledgment that transmission is important does not immediately identify just one school of thought, because a variety of theoretical perspectives exists for dealing with transmission processes. Leading contenders for the job of explaining transmission include mathematical communication theory, as developed at the Bell telephone labs by Claude Shannon and his colleagues, and cybernetics, the brainchild of Norbert Wiener and others. Both of these theories are about the process of getting information from one place to another. However, in the course of being transmitted, cultural information is typically also reproduced rather than merely being passed along: Brains cannot transmit a belief without learning it first; a copy is thus left behind. The transmission of cultural beliefs therefore not only involves the manifestation of information in physical form (for example, as sound waves in speech) but also requires the duplication of the information packet itself. In this respect, it is akin to biological reproduction.

A well-developed conceptual framework for this task is ready-made: epidemiology, the biological study of how pathogens disperse through populations. The question is how to extend this paradigm to the explanation of culture. From an epidemiological perspective, the central question for cultural evolution is why certain kinds of traits, among all of those that individual creativity tosses up, spread rapidly to others, becoming endemic in a population (and hence becoming "cultural"), while others remain idiosyncratic, characteristic only of scattered individuals. The focus then is determining what qualities cultural beliefs must exhibit to succeed in reappearing across generations, and what factors determine that tiny fraction of ideas or beliefs that come to dominate, through the networks of social communication. "Informational epidemiology" is thus meant to explain the distribution of beliefs in a population as the cumulative result of a history of diffusion events.

What we have here, in effect, is the cultural trait seen as a virus, an analogy already seen in the memetics literature. As I've argued, this analogy is weaker than one that treats cultural evolution as analogous to genetic evolution. The problem with this alternative is that we don't know whether a replicator-based approach to culture is appropriate either.

There is a happy compromise position, however. A group of contemporary theorists seeks to formalize an evolutionary perspective on culture through the adaptation of population genetic models to the peculiarities of cultural transmission. Although a generally accepted rubric for this group has not reached wide agreement, a good description is "cultural selectionism." This is appropriate because cultural selectionists require only that cultural evolution rely on what John Maynard Smith has called "units of selection"—entities such as organisms with the characteristics of multiplication and heredity. This is their main sticking point with memetics (discussed next), which argues there is a cultural replicator. Replicators are an example of what Maynard Smith would call a "unit of evolution"; they differ from units of selection in being capable of variability as well.

A moniker emphasizing selection also distinguishes this modern school of evolutionists from earlier applications of evolutionary theory to social phenomena, which tended to see the long history of life on Earth as a forced march up the ladder of improvement reaching toward Heaven, through higher and higher manifestations of God's Great Design. This was a form of manifest destiny applied to the biological realm, which ignored the possibility of reversals in complexity (if not of fortunes) within evolutionary lineages, such as the loss of eyes when organisms move to an underground lifestyle. As a result, the earlier thinkers who applied Darwinian ideas to the sphere of human social life (such as the Victorian sociologist and philosopher Herbert Spencer) were inclined to think of all evolutionary change in terms of progress, a trend that culminated in the much-derided Social Darwinism of the 1920s. Selection is not a process that heads toward any predetermined goal over the long term but rather reflects local conditions at each moment.

Cultural selectionists define culture as information capable of affecting individuals' phenotypes that they acquire from others by teaching or imitation. This definition obviously emphasizes transmission, as I have argued it should, and through a particular mechanism of heredity: social learning. At the population level, culture can be considered a pool of information that becomes modified as people learn new things. If culture must be learned, and learned anew each generation, then cultural knowledge must obviously be transmitted to people after they are born. Presumably transmission is never perfect, so variation between the cultural repertoires acquired by individuals of a given generation can arise. And since only a part of a group's cultural knowledge can be housed in any given mind (our mental "jukeboxes" are too small to hold everything, remember), culture is necessarily distributed across individuals. This variation in knowledge is the basis on which various kinds of selection processes, ranging from the physical to the psychological, can operate.

Since culture exhibits, by this definition, the three characteristics required for evolution through selection—variation, heredity, and selection—cultural evolution can be analyzed using Darwinian methods. This makes culture a part of biology for cultural selectionists. Culture is the quintessential human adaptation, which a major part of the human brain has evolved to express. What distinguishes cultural selectionism from sociobiology and evolutionary psychology, then, is its emphasis on traditions of information learned through social transmission, supported by social learning abilities that have evolved to support such traditions. The basic claim is that *to explain variation in the behaviors and beliefs of humans, you need to add culture to the standard equation of genes plus environment.*

The formal approach associated with this school (based, as noted, on population genetics) is generally called "gene-culture coevolutionary modeling." These models emphasize the novelties that must be introduced in the standard genetic machinery to account for the unique features of cultural as opposed to genetic inheritance. Examples of these new modes of transmission include multiple parentage (you can learn aspects of some cultural trait from a variety of people), transmission within peer groups (rather than inter-generationally), and psychological tendencies to acquire particular values of a given cultural trait (rather than the 50/50 chance of getting either the mother's or the father's type, as in the case of genes). Thus we might have a model in which an offspring has varying probabilities of adopting Protestantism from his mother, Catholicism from his father, or a New Age religion from the culture at large. The models' job is to keep track of how the relative frequencies of alternative values of the "religion" trait change over time in a group.

Cultural selectionists also make much of the fact that the standard distinction in genetics between genotypes (as replicators) and phenotypes (the qualities of organisms) has a cultural equivalent: Ideas are the replicator ("blueprint") of culture, and behavior its expression, or phenotype. Mental planning, as the intermediate step between an idea and its manifestation, can also be treated as something like *its* biological analogue: the development of an organism from its genetic sources. And just as in genetic inheritance, variation is introduced through errors in the transmission of information, leading to the diversification of cultural forms.

Cultural selectionists emphasize "population thinking"—the idea, first articulated by the eminent biologist Ernst Mayr of Harvard University, that the Darwinian "revolution" was more about the transmission of a pool of information through time than about particular mechanisms for that transmission, such as natural selection. In the selectionist view, evolutionary theory "is [simply] an accounting system for keeping track of information as it is propagated forward through time by whatever means, carried by whatever vehicles carry it, interacting with the world and each other in whatever ways." This is a very general formulation of the evolutionary process, designed to make it applicable to culture, whatever mechanism turns out to cause change in cultural traits over time. The focus on population thinking serves, in effect, as the selectionists' mode of generalization. This degree of abstraction is taken as a virtue, because it allows a multitude of models to be used in the face of ignorance about what's really happening inside the explanatory black box. Cultural selectionists want to be agnostic about the process underlying cultural evolution.

What can be treated as a population in the case of culture? One good answer is the set of ideas in people's heads that are acquired from others. So cultural selectionists hold to the standard interpretation of culture as being something essentially cognitive in nature. Culture is just the population-level consequence of the evolved psychological ability to imitate. As with evolutionary psychology, selectionists see individual cognitive abilities as the locus of selection. Selective forces in cultural evolution are psychological biases, which lead to a preference for Protestantism rather than Catholicism, for example. So cultural selection is a process analogous to natural selection, but it works on variation in cultural traits rather than biological ones. What selectionists emphasize (in contrast to evolutionary psychology) is that in both cases the type of trait favored by the selective force will increase in frequency in the population. It's just that the location of the selective force happens to be different: With culture, it's in the head as well as in the environment. So cultural selection is a Darwinian process by which particular *traits* increase or decrease in frequency due to their differential probability of being adopted by individuals. In contrast, natural selection is specific to the differential survival or fertility of *individuals* expressing different types of traits.

Of course, natural selection can also work on cultural variation. If a socially learned idea causes you to jump off a cliff, then the idea will die right along with you—and never get learned by anyone else. That's an example of natural selection on a host due to a behavior caused by a cultural trait. One consequence is that the cultural trait has also compromised its own likelihood of further transmission, so this is also an example of natural selection acting on the reproduction of a cultural trait. This can happen because the trait causes its host to engage in some behavior that reduces the host's longevity (as in our example), or because it reduces the likelihood that its host will pass that trait on to others—for example, a belief in the virtue of celibacy can leave someone with no offspring to be impressed by that virtue. The fate of cultural variants therefore depends both on mental biases (which influence whether they get adopted in the first place) and natural selection on people as "hosts" of that cultural variation. It's a question of a trait getting past psychological filters, and of particular host organisms surviving to spread that trait further.

In an effort to keep things simple, the typical practice among gene-culture coevolutionary modelers is to focus on the dynamics of cultural traits alone, unless it's important to track the relationships between these traits and genes. If cultural inheritance is completely independent of genetic inheritance, then cultural evolution can be modeled separately, as a process happening by itself. These models then trace changes in the frequency of cultural traits in a population through cycles of social interaction (leading to the possible exchange of cultural information) and differential social transmission of the trait (cultural selection).

On the other hand, there are situations in which an individual's propensity to adopt a particular cultural trait depends on his or her genetic makeup. In this case, there is an interaction that can be included in the model as a parameter which determines the probability that an individual will acquire a particular trait according to which genes they have.

This kind of interaction between genetic and cultural inheritance is more likely to occur at longer temporal scales. A good example is the evolution of lactose intolerance. Some people cannot properly digest dairy products because they lack an enzyme (lactase) for processing the complex sugar lactose. By describing individuals in terms of the presence or absence of the lactase-producing gene and a culturally acquired tendency to drink milk or not, we can model how milk consumption coevolves with the gene allowing individuals to absorb lactose. The analysis of this model suggests that the genetic variant for absorbing this sugar can reach a high frequency only if the children of milk drinkers themselves drink milk. There is even a broad range of conditions under which the lactase-producing gene cannot spread, despite conferring a significant biological advantage on individuals, in the absence of this cultural preference. So this example demonstrates that genetic change can strongly depend on cultural preferences. It isn't just a one-way road from genes to culture!

A mechanism for producing much quicker change in cultural traits than interaction with genes is, of course, learning. Learning is a means by which information about local conditions can be acquired through sense organs, allowing organisms to adjust their behavior on a temporal scale faster than is possible through the induction of changes in "hardwired" routines through genetic mutations between generations. In particular, social learning—acquiring information from the other individuals in your environment—has two major advantages as a mechanism for the acquisition of information. First, you can avoid the costs of going out and engaging in potentially dangerous trial-and-error learning for yourself, like figuring out the best way to kill a large animal for supper. Instead you can just ask somebody or, better yet, watch them try it first. Second, you can selectively adopt information from only those people with little interest in deceiving you (such as your parents, with whom your genetic interests overlap significantly). This helps to avoid engaging in behaviors that run counter to your better interests and ensures that what you learn is "good" data about the way the world works. (Just as in genetics there is the danger that mutations or lethal recessive genes will emerge to cause crippling or fatal diseases, so too the danger of copying other people's cultural information is that it may be out of date or inappropriate to one's own circumstances.)

Even evolutionary psychologists, despite their general disdain for transmission, acknowledge the benefits of social learning. Steven Pinker, for instance, sees culture as

> a means of exchanging knowledge. It multiplies the benefit of knowledge, which can not only be used but exchanged for other resources, and lowers its cost, because knowledge can be acquired from the hard-won wisdom, strokes of genius, and trial and error of others rather than only from risky exploration and experimentation.

Learning socially, once the ability has evolved, may represent a more efficient means of acquiring important information about the local environment than through direct interaction with that environment. Selectionist models have indicated that this is especially true if aspects of the environment important for survival and reproduction change very rapidly, or if individuals tend to migrate to ecological niches different from those in which their parents lived. In these conditions, the degree of reliance on personal experience through individual learning from the environment should be high. On the other hand, if the resemblance between parental and offspring environments is good, due either to low mobility or to little change in the ecological context of life, then individuals can safely rely on the information acquired by others as being relevant and appropriate. Doing a bit of social learning should be favored because it will still produce the right kinds of responses without the cost of trial and error.

So cultural evolution is embedded in, and constrained by, genetically evolved psychological propensities. At the same time, these biases can produce variability in cultural traits that is not directly tied to ecological differences. Universal psychological biases working on slight differences in initial trait values can lead groups to quite disparate cultural equilibria over time, thanks to dynamics internal to the cultural realm. For example, selectionist models show that cultural "badges," such as distinct dialects, styles of dress, and forms of religious practice, can erect barriers to the free flow of ideas in separate environments, like, say, Germany and France. These traits can be completely inconsequential from a genetic viewpoint (that is, have no direct effect on reproduction or survival) but nevertheless evolve as ways of identifying membership in different cultural groups. Ethnic groups thus form the cultural analogues of reproductively isolated species. The main difference is that the barriers to transmission are more permeable and the rate of cultural evolution much higher than in the case of genes. As suggested above, evolutionary psychology has a difficult time explaining stylistic differences between groups in things like aesthetic productions (like the American Hollywood versus Indian Bollywood styles of motion picture–making) based solely on universal psychological propensities and their survival value, even when ecological conditions vary.

Cultural selectionism holds that the differences between genetic and cultural transmission introduce important changes in evolutionary dynamics. In particular, it can produce different kinds of outcomes from those a purely genetic system would deliver, because information acquired socially constitutes a second, non-genetic system of inheritance. Selectionists point out, for instance, that cultural information can be transmitted in ways impossible for genes: Parents can pass their genes on to their offspring, but cultural ideas can flow in the opposite direction as well—from offspring to parents, for example. Certainly kids make their parents aware of new fads by pestering them to buy the associated paraphernalia—for example, toys tied to the latest movies. Further, individually learned skills or ideas can subsequently be transmitted to others, making culture a system for the inheritance of acquired variation.

Evolutionary dynamics become more complicated with culture, thanks to new transmission-based evolutionary "forces" based on social learning abilities. And differences in the spread and diffusion of cultural information account for the novel dynamic properties of cultural evolution, such as the fact that it can produce much more rapid change than genetic evolution, or keep maladaptive cultural practices around even in the face of natural selection against the individuals who engage in such practices.

This is true even though the capacity for culture, including psychological abilities for social learning, may itself have been the product of natural selection acting on genes in an earlier time. In perhaps the most famous book on cultural selectionism, *Culture and the Evolutionary Process,* Robert Boyd and Peter Richerson, both biologists at the University of California, showed that a reliance on social learning, and all that entails, can arise through standard evolutionary processes, with this psychological ability then transforming the world through new dynamics of information transfer. This purely biological theory for the origin of cultural inheritance through the action of natural selection effectively defeats an important claim of sociobiology: All human activity cannot be reduced to biological fitness maximization by natural selection acting directly on genes.

So genetically non-adaptive cultural evolution is possible and is more likely when the differences between genetic and cultural inheritance are most marked. Indeed one of the earliest findings to emerge from cultural selectionist analyses was that a variety of mechanisms exist through which traits with high cultural fitness can increase in frequency despite being maladaptive from a genetic perspective. This is *not* something that adaptationist positions like sociobiology and evolutionary psychology would predict.

So the selectionist framework embraces the full range of possibilities for the evolution of cultural traits: Cultural traits may influence the fitness of genes (as in the lactase example above), or genes may determine what direction cultural traits will move in (as when a sect dies out because its rule of celibacy makes it unable to attract recruits), or genes and cultural traits may each go their own way quite independently (when a cultural trait has no biological consequences, it can drift through minds and populations at whim). There is no overarching answer to the question of who benefits more, genes or culture; rather the answer must be sought for each case independently.

The cultural selectionists' strongest claim is that cultural evolution doesn't *necessarily* depend on replicators. For them, entities with fidelity, longevity, and fecundity (the ability to multiply)—the triad of features that good replicators exhibit—do not necessarily populate the cultural world. Cultural selectionism admits there is a second inheritance system operating in human society, that of culture. At the same time, however, proponents of this view argue that culture need not reflect the activity of a replicator operating at the social level to produce this cultural heritability. How can cultural evolution be called a "dual inheritance" system (by Boyd and Richerson in their book *Culture and the Evolutionary Process*) when it doesn't have dual replicators? This is possible for exactly the reason we identified earlier: Evolution depends on inheritance but not replication. Cultural selectionists argue that all we require for the cumulative evolution of complexity in culture is a system for maintaining heritable variation—that is, for a correlation between generations in phenotypic values. They point out that the failure of memetics to come up with a reasonable theory of meme replication has been taken to herald the death of cultural evolutionary theory. But this need not be the case, for a Darwinian approach to culture can survive the death of the meme. It can live on in the form of other mechanisms of inheritance.

What might these mechanisms look like? For one thing, in the cultural selectionist view, cultural inheritance need not depend on "atoms" of information like memes being faithfully duplicated in each generation of offspring. Further, selectionists argue there need be no cultural communication chains tied together by replication events. Instead the same traits may continually reappear through other causal routes. Some kind of gene-environment interaction—like the evolutionary psychologists' evoked culture—may be responsible. Alternatively, many organisms modify their environments in significant ways (the classic beaver dam example springs to mind). These modifications, sometimes in the form of artifacts, can serve as educators for future generations. These artifacts persist in the environment, so people in the next generation can learn directly from them.

Information can thus be inherited by each individual in the form of structures in their own brain or through artifacts left behind by their ancestors. In either case, there is no direct mind-to-mind link, no cultural replication. But that's okay with the cultural selectionists because the conditions of their population genetics models are still satisfied: There is a cross-generational correlation in cultural traits, even if the correlation is not due to the operation of a replicator at the cultural level. These are still cases in which culture evolves, because evolution is not dependent on the existence of a replicator working at the individual level, diffusing itself through the population. All the replication producing inter-generational correlations may be happening through genes, or even thanks to the inheritance of aspects of the environment. Memes may not play a role at all.

WHO'S IN CHARGE?

Major schools of thought deny that culture replicates. Some even reject the apparently commonsensical impression that most cultural knowledge is socially transmitted. But denying the importance of transmission is illogical because cultural variation does not mirror ecological or environmental variation. Instead culture appears to have a dynamic of its own, due to histories of information transmission independent of the genetic channel.

Cultural selectionism, which recognizes this central fact, thus has considerable strengths and seems a viable explanation of our target phenomenon: culture. It certainly seems to avoid the pitfalls we identified with sociobiology and evolutionary psychology. But there is a second school of evolutionary thought that also emphasizes the transmission of information: memetics. This leaves us with two candidates still standing for the office of Culture Explainer. Both memetics and cultural selectionism admit that transmitted information is central to the phenomenon we call culture. Where these two groups differ is on the mechanism by which culture is reproduced. Selectionism requires only heredity, while memetics requires replication. This makes memetics a special brand of cultural selectionism.

This contretemps between cultural selectionism and memetics makes it clear that earlier views on memetics have been framed inappropriately. And asking the right question is the first and perhaps most important step to making progress in our understanding. But if we look at the question that each of the major defenders of memes finds to be the most fundamental, we find they do not identify the issue that appears crucial from our review of the situation.

A major proponent of memes these days is Susan Blackmore. The central issue in her book *The Meme Machine* is stated as follows: "Imagine a world full of brains, and far more memes than can possibly find homes. Which memes are more likely to find a safe home and get passed on again?" This question not only presumes the existence of memes but also is concerned solely with selection among them. In essence, her question is *which* memes benefit from the biases people might have about learning them? Because Blackmore takes so much about memes for granted, she feels it is legitimate to move on to this secondary topic. Here we will take nothing for granted and start from ground zero.

Dan Dennett, author of *Darwin's Dangerous Idea,* is perhaps the best-known contemporary defender of memes. When he talks about memes, he perennially asks the question *"cui bono?"* (Latin for "who benefits?"), implying that memes rather than genes can be the winner when it comes to contests involving cultural traits. This question is taken from Richard Dawkins, who originally asked it in the context of parasitism. Dennett even adopts the same example to illustrate the basic principle: the story of the ant and the "brainworm" fluke. The parasitic fluke (a small worm) needs to get into a sheep to reproduce and uses the ant as an intermediary host. It infects the ant's brain, induces the ant to climb a grass stalk (which it otherwise wouldn't do), where the ant, and hence the fluke, are more likely to be gobbled up by a passing sheep. Such a behavior is quite stupid from the perspective of the ant, because natural selection is unlikely to favor ant genes that regularly leave ants dead. So who can be benefiting from this behavior, if not the ant? The answer rapidly comes back: a parasite that short-circuits the ant's brain to make it do counter-productive things.

So far, so good. Parasites are biological entities like their hosts. Many of the same genes are in both hosts and parasites, so the battle is one *between* genes. Even if these genes happen to be located in organisms of different species, they are still part of a single system of replicators.

The problem is that Dennett takes his "who benefits?" question and applies it to a contest between different types of replicator—memes versus genes. But this is not the same situation as the ant and the fluke, where only genes are involved. What genes can contend among themselves is not necessarily the same as what genes and memes can fight over. Copies of the same gene in different bodies typically do not compete over control of one of those bodies. And if genes are in organisms from distinct species, they don't even compete for the same set of environmental resources, because species tend to inhabit different ecological niches. For example, despite their considerable overall similarity, each of the many different species of finch on the Galápagos Islands has a different diet, thanks to adaptations in the shape of their beaks. Some types of competition are thus excluded to genes. But a meme and a gene *can* clash over just about anything, including the direction of a particular behavior, even in the very same organism, because they are not part of the same replicator system.

Memes and genes may also have varying stakes in their various rivalries. For example, memes depend on behaviors such as social communication to get themselves transmitted. So passing on a particular bit of gossip may mean life or death to a meme. On the other hand, gossiping is a fleeting moment of idle behavior having little consequence to the genes inside the gossiper. Genes simply want to keep their rumormonger alive and well, and the latest slander about Mary's infidelity isn't going to influence their individual's health and well-being very much. This means that memes will invest more of their energy than genes to ensure they win the "to gossip or not to gossip" fight.

This doesn't imply that the meme will always succeed, however. Winning a military battle may matter more to the Abbedabbies than the

Zoobedoobies, but if the Abbedabbies are stuck with sticks while the Zoobedoobies have guns, then "mattering" doesn't matter much: The Zoobedoobies will win, even if winning is only of marginal importance to them. Similarly, the *effective* power of these dueling replicators (memes and genes) depends on their relative strength, their "technologies" of control over the behavior of the organism that hosts them. The brain must decide what the organism is to do next: Should it be the gossiping option that memes hope the brain will find appealing, or the foraging for food alternative that the genes would prefer? On the other hand, both memes and genes might benefit from that single behavior, or both might suffer from it. Which replicator proves more persuasive at the moment of decision is not the same as Dennett's question of "who benefits?" The outcome of a conflict alone cannot tell you whom the adaptation has evolved for. This means that, once we switch to a question of coevolution between different replicators, Dennett's question doesn't pick out the crucial factor for determining what course evolution is following.

Dennett's perspective also implies that maladaptation must be due to the activity of a parasitic meme. What's a maladaptation? It's a design feature or behavior of an organism that makes no sense from the point of view of that organism's genes. Cases of maladaptation are certainly a major fund of information about relationships between parasites and hosts, as in the case of the ant and the fluke recounted above. Dennett's question implies that you need only to look for memes when someone is doing something maladaptive, because it must mean a parasitic replicator with different interests is at work. But Dennett's perspective is too narrow here. Why? First, some parasites don't inflict penalties on their hosts. The bacteria in our gut are also parasites, but their relationship with us is one of mutual benefit, or symbiosis: We get to eat foods we couldn't otherwise digest, while the bacteria get a nice home and a constant food supply. If biological parasites don't necessarily manifest themselves through maladaptations, cultural parasites such as memetic ideas certainly need not. A meme may be in the driver's seat but nevertheless produce an outcome that favors the host. In fact, favorable outcomes should happen whenever the meme's interest coincides with that of the person it uses as an instrument. We may find that memes often foster the standard goals of genes, like the maintenance and reproduction of the host organism.

Second, some maladaptive behaviors will prove to be the result of conflict between the genes within an organism or between a gene's expectations and the creature's current conditions. These would constitute maladaptations in which memes play no role. For example, pregnant women may reject a variety of foods because they are novel or exotic, even though their caloric requirements have ballooned (along with their bellies), due to an evolved fear of ingesting unfamiliar substances. So the interests of genes and memes are likely to converge in some respects but diverge in others. The presence of maladaptive behaviors will not be simply related to the activity of memes.

The more appropriate question to ask, one which will be central to this book, is *cui impello?* Who drives the cultural evolutionary process or sets it in motion? In other words, Who's in control? It's not who benefits but *whose interest is being expressed* that needs to be asked when you're attempting to explain what happens in a *coevolutionary* system, as when the dynamics of genes and memes interact in ways that determine what happens to either of them. It is a question of causes, not consequences, of who is responsible, not who is rewarded.

How might such a difference in perspective manifest itself? If genes are in control, then the duplication of cultural information takes place passively, at the behest of gene-produced brains. If the memetic perspective is correct, then at least some bits of cultural information have the ability to influence the probability of their own duplication. If memes are causes, then they can bias the course and tempo of evolution. If they are the mere products of someone else's activity, they cannot. Control makes all the difference. It also makes the field of memetics essentially about power: the relative power of replicators dueling for control over the determination of our thoughts and behavior. The crucial issue that distinguishes cultural selectionism from memetics is how cultural reproduction is achieved, and whether this involves the special kind of thing called a replicator. The French poet Paul Valéry claims that "poetry can be recognized by its ability to get us to reproduce it in its own form: it stimulates us to reconstruct it identically." Is culture like poetry in requiring replication? Whether information evolved from being passively reproduced to actively replicating during social communication thereby becomes the deciding issue for whether memetics should be the favored evolutionary approach to culture.

At this point, we have come full circle, to the question we raised at the beginning: Does replication occur in the context of communication? Now we can reframe the question as follows: Is it to be heredity or replication in cultural evolution? Is Prusiner correct about the need to invoke a new kind of replicator to explain the case? Shall we vote for cultural selectionism or memetics? These are all questions that dance around each other, as somewhat different ways of looking at the same basic issue: How are we to account for communicative events that spread cultural information though social groups? Are genes enough or do we add memes to the mix: *Cui impello?*

The identification of new replicators like memes would fundamentally change how we view the world. Whole new arenas of evolutionary dynamics open up if a novel replicator is at work. And consider what it would mean for our view of communication: Are we in control of what we say and do, or are we victims of brain parasites? Who, or what, benefits from our own thoughts? The implications of the meme question for human agency and free will are profound indeed.

RESCUING MEMES

How then can we make any further progress on our quest to find memes? First, memetics will prove to be the best explanation for culture if it leads to novel predictions that are proven to be correct. The very specific claim that cultural similarity is due to replication is a hypothesis from which novel, testable predictions can be made. As yet, however, memeticists can point to very few, if any, established or confirmed results unique to their explanatory approach. But this situation could change if the model of meme replication I provide later proves scientifically fruitful.

Second, memetics would be the best explanation of culture if it could be shown to account for a wider variety of phenomena than other theories of culture, even though that explanation involved a somewhat larger number of principles. This would render moot the parsimony argument—the criticism that memetics is just too complicated to be right. The view of memes I develop in this book has this quality because it also has implications for psychology and communication.

Third, memetics would be the best explanation of culture if cultural entities having the characteristics of replicators could be shown to exist. Such a demonstration would dominate any other consideration. Unfortunately this knock-'em-dead argument seems the least likely prospect at present, simply because we don't currently have a testable model of what a meme might be.

Still, we can ask whether things like memes are really plausible. Could there be something acting independently of genes that can influence the likelihood they will be duplicated, operating through social communication? If we find such a possibility is not very likely, then we have effectively answered our question in the negative: Memes probably don't exist. In fact, as we will see, finding a model of communication that is compatible with a replication event is not easy. However, it is possible, and proves to have a number of important repercussions of its own, although it requires some new thinking about how communication occurs.

Memes *can* be rescued from the Airy-Fairy-Land in which they now exist. Indeed I argue that they have a very particular kind of real-world existence. This conception of memes also happens to be at variance with previous views, as it probably has to be to open up new avenues for exploration. I can't *prove* my claims here, since they are empirical ones that will require support from actual investigation. But, as the Chinese say, the longest journey begins with but a single step.

The task at hand is to clarify what a meme might be and how it works. If we can determine how memes must perform the crucial evolutionary tasks of replicating, mutating, and being selected, we will have struck gold. Then we can really begin to harness the power of memes—and perhaps take back control over our own thought processes.

Chapter Three

ADDING ROOMS TO DARWIN'S HOUSE

All life evolves by the differential survival of replicating entities.
— *Richard Dawkins*

On his return to his native England from a round-the-world trip aboard the *Beagle,* Charles Darwin settled at Down House, not far from London. There, he lived for forty years as a Victorian country gentleman, writing the works that would make him famous. As the site for such important events, the house is a national treasure, and has been recently restored by English Heritage to something like the condition Darwin left it in. In this house, his wife, the former Emma Wedgwood, bore Darwin ten children. This made daily life somewhat cramped, but no significant extension was made to the family estate.

Since Darwin's death, however, quite another kind of fate has befallen his *intellectual* property. Theoretical carpenters have been tacking on rooms left and right to Darwin's scholarly mansion. Whole new wings have been added to the main "biology" hall, as evolutionary approaches to chemistry, economics, and computer science have become popular. An entire "upstairs" has also sprung up in the form of Darwinian humanities. Even cosmologists have found inspiration in Darwinian thinking, although making cosmology evolutionary has meant not just expanding the Darwinian architecture to the entire universe, but postulating the existence of multiple universes, competing with one another for survival on a cosmic scale.

What I have in mind here entails a much more modest aim: to add a "culture room" to the house that Darwin built. But if we're going to explain cultural change as a process analogous to biological evolution, complete with replicators, we had better be sure that Darwinian principles apply to more than just genes. At the same time, as I intend to show in this chapter, Darwinism is too widely invoked nowadays. Before I elucidate the grounds on which Darwinism can properly be broadened to include phenomena like memes, a bit of history is warranted to help us understand just where Darwinism stands right now.

EXPANDING THE LIVING ROOM

By the 1870s, Gregor Mendel, the curious monk from Moravia, had conclusively shown through crossbreeding how the pea plants in his garden (and by extension any other organism) inherit certain characteristics. A single genetic mutation could make a visible change in the resulting organism: Smooth-skinned peas became wrinkled, green peas yellow. But in the first decades of the twentieth century, the statistical work of Francis Galton, the gentleman scientist and explorer (and Darwin's nephew), indicated that tiny variations in a trait such as height could also be produced through genetic inheritance. In such cases, mutations produced only incremental changes in an individual's observable qualities—in their phenotype—not the rather manifest categorical changes in character that Mendel's work suggested. Mendelism had to be reconciled with so-called quantitative or Galtonian inheritance before evolutionary theory could be applied to both classes of phenotypic traits—to both the colors of peas and the height of humans. This unification was achieved in the 1930s when it was realized that there could be multiple genes underlying the manifestation of a given trait, each of which make an additional contribution. So two height-inducing genes might make you two inches taller, three such genes three inches taller. In this fashion, microevolutionary changes in phenotypes could be produced, resulting in the continuous distribution of height seen in a population. This continuous "blending" of inherited attributes was produced by variation in genes, each of which was nevertheless particulate, just as Mendel had imagined. This insight, due to Ronald Fisher, a biologist at Cambridge, and developed into the formalized models of population genetics by J.B.S. Haldane and Sewall Wright, has been called the Neo-Darwinian Synthesis. This revitalized Darwinism, now generalized to include both qualitative and quantitative characters, has formed the cornerstone of contemporary evolutionary theory for the last half century and has stood firmly against many assaults over that time, from both inside and outside the academy.

Other, more recent, attacks on Darwinism have led to less important generalizations of the theory. For example, in 1972, two eminent paleontologists, Niles Eldredge and Stephen Jay Gould, came up with the idea that the pace of evolution exhibited a large-scale pattern they called "punctuated equilibrium." Inspired by the fossil record, which showed long periods of stagnation in what species looked like, they suggested that the formation of new species usually took place rapidly, and then genetic evolution would settle down again for a restful period until some new challenge spurred another burst of change. This view was widely touted, especially by creationists intent on showing that famous Harvard-based evolutionary biologists didn't believe in Darwinism anymore. But, of course, Gould was quick to quash this mistaken impression by affirming his allegiance to Darwin. Darwin's own emphasis on gradual phenotypic change in the evolutionary record was largely a historical peccadillo, a response to his own historical circumstances. Darwin needed to distance himself from the predominant ecclesiastical opinion of his day that the universe was created in six days and so emphasized that evolution took a long time to occur. But gradualism is not a fundamental characteristic of the evolutionary process, so the suggestion that sometimes evolution may proceed in jerks does not affect the utility of Darwin's basic ideas.

More recently, Stuart Kaufmann, a fecund thinker with primary allegiance to complexity theory, has championed the notion that natural selection is not the only cause of adaptive design. Perhaps equally important, in his mind, is the order or structure that consistently appears each generation through the operation of physical and chemical constraints on what can develop. As a result, many aspects of an organism's form don't have to be coded for by genes but can arise through more fundamental principles that apply to the construction of any physical object, not just living things. In particular, the rows of seeds spiraling up the sides of pinecones, to left and right, will always come in adjacent numbers from what mathematicians call the Fibonnacci series. In this series, each term is the sum of the previous two in the sequence (that is, 1, 1, 2, 3, 5, 8, 13, 21, and so on). So a pinecone may have 8 rows spiraling left and 13 going right. It is believed that this pattern is generated by microscopic events at the tips of the growing tissues that eventually form these seeds. Many of the beauties visible in the complex forms of organisms may, in fact, arise as the coincidental consequence of many bits and pieces using simple rules for interaction. Such order essentially comes "for free" from a biological perspective. Kaufmann saw this kind of unprescribed emergence of global structure through the self-organization of parts as contrary to Darwinian principles—at first. However, he later came to see his work as falling within the Darwinian fold as well. Even if natural selection is not solely responsible for the form organisms take at the end of development, it is still the primary determinant of the direction lineages of organisms trace over time. Since these chains of descent are the true object of evolutionary change, natural selection can still be called the most important *biological* agent, just as Darwin expected.

These episodes were not fundamental attacks on Darwinism because they didn't threaten the key claims of his theory. Darwin himself recognized what was crucial to its survival and eventual establishment. He emphasized that "if it could be proved that any part of the structure of any one species had been formed for the exclusive good of another species, it would annihilate my theory, for such could not have been produced through natural selection." So Darwin was well aware of the problem of altruism—one organism coming to the aid of another at some cost to itself—and described it as that "one special difficulty, which at first appeared to me insuperable, and actually fatal to the whole theory." Darwin was referring in particular to the social insects, which exhibit an extreme form of altruism. Already in Darwin's day it was common knowledge that hymenopteran colonies (honeybees, wasps, bumblebees, and ants) usually consist of one reproducing queen and a multitude of sterile workers. This means that most of the members of the group forgo completely the genetic imperative to reproduce; instead these sterile individuals cooperatively help the fertile animals to raise the group's offspring. This sort of altruistic social behavior is also characterized by another trait: At least two generations overlap in life stages capable of contributing to colony labor, so that the offspring can assist their parents during part of their life cycle. Explaining the abandonment of reproduction by worker castes in some social insects was a crucial dilemma to which Darwin devoted an entire chapter in his *Origin of Species*.

It was not until the 1960s, however, that evolutionary theory could properly explain a significant part of all the cooperation occurring in nature. A revolutionary insight by William D. Hamilton (the recently lamented Royal Society Research Professor of Zoology at Oxford) was so important that it spawned an entire field of research in biology about social life—sociobiology (discussed in Chapter 2). Sociobiology represents, in effect, a major generalization of Darwinism in terms of expanding its scope to encompass a whole new range of phenomena. Why did Hamilton's theory have such an impact on modern evolutionary biology?

Hamilton showed that altruism, the conundrum dreaded by Darwin, could be explained, and quite simply, if we would just shift our point of view down from the level of organisms to that of the genes. By so doing, we could recognize that related animals are likely to share many genes. Not only that, but if the gene copies could evolve ways to help each other out, they might benefit in the only terms of relevance to evolution, that of gaining representation in the next generation. Although one copy of a gene responsible for some altruistic act might suffer if the organism sheltering it dies as a consequence, offsetting reproductive benefits might be conferred on other copies of the gene in the organisms toward which the act is directed. The result could still be a net increase in the number of such genes surviving into the future.

In retrospect, of course, Hamilton's insight seems obvious enough: He simply recognized that genes exist in multiple copies within populations and should evolve to exploit the possibility of collaborating whenever circumstances permit. Hamilton simply looked past the skin (or exoskeleton) of the individual organism to see that natural selection would favor strategies of mutual assistance between related genes, regardless of where these brethren were located.

So even though most of the ants in some colonies are not themselves reproductive but provide defensive services to the group, this makes evolutionary sense. This is because the soldiers are more related to their siblings than they would be to their own offspring (if they had them). The gene that makes soldiers into soldiers rather than reproductives can survive because groups with this social structure will out-reproduce their competitors. "Hamilton's rule," which summarizes his insight formally, suggests that an individual that profits from the altruistic act must share a higher proportion of genes with the benefactor than the cost-benefit ratio this act imposes on them. If this inequality holds, then the genes behind such altruism can evolve.

Of course, to make this all work, several tricks must accompany the desire to assist related genes: The genes must be able to recognize each other reliably (which can be difficult because they are typically packed deep inside these sacks of substance called organisms) and simultaneously spawn the appropriate behavioral response when the recognition bell goes off. This can be a lot for one gene to manage, so it may be necessary for cooperative gene-complexes to evolve before cooperative behavior will be seen in a species. But Hamilton's insight—the "gene's-eye view" of nature—led to the eventual formulation of Darwinism that now prevails.

A DARWINIAN UNIVERSE?

The successful application of this newly generalized Darwinism to social problems spawned a massive outbreak of interest in evolutionary thought. In the last 25 years or so, Darwinian theory has been applied to a dazzling array of problems. Testimony to the apparent utility of this theory comes from the recent burgeoning of fields such as evolutionary ecology, evolutionary economics, evolutionary psychology, evolutionary linguistics and literary theory, evolutionary epistemology, evolutionary computational science, evolutionary medicine and psychiatry—even evolutionary chemistry and evolutionary cosmology.

What unifies these approaches, despite being concerned with such varied topics? What justifies calling all of these theories Darwinian? In fact, Darwinian theory—both in its original formulation by Darwin himself and in its twentieth-century "synthetic" guise—is limited in several respects and hence not really fit to serve as the foundation for such varied use. Why is this?

First, because Darwin's theory, as the title of his original work suggests, is a theory of evolution *by means of natural selection*. Natural selection is recognized as the driving force behind directional change in evolution, the process central to the production of adaptations. But it doesn't adequately cover changes in the frequencies of genes by other means, such as randomizing factors like genetic drift. However, this deficit has been remedied in large part by the Neo-Darwinian Synthesis of the 1930s, which formalized the relationship between the various kinds of influences on genetic change, both random and directional. So this criticism no longer troubles evolutionists.

Second, Darwin's is a theory about the evolution of *organisms*. However, his work doesn't tell us how to identify an organism. Organisms may appear obvious, but that is only because the common bias has been to treat the higher vertebrates and especially mammals as the paradigm organisms, despite their rarity and atypicality in the kingdom of living things. Mammals like us are well-organized, complex conglomerations of cells contained in a sack of skin; death causes the disintegration of this composite. But much more common in the kingdom of living things are "lower" animals like clonal organisms, plants, bacteria, and viruses. These organisms have very different ways of going about evolutionary processes such as reproducing, finding the material necessities for continued existence, and so on. Evolution has even produced "in-between" kinds of creatures like colonial protists, cells that sometimes aggregate but other times disassemble and go about their business individually. Darwin presented exemplary cases of organismal evolutionary analysis in *The Origin of Species,* but he left vague the intended, let alone actual, scope of the theory. It remains uncertain whether neo-Darwinian theory—even as developed during the twentieth century to include population genetics and the conceptual apparatus of the evolutionary synthesis—fully applies to all branches of the tree of life.

Third, Darwin's theory is not only about organisms but suggests that selection operates *only at the organismal level.* However, biology nowadays concerns many levels of organization—from molecules to cells, organisms, populations, species, and even ecological communities. As noted, the recent revolution in evolutionary thought, centered around sociobiology, can be seen as a switch to the view that the gene, not the organism, is the fundamental level of organization and selection. But many evolutionists suspect that selection may occur at more than one level of organization simultaneously.

Thus, while we see that applications of "Darwinian" thought abound, it remains questionable whether these applications are really

Darwinian at all. How can a theory devised to deal with organisms be legitimately extended to the immune system, the competitive strategies of corporations, or the life history of stars, as recent developments in the Darwinian "movement" have alleged? This question is central to enlarging the scope of evolutionary theory to the problem of human culture and communication—and ultimately of how we think. For a general *evolutionary* science of culture and psychology, it is crucial that we establish a solid foundation for the jump from biological replicators to something as controversial as memes.

REPLICATORS, INTERACTORS, AND LINEAGES

How might evolutionary theory be legitimately generalized from the case of biology to culture? The "gene's-eye view" of evolution described earlier isn't a good enough foundation for understanding cultural evolution because it is still limited to one level—the genes. So a straightforward application to culture is legitimate only if cultural change is reducible to evolution in genes, in which case the social sciences become superfluous. This hardly seems likely.

One way of making evolutionary theory more comprehensive, which Richard Dawkins has dubbed Universal Darwinism, comes from a recognition that there are two functions that agents of the evolutionary process can play. The Universal Darwinist strategy looks at the roles that genotypes and phenotypes (the observable expressions of genotypes) play in the evolutionary process. It then subtracts the features of these agents that appear peculiar to genes and organisms. Such features tend to be associated with the particular material from which genes and organisms are constructed and the particular mechanisms by which DNA replicates or by which sexual or asexual organisms reproduce. From this now more abstract perspective, the evolutionary process is regarded as having two components, replication and selective interaction, and these are defined for any entities performing those functions. Dawkins's strategy, in effect, is to abstract from the matter and structure of the concrete mechanisms of genes and organisms to yield a theory specified solely in terms of functions.

Richard Dawkins has given the name "replicator" to any type of information with the characteristics required to form the foundation of an evolving system. He suggests that examples include not only Mendelian genes but also the computer programs called viruses, and memes. Replicators are thus the basic units of evolutionary processes, and their basic function is to pass on their structure largely intact through copying.

With this identification of replicators, Dawkins essentially argues that Darwinian evolution can occur anywhere—in genes, in culture, on Mars, and all at the same time—as long as replicators are at work in each evolutionary process. For Dawkins (as the quote at the head of this chapter suggests), evolution is all about *replication.* In effect, he asserts that *Darwinism can be well and truly generalized—indeed universalized—by identifying replication as the central phenomenon underlying any evolutionary process.* It is based on this argument that Dawkins originally introduced the notion of a meme, as a hypothetical example of a different replicator associated with culture that could evolve just like a gene. It is through the concept of replication that Dawkins believes evolutionary theory can be properly extended to cover cultural evolution.

Despite the central role of the replicator concept in evolutionary biology, insufficient attention has been devoted to defining just what replication is. Dawkins himself is not explicit on this point, and philosophers of biology have yet to reach a consensus. It seems that replication is just what replicators *do.* In case of any uncertainty, the tendency is just to point to DNA as an example of how it is done. Anything else worthy of the name replicator must be "like" DNA.

This isn't good enough, of course, if one's goal is to come up with more broadly applicable criteria that can be rigorously applied to candidate replicators. We will have to do considerable work in the rest of this chapter to lay out exactly how replication works, so we can then apply this knowledge to the novel case of memes.

Let me get a start on this problem by defining four minimal conditions for replication. Generally conceived, replication is a relationship between a copy and some source exhibiting the following characteristics:

— *Causation:* The source must be causally involved in the production of the copy
— *Similarity:* The copy must be like its source in relevant respects
— *Information transfer:* The process that generates the copy must

obtain the information that makes the copy similar to its source from that same source, and
— *Duplication:* During the process, one entity must give rise to two (or more)

The first condition says that the original replicator must participate in the process that results in the appearance of its copy, a second replicator. This concept of replication is neutral about how much control a replicator has over the replication process. A replicator must be responsible, given the context, for making the process happen in some sense. But the replicator need not be *entirely* responsible. Many other factors can be involved, and conditions may have to be "just right."

The second condition is simply that the copy and its original must be of the same class of entities, so that they can rightfully be called similar to one another. The reference to "relevant respects" generally means with respect to those qualities that allow the replicator to maintain its generative ability and those aspects that produce the similarity of the copy to its source. At the same time, this clause allows for the possibility, indeed the necessity, of occasional mutations in what gets copied. Such variations are the fuel on which the evolutionary process works.

This leads to the third condition, which is necessary in order to exclude from our definition of replication the possibility that copies of a source appear over and over but as the result of activity by something besides a replicator. In a way, the "information transfer" requirement is that everything necessary for replication is present at the point where replication is taking place. Replication must not refer to, or rely upon, information brought in from outside. The process has to be "local," or encapsulated in an informational sense. No instructions can fly in to rescue the situation, like a hand of God descending from the heavens. Most important, it is this transfer of information from one generation of replicators to the next that defines the relationship of descent between them. The encapsulation of a replication event as a node in a chain of similar events also establishes the existence of an evolutionary lineage.

Duplication, the fourth condition, is also an intrinsic part of what we mean by replication. The word "replication" has a sense in which something must be repeated. Replicators multiply. They can do this only if their number increases. So duplication results in a changing number of the relevant kind of entity being counted. The source and copy are individuals that form part of a population of alternative forms, so that by being duplicated, the relative frequency of the type they both represent in that population changes. Such changes in relative frequency are the substance of evolutionary change.

The duplication criterion also has the salutary effect of taking the replicator notion out of the realm of the abstract. The first three conditions merely identify replicators by their primary function: copying themselves. Duplication, on the other hand, forces investigators to find two physical entities and determine how they got there.

The process by which the quintessential replicator, DNA, accomplishes the feat of producing copies of itself obviously fulfills all four of these criteria in spades, as can be seen from even the simplest description. Gene products read off from DNA spur the molecule to unfold, allowing replication to begin. Just before cell division, enzymes unzip the two strands of the molecule, exposing the bases to the cell's internal soup of raw materials. Another enzyme, also produced by DNA and known as a DNA polymerase, then marches along each of the two strands, triggering each base to pair up with a complementary base from the soup. This is how DNA *causes* itself to be duplicated. Through this step-by-step process, the polymerase copies the genetic information and creates two new, double-stranded DNA molecules. Since both molecules have the same structure and sequence as the original, the similarity condition is fulfilled. Since the chemical bits of DNA that can bond together are specific pairs—with every cytosine bound only to a complementary guanine, and every adenine only to a thymine—the sequence of bases attached to the newly exposed strand depends on which ones are present on the other side of the chain. This is information being passed—inherited—from the source strand to the copy being made. And finally, at the end of the process, there are two molecules where before there was only one: duplication.

What other kinds of phenomena fall into this category? Some examples will help us to firm up the four conditions for replication, which remain somewhat abstract at the moment.

You might think, for instance, that the first-class British postage stamp, with its picture of the Queen on the front side, qualifies as a replication of the Queen herself according to our criteria: Elizabeth Windsor caused the stamp to have that particular face on it by being a queen rather than a commoner (information transfer); there is a resemblance between the face on the stamp and that of the person (similarity); and the Queen perhaps signed a decree to the effect that such a stamp should be manufactured (causation). And there is duplication as well: The stamp and Queen exist side by side; the Queen's visage is not harmed by the taking of the photo that led to the stamp's creation.

But this isn't really an example of something we would want to call replication because stamps don't go on to make other stamps. There's no evolutionary lineage here. So what's wrong with our analysis of Queen "replication" so far? Perhaps the conditions are too general, only necessary but not sufficient to distinguish true instances of what we are after? At minimum, we need to firm up what we mean by similarity. Resemblance isn't enough. A three-dimensional object such as the Queen and a two-dimensional object such as a picture of her are not really the same in the sense that we mean. Similarity requires a closer physical relationship. An important implication of this perspective is that replication is not founded on a notion of similarity, but of causal linkages: The notion of linkage, of *lineage,* becomes of paramount importance.

In addition, the notion of cause is too broadly interpreted in this example as well: The Queen, even if she did sign an edict, didn't really *cause* the stamp to be made in the sense we are using the term, in that she didn't manufacture it herself in any meaningful sense—she didn't turn the handle of the printing press, for example. No lineage of replication events follows from the decree because a causal process involving paper and ink and machines produces each stamp. So the notion of *proximate* cause gets closer to what true replication involves. The face information doesn't go from one stamp to the other, nor does it go directly from the Queen's face to the stamp either. So on all of these counts, the Queen does not replicate every time the Royal Mail emblazons her face on a sheet of paper.

Let's investigate another technological example involving a piece of paper, this time sitting on an office worker's desk. On the paper is a graph of third-quarter sales figures. Say the department head calls for a meeting about sales. The paper must be copied but is in no sense able to do all the work of copying itself. Rather the worker must bring the piece of paper to the copier and place it inside the machine. The paper itself does have some limited power to make this happen by virtue of what it has printed on it: The graph is much more likely to get photocopied in this situation than a poem by a third-grader. Then, given the existence of the surrounding machinery—and the obliging human touch of punching the button—the piece of paper is copied.

Is the piece of paper a replicator, then? It seems to fulfill all of the criteria. The printing on the paper serves as a template of information—the relevant information—for what should appear in the duplicate (causation). The copy, if the copier is any good, should also be very similar to the original (similarity). What is impressed upon the blank sheet should be the information derived from the original (our inheritance relationship fulfilled). And at the end of the process, both original and copy are there for inspection, side by side (duplication). The printed information can also be further reproduced, becoming a lineage, as required of an evolutionary replication process. Any mutations, such as a smudge on the glass, will be faithfully reproduced in subsequent copies. So evolution in the message can occur. For example, if the smudge appears in a spot that leads to misinterpretation of a word, it may reduce the likelihood of further reproduction—a form of selection.

It's true that we still have to take the notion of causation a bit loosely: The paper, for example, isn't the proximate cause—the person punching the copier button is that. This is why some say that replication should be limited to "direct" causation, as we noted in the example of the stamp above. Still, the paper is relatively inert, a good store of information; the copier is a protected environment for replication; there is good similarity of content and substrate between the source and copy. Dawkins himself considers this case and concludes that the piece of paper—or more precisely, the information printed on it—*is* a replicator. So replicators can be artifactual in nature, not just biological.

Perhaps the most important quality a replicator has, from an evolutionary point of view, is its "power"—a quality originally identified, again, by Richard Dawkins. He argued that replicators, besides merely copying themselves, also exert some additional positive influence on the probability of their own replication. Those replicators with greater power will be favored over those that have less of a positive effect on this probability. More copies of these replicators get made compared to those without such power and thus bias the statistics of reproduction in their favor. Power is what determines who benefits—or which replicator wins the battle to control the events over which different replicators, like genes and memes, might be competing.

There are many ways in which replicators can ensure that they not only survive but prosper in the next generation. Viruses produce structures that fool other organisms into treating the viral genes as their own and make them copy them willy-nilly; genes in multicellular bodies produce structures that increase the probability the organism will reproduce and thus likely pass on copies of these genes; and organisms can engage in behaviors that increase the probability that a sibling will reproduce and so pass on copies of the gene sequences that both organisms inherited from their parents. Indeed the ingenious ways in which replicators ensure they get copied again and again is one of the best signs of natural selection at work.

Dawkins also identified a second class of agent important to thinking about a universal evolutionary process. He originally called them "vehicles," but most people nowadays prefer to call them "interactors." An interactor is another technical but powerful concept in contemporary evolutionary biology. Interactors are paired with replicators, both conceptually and causally, because replicators *produce* interactors. For example, genes (replicators) produce organisms (interactors). A person, like any other organism, is the temporary storehouse for an aggregate of genes that will come together only once. That's what makes a person an individual. So interactors, unlike replicators, don't repeat themselves; they don't form lineages. Interactors such as organisms multiply the collective capabilities of genes (such as providing the ability to get around in the environment) and obey the genes' will—a will that primarily consists of getting reproduced as often as possible. These qualities make it worthwhile for genes to collaborate in the construction of interactors.

Interactors, by virtue of interacting with the environment, can either foster or cause the demise of the replicators that are housed inside them. Interactors are what get selected, by virtue of what they do—and what they do replicators had a hand in producing. Interactors can be distinguished from replicators (of any sort) by their inability to pass on their structure directly. People can't sprout clones of themselves from out of their foreheads like Zeus, for instance. The structure of organisms is only indirectly transmitted to future generations through their role in perpetuating the genes that caused their construction in the first place. Replicators are directly involved in replicating themselves but interact with their environments only indirectly, through the interactors they construct.

Of these two basic evolutionary concepts, the idea of a replicator is more fundamental, because interactors are not necessary components of reproduction. At the beginning of life on Earth, for example, replicators were very basic and had to interact with the environment on their own behalf. In effect, they played both major evolutionary roles, of replicator *and* interactor. It's likely that specialized structures for interaction simply hadn't been invented yet. But since the job of replicators is to hold on tight to a message and that of interactors is to interact and perhaps be changed by that interaction, the two jobs are in conflict. It's therefore going to be to a replicator's advantage to evolve the ability to produce specialized interactors. Then the replicator can remain relatively inert and protected, while its minion, the interactor, undertakes the dirty, dangerous job of getting around in the world while staying in one piece. So, as material entities, replicators are also the driving force behind the evolutionary process, generally conceived.

It is important to remember that evolution is a process, not just a collection of things like replicators and interactors. From this perspective, perhaps the most important evolutionary concept is *lineage,* an identity that persists indefinitely through time—either in the same or an altered state (because mutations can occur)—as a result of a replication sequence.

A chain of replication events constitutes a lineage. These events are causally and physically linked by the replication relationship in each instance. Replication leads to chains of similar events in which those participating in replication do all the work themselves, rather than being caused by processes operating at other levels of organization. In a lineage, each instance is related by the fact that a replicator of the same type produced it. The events can thus be linked together as representatives of the same thing. This relationship between instances or events is what biologists call *descent.*

Lineages cannot simply reflect the persistence of something that happened long before. The recurring features defining the lineage must be reproduced from scratch each time they reappear. So, for example, the younger generation cannot wear clothes similar to those of their parents simply because the clothes themselves have been passed down. Rather, to constitute a case of cultural replication and another link in a true lineage, similar pieces of apparel have to be manufactured and chosen as a cultural "statement" by the young.

Another example makes this point about informational inheritance clear. Consider a number of tape recorders, placed in a line at some distance from one another. Each recorder has been set up so that "hearing" a particular tune play will cause it to play the same tune, pre-recorded on its tape. Someone punches the play button of the first machine. The sound it makes is sufficiently loud so that the next one in line can "hear" it and is stimulated to reproduce the tune. This, of course, causes the next one in line to play the same sequence of sounds as well, and so on. Without knowledge that these machines include stored information in the form of pre-recorded music, we might be led to the conclusion that this tune has replicated itself through a chain of replication events involving the different machines. But, in fact, information is not inherited from machine to machine: What makes the tune "survive" down the line is a set of conditions that held true before the first recorder ever disturbed the air. These conditions persist throughout the experiment, and nothing is created except a set of signals produced by whatever program has been attached to the recorders to make them sensitive to this stimulus.

Lineages are continuous, unbroken sequences linking replicators past to replicators of the future. Your family tree is underpinned by a number of gene lineages, for instance. Lineages are the true substantive focus of evolution because it is lineages that allow the cumulative effects of selection to produce adaptations over time. Notice that a lineage is restricted to the case of replication. Interactors like organisms do not constitute elements of lineages; even though they may be the bearers of adaptations, those adaptations did not evolve to benefit organisms but rather the replicators that make them. It is consistent pressure on similar interactors that allows small modifications to accumulate in the suite of replicators responsible for those modifications. Each new interactor is not only viable, but a superior type for that environment. By making use of the material generated by previous generations of selection, increasingly complex interactors can be produced through the accumulation of adaptations.

Universal Darwinism thus suggests there are only two major agents for evolutionary theory to consider. These actors, replicators and interactors, are identified by their functions, regardless of whether we are talking about the evolution of molecules or universes. These agents dance back and forth between replicator and interactor forms through generations to produce lineages, which are the record of their activity over time.

Richard Dawkins argues that "evolution is the external and visible manifestation of the differential survival of alternative *replicators*." But, in fact, as we saw in Chapter 2, you can have evolution without replicators. Heredity alone is sufficient to generate an evolutionary process. The last several sections of this chapter have thus far been devoted to replication as the foundation of an evolutionary process. However, the scope of evolution is broader than envisioned by Dawkins, whose amendment of Darwin does not produce a truly universal Darwinism, as he hoped. Everything Dawkins claimed is still valid; it's just that the reach of the Dawkinsian program is more limited than its author believes.

Dawkins's Universal Darwinism is not the only scheme that has been developed for generalizing evolutionary theory beyond the classic Darwinism of natural selection among well-defined organisms. Richard Lewontin of Harvard, another of the most distinguished and productive of contemporary evolutionary biologists, suggested a second strategy for generalization early in his career. In a nutshell, his idea is simply to replace the word "organism" wherever it appears in Darwin's *Origin of Species* with the word "individual." If Darwin's book still makes sense, then the theory is not specific to organisms but can refer to anything that can properly be called an individual. In effect, although Darwin stated his principles in terms of organisms, they can be interpreted more generally than the theory's originator intended. By rereading Darwin in this way, Lewontin formalized his insight in terms of three principles, which he claimed are necessary and sufficient conditions for evolution to occur. This led him to the criteria we presented earlier—heredity, variation, and fitness—as those that successfully identify any kind of individual capable of evolving.

Lewontin's strategy for generalizing Darwinism can be characterized as "levels abstraction" because it identifies different levels of organization at which individuals—the locus of selection—can be identified. Only those properties of organisms *essential* to their role in evolution are used to define the entities ("individuals") that satisfy that role. It is interesting to note that individuals defined at levels *above* the organism, such as species, will not have many of the qualities of organisms, which are specific to that physical scale. This is because organisms are more tightly organized—and the fates of the genes present in them more interdependent—than is the case for groups, whose membership can fluctuate. Genes simply don't sweep in and out of an organism like organisms can from a group. As a result, adaptations have evolved that reflect the long-term selection pressure on cooperating groups of genes.

Lewontin's three principles crosscut Dawkins's Universal Darwinism: Where Lewontin generalizes over levels, Dawkins generalizes over roles. Still, the two methods of abstracting Darwinism are compatible. You can think of Lewontin's conditions for individuality as the qualities that Dawkins's interactors—the individual "units" of selection—must have to count as the foundation of an evolutionary process. However, these two perspectives remain incompatible because Dawkins's view is restricted to replication, while Lewontin's is based only on the principle of heredity. They are therefore different perspectives one can take on the generalized evolutionary process, with Lewontin's being "more universal" than Dawkins's.

Lewontin's theory is certainly the most general one yet developed, and one that can be fruitfully applied to culture. As we have seen, cultural selectionists suggest we should take Lewontin's line on culture, that cultural evolution need only exhibit heredity. Others—memeticists, essentially—claim the Dawkinsian approach to Darwinism is required (because memes are replicators).

What distinguishes memetics from other evolutionary approaches to culture is its bold central claim: Memetics argues that cultural evolution is the result of a replicator called the meme. As discussed earlier, the prominent contemporary memeticists Dawkins, Dennett, and Blackmore identify the issue of replication as central to memetics, which they should do to be consistently different from cultural selectionists. So we can legitimately say that the difference between memetics and the other schools of evolutionary thought on culture boils down to the distinction between replication and inheritance. Inheritance alone is *not,* in fact, the issue dividing memetics from alternative explanations of cultural evolution, including cultural selectionism. The specific claim of memetics is that an entity, called the meme, exists that fulfills all the criteria of a cultural replicator: causation, similarity, information transfer, and duplication. This is, in fact, the only claim that is unique to memetics—the claim that defines it and separates it from its rivals.

We are therefore back to our problem of identifying how such a replicator would work. Luckily there is yet more we can learn about replication—and hence about the nature of memes as replicators—by looking at another area of research on this crucial phenomenon.

THE REPLICATION REACTION

Defining the general category of "replicator," as Dawkins's Universal Darwinism has done, permits the creation of a formal theory of replicators. This theory is another important source of information about the nature of replication that can influence our search for memes. Memes will have to satisfy the constraints of any such general theory to be considered legitimate members of the replicator "club."

Constructing a general theory of replication has been the project of what can be called the Vienna School of evolutionary theorists—primarily people surrounding the Nobel laureate Manfred Eigen, a physical chemist at the University of Vienna. Replicator Theory was developed by this group largely to model the origin of life and hence the emergence of the first replicators from a dead universe. It isn't surprising that the minimum—and therefore most general—requirements for defining replicators should come from a research program addressing the problem of how to recognize the first primitive signs of self-reproduction. However, as the developers of Replicator Theory themselves emphasize, the "chemistry" of replicator "reactions" in the theory is really quite stylized. Replicator Theory has been applied to all kinds of phenomena with equal alacrity and ease. Indeed it has already been adapted to deal with the combination of symbol strings in a kind of artificial chemistry. Replicator Theory is, in fact, the most general theory of the evolutionary process we have: It can account for predator-prey interactions and purely symbolic intercourse. Its application to super-molecular replicators like memes should therefore be assured.

Replicator Theory begins by noting that when replication occurs, two bits of organized matter appear where before there was only one—an example of the duplication principle in action. This is an event that typically must be accompanied by the input of energy, either because the original bit of matter is not consumed by the reaction and remains intact or simply because a second novel bit of structure gets added to the mix. To accomplish duplication in either of these ways, we need to add some energy to the system to get things going. So replication is almost by definition a negentropic event. That is, it involves a local reduction in the amount of disorder in physical matter. After all, something new—another replicator, which almost certainly entails some additional complexity of structure—is being created.

Often, however, replicators won't have the ability to reproduce and simultaneously manage the allocation of energy during the reaction. This is especially true if they are rather simple. What this means is that the replication event will often require the assistance of a catalyst. Catalysts are a special class of molecules with the ability to speed up a chemical reaction without themselves being consumed or modified by it.

Further, if all of these events are to come off efficiently and repeatedly in reasonable time, they will also require resources and replicators to be locally concentrated—that is, the participants in the reaction must be together in some kind of container to protect them from environmental distractions and noise. For example, DNA has secluded itself inside a double envelope of nucleus and cell protoplasm, both of which enclose these highly stable molecules behind walls that shield the replication reaction from interference. DNA even recruits assistants such as messenger RNA and various proteins into this safeguarded domain at the crucial moments to midwife its birthing labors.

Generally speaking, a replication reaction begins when an enzyme and its substrate (the specific chemical with which it interacts) come into contact. They may bond together to form a temporarily more complex whole before dissociating again to produce the end products, which are typically different combinations of atoms than went into the process. This reaction generally proceeds at a constant rate, converting new bits of raw materials into reaction products. As the substrate is used up, these products may accumulate locally. In some cases, this buildup of by-products (waste) or end products (the desired ones) will begin to cause the reaction to reverse itself or at least to slow down, perhaps due to a paucity of freely circulating components that haven't already been bound into new replicators. Many such reactions are also sensitive to various local conditions, such as the acidity and temperature of the microenvironment.

Small structural differences in chemicals can cause gross differences in levels or kinds of activity. So several classes of catalytic events resulting in replication can be distinguished on organizational grounds. Each of these types of reaction has been identified in nature.

The first and simplest form is autocatalysis. This happens when a replicator serves as its own catalyst. In this kind of reaction, one replicator, thanks to the availability of an energy supply, can convert some raw materials into a second copy of itself, perhaps together with some by-products, as waste. So the simplest form of the replicator equation can be expressed as follows:

$$1 \text{ replicator} + \text{materials} + \text{energy} \rightarrow 2 \text{ replicators} + \text{waste}$$

Such a reaction can realistically be called self-replication because it is the closest approximation to a case of "going it alone." This is typically a single-step operation, and its dynamics can be described by what is called a "first-order replicator equation." The *kinetic* equation associated with this dynamic—which specifies the actual levels of energy and real timing of the event—describes a Malthusian growth process. In Malthusian or exponential growth, births are a linear function of density (or the existing number of similar replicators). Each existing replicator gives birth to just one other one each time the process is iterated. Thus the total number of replicators grows at a constant rate over time. In effect, the per-capita birth rate for new replicators remains constant, as long as all the necessary materials are kept nearby in quantity.

This regime can be called "survival of the fittest," because when two replicators are at work in the same microenvironment, the one with the faster growth rate will come to dominate the slower-growing replicator, even in a system with freely available resources. In effect, the slowpoke is selected against by time itself. It simply gets out-duplicated. The pattern of evolution in such a case is simple. Everything depends on the initial proportion of each replicator in the population and its rate of reproduction relative to its competitors. We can think of this as a race between two replicators: one that is slow and steady, like the turtle in the child's fairy tale, and another that is fast but tires rapidly, like the hare. Unless the turtle has a great advantage in terms of a lead at the starting block, it will lose the race to the faster-reproducing rabbit.

Replicators that use templates as the means to duplication tend to follow this pattern of growth. This is, of course, the method used by DNA to achieve replication: Each strand of the double helix serves as a kind of template for the other. Having split lengthwise, the complementary strands of DNA unwind from their three-dimensional form into two more accessible straight lines. At this point, the message of DNA on one side can be read off by ribosomes to produce its complement, which is then tacked on to form a new stretch of DNA.

Another example is the operation of ribosomes themselves. Ribosomes are a peculiar type of RNA that manufacture the proteins used in all living things. A ribosome can string together amino acids in a precise linear sequence, following instructions provided by the messenger form of RNA. It grasps a specific amino acid, brings it together with the growing polypeptide, and thus causes a catalytic reaction that results in the amino acid being added to the end of the polypeptide.

These examples are very familiar to any biology student, and survival of the fittest is, of course, the idea in the back of everyone's mind when they think of evolution. Indeed survival of the fittest has been linked to natural selection ever since Darwin himself adopted the catchphrase (which originated with Herbert Spencer) in a late edition of his *Origin of Species*. But there are other kinds of replication processes that lead to very different kinds of dynamics. Darwinism is not limited, as Darwin himself thought, to an optimizing process, the best possible outcome. The alternative evolutionary regimes are much less familiar because they have been derived more recently. But they turn out to be just as important in understanding the world around us.

The first of these other types of evolutionary dynamic is based on a second, more complicated type of replication reaction. It occurs when an independent catalyst is required to achieve the replication. Here, a replicator must combine with the catalyst, an energy supply, and source materials to produce the replicator copy and some waste products. This reaction process may also involve intermediate states in which the catalyst is first bound to one or more of the other participants in the reaction, which creates a complex but temporary reaction product that separates again to reveal the final form of the duplicated replicator. This general situation is covered analytically by the second-order replicator equation.

In this case, the more replicators there are, the faster their population grows. Replication can now occur even faster than in the earlier case of autocatalytic replication, because the number of births at any given moment is less dependent on the number of replicators that has already been produced. What becomes more crucial is the availability of the catalyst. If this is in abundant supply, then the births of new replicators can increase in nonlinear fashion over time. The increase in births can be hyperbolic or super-exponential—effectively much faster than the Malthusian growth of the previous dynamical regime. (Mathematically this is represented as an exponent on the replicator's growth rate of greater than one.)

When the rate of growth is this fast, the population of replicators rapidly satiates the environment. So in a competition, the replicator with the highest initial concentration, when *multiplied* by its growth rate, wins the race. Thus initial conditions have a large impact on the final composition of the population. In effect, this kind of growth gives an evolutionary advantage to those first on the scene. Once a solution is found, it becomes difficult for any later arrival to invade the population and dominate it, regardless of how good it is. So in our fictive race, no matter how much of a head start the turtle replicator has, it's unlikely to beat the hare, regardless of how lazy and unworthy the hare is. This regime is therefore called "survival of the *first*." It is a winner-take-all situation in which, given a particular path to follow, the first to finish the course effectively finishes it for everyone else as well. If the hare makes it first to the finish line, it keeps the winner's award forever, whether or not the turtle has characteristics that make it more fit for living in that environment. Survival of the first is not necessarily survival of the best.

Examples of this evolutionary regime, in which being common is an advantage, are protein-catalyzed replication and sexual reproduction. With sexual reproduction, the complication is that two organisms at minimum are required to produce a new one. The male and female are again complements, but now both must "come together" before replication can occur (unlike DNA, where one complement already exists and produces the other). The reason species that evolve the ability to reproduce sexually are at an advantage when common is because it then becomes easier to locate an appropriate partner. There are simply more partners around to meet, and individuals are not restricted, due to lack of choice, to potentially unsuccessful matings with partners having incompatibilities or defects.

Such a dynamic also characterizes hypercycles, a quite different kind of replicator that may have preceded template-based replicators like RNA and DNA in the history of life on Earth. Hypercycles are a mutualistic group of molecules that catalyze one another in a ring-like fashion, such that A produces B, which produces C, which produces D, which produces A again. It is the group as a whole that constitutes a replicator in this case. Replication is expected to be relatively slow in a hypercycle because it requires that every member of the circle be present in sufficient quantity to complete the causal sequence (there are no shortcuts). But the cycle that works the fastest can outcompete any others—especially if the competitor cycles utilize any of the same molecules, because the speedier cycle will get to those crucial molecules faster. There are simply more of them already around, so one of the partners of that cycle is more likely to bump into the next guy in the chain than a member of a rare cycle. When it is common, a hypercycle has the ecological advantage of being everywhere.

On the other hand, larger hypercycles (those with more members— say, running through F rather than D before making the circle back to A) are at a disadvantage because it becomes more probable that one of the member molecules will mutate. As a result, that member will no longer produce the next molecule in the causal ring, meaning that the cycle is no longer completed and the replication of that cycle ceases. If any of the constituent reactions has a high energetic cost or involves relatively rare components to assist in the production of the next member, then these too are disadvantages to that cycle.

The opposite of superfast or hyperbolic growth is parabolic or subexponential growth. This kind of rate characterizes yet a third evolutionary regime. Here the higher the concentration of the replicator population, the slower the rate of overall growth: Self-limitation becomes stronger over time until no further replication takes place, and the replicator's population size levels out. This regime can occur when birth is inhibited by the fact that each replication takes two replicators out of play. Examples of this include oligonucleotide analogues and many of the synthetic replicator proteins produced artificially in laboratories. In oligonucleotides, the source replicator is a template for the production of a copy from raw materials, but once the copy is produced, it remains bound to the source molecule, such that the two now function as one. Further, some of the substrate needed to make more replicators has been used up without an effective increase in the number of replicators roaming about looking for new victims. So each instance of replication further limits the ability to do it again.

This third regime can be called the "survival of everybody" because multiple types of replicators, with different rates of growth, can coexist at the end of the race. In effect, both the hare and the turtle wind up at the finish line at the same time, because the closer the hare gets, the more tired it becomes, while the turtle, moving slower, is not so incapacitated and edges ahead until it too faces the same constraints. There is a cost to being common in this regime, such that any new variant replicator that comes along is free to try its hand at surviving in the soup. This is just the opposite of "survival of the first," where there is a cost to currently being the rare type in the population. With the survival of everybody, rare replicator types can increase in frequency, regardless of their abilities.

Such a regime may be due to standard ecological interactions, such as the tendency of predators to focus on those prey species (or parasites on hosts) that are relatively more abundant, leaving the rarer species with the advantage of evolving less controlled by that pressure. Replication may also simply exhaust the necessary resources on which replication is based, with the replicator being unable to substitute somewhat different but more freely available resources. A more social reason for this kind of evolutionary dynamic is that females may prefer rare types of males, which leads to an equilibrium in which different kinds of males can persist.

Thus, depending on how the present density of a replicator influences its rate of growth over the next interval, different evolutionary regimes can come to pass. (The table below provides an overview of the differences between evolutionary regimes.) If, on average, each replicator produces another one like itself each time around, then survival of the fittest will result, because there is no intrinsic advantage with time: The replicators are competing on equal terms. The race then becomes a question of survival, and those replicators—or, more appropriately, the interactors they produce—with the best fit to environmental conditions will win. However, if a replicator achieves a less good record of reproduction, survival of everyone is possible. This is because as any candidate gets slightly ahead, it loses steam, giving the others a chance. So it will often be neck and neck at the finish line, with only relatively random events giving the nod to any one competitor. On the other hand, if a replicator can produce more than one copy of itself each time the starting gun goes off, then the fastest one will win the race—by dint of sheer numbers.

The two less familiar kinds of evolutionary process—in which rates of growth are either greater or less than one—are different in that the best "runner" doesn't necessarily win the race. With these alternatives, there is no optimization of biological fitness, and adaptations need not accumulate over the long term. The heritable features of an interactor are no longer the primary determinants of success in the evolutionary race; instead an element of temporal chance enters the picture. So replicator types with superior abilities do not necessarily increase in frequency over time. The ideal outcome arises only when replicator birth rates are Malthusian. But this is just as expected, since it was a reading of Malthus that inspired Darwin's conception of survival of the fittest in the first place.

REPLICATION REGIMES

Selection Regime	Replication Type	Mechanism	Growth Rate	Characteristics
survival of the fittest	first-order replication (autocatalysis)	template (ribosome)	Malthusian (exponential)	success depends on own characteristics only
survival of everybody	first-order replication (autocatalysis)	complement-matching (DNA)	parabolic (sub-exponential)	selection is not based on own qualities
survival of the first	second-order replication (catalysis)	sexual reproduction	hyperbolic (super-exponential)	invasion is difficult, regardless of individual capabilities

So here's the bottom line from Replicator Theory: *Replicators use different sorts of mechanisms to make copies of themselves. The type of mechanism they use determines the speed at which those copies can be made. These growth rates in turn determine very different classes of evolutionary dynamics.*

Thus replication proceeds at a pace dependent on the concentration of existing replicators of the same type or the availability of other resources, and on whether the replicator or an assistant serves as the catalyst to the reaction. These different mechanisms of replication produce distinct rates of increase over time. Each replication regime is consistent with basic Darwinian principles, but the path evolution follows—and who wins the race—turns out to depend on the distinctive means by which a particular replicator accomplishes the vital task of replication. Different replicators, because they are physical things, will accomplish their evolutionary goals through different mechanisms. And this means that the outcomes will vary because the dynamics of selection depend on the rate at which a population grows, which in turn critically influences the kind of equilibrium one can expect at the end.

Different kinds of evolutionary regimes are possible even with the first-order replicator dynamic, because growth rates can range from parabolic (or sub-exponential) to exponential. So one class of replicators, the self-catalyzers, can themselves produce qualitatively different kinds of evolutionary dynamics. What varies within this class is the kinetics of the replication reaction, or the steps through which each replicator must go to produce a copy. In effect, which molecules can be found bound together varies at each point in the reaction. It is these *intermediate* combinations of elements that are key to the kind of growth rates that can be achieved. An appreciation of mechanisms is therefore absolutely crucial to understand evolution at a given level of organization. Abstract models are simply not enough.

As noted earlier, the first three criteria of replication—causation, similarity, and information transfer—would leave it a purely functional notion, while the duplication criterion tended to make the definition more mechanistic in nature. This trend toward greater physical specificity is furthered by the need to identify a replication reaction for each kind of replicator. Where, how, and under what conditions does replication occur? Answering such questions goes a long way toward eliminating the confusions, which abound in the philosophical literature, about what can be classified as a replicator (suggestions have ranged widely, from photocopies to bird nests to the life cycles of organisms to species). Replication reactions ground the replication process in a time, place, and sequence of events that result in the duplication of some physical structure.

The replication reaction concept also makes it clear that, despite the often-repeated phrase "self-replication"—and the idea that replicators are omnipotent agents able to achieve their goals at will, without help— no replicator really ever "goes it alone." How could it when it must by definition produce two out of one? At minimum, it will require raw materials out of which to construct the copy. DNA requires an incredibly elaborate group of co-participants to reproduce itself and must also find itself in a very special location: the nucleus of a living cell. Take DNA out of this context—place it in a test tube, for example—and the molecules just sit there in gooey clumps, uselessly tangled up like submicroscopic wads of genetic spaghetti.

So replication appears to be a highly complex, specialized kind of event that typically takes place within a close confine. A replication reaction is the cauldron within which replication can take place. To find the traces of a replicator (such as a meme), then, we need to look out for a chain of specific events—a chain of replication reactions.

References to theory, however, do not exhaust the evidence from which we can draw to make claims about the nature of memes. There are also other replicators—either recently discovered or recently invented—that provide a much broader perspective on what a replicator can be like than the standard example, genes, provides. Therefore the next chapter explores what we can learn from the ways in which these other replicators work. While this exercise has its own intrinsic interest, it will also prove crucial to limiting the kinds of approaches we can take when building our model of meme replication.

Chapter Four

THE REPLICATOR ZOO

We can see why there should be so striking a parallelism in the distribution of organic beings throughout space, and in their geological succession through-out time; for in both cases the beings have been connected by the bond of ordinary generation, and the means of modification have been the same.

— Charles Darwin

Nowadays it's not just genes anymore. Replicators abound. Indeed it is now widely argued that DNA was probably not the first replicator to arise on planet Earth, being rather too sophisticated in its repair abilities and copying fidelity to have sprung up from nothing. Many kinds of replicators now exist or have existed, on many substrates. For insight into the general prospect of replication as a phenomenon, we need not rely just on theory, which we've just discussed, but can learn from the exist-ing *examples* of replicators. All these mates-to-memes should provide considerable heuristic inspiration for anyone attempting to delineate the characteristics of a cultural replicator. This intellectual wellspring has gone largely untapped by memeticists, but we can use it now to help us determine what a meme's qualities must be. I therefore propose that we look closely at two well-documented cases of alternatives to genes: pri-ons, a biological replicator independent of genes, and the replicators based inside artifacts called computer viruses.

SPONGY GRAY MATTER

In 1957, a few years after he co-discovered the double helix, Francis Crick proposed a very famous hypothesis. It states that "once 'information' has passed into protein *it cannot get out again.* In more detail, the transfer of information from nucleic acid to nucleic acid, or from nucleic acid to protein may be possible, but transfer from protein to protein, or from protein to nucleic acid is impossible." Later, after it had proven to form the foundation of molecular biology, Crick called this hypothesis the "central dogma" of biology. Crick's dogma defined the difference between two kinds of biological specificity, or the transfer of informational constraints—that of each DNA sequence for its complementary strand, and that between DNA and protein—in terms of information flow.

But in the final years of the last millennium, Crick's central dogma fell. The reason? Direct protein-to-protein information transfer was found to be possible in prions, the infamous class of proteins we met in the first chapter of this book. They are a new class of replicators with many features distinct from DNA- or RNA-based replicators, thanks largely to the fact that they have a different substrate than genes or viruses. How are they capable of replicating themselves without relying on DNA? With the aid of a catalyst, prions cause another molecule of the same class to adopt an infectious shape like their own simply through contact. Thus prions are an important and only recently discovered mechanism for the inheritance of information through means other than genes.

This is an exceptionally profound development for biology. As the first instance of a very different mode of replication from genes, but within the biological realm, prions have generated considerable excitement. In fact, they have had the scientific community in an uproar for years because they cause infectious diseases strictly through the transmission of protein particles. The discovery merited a Nobel Prize for Stanley Prusiner, who, against a backdrop of resistance and disbelief among most of his colleagues, doggedly pursued the possibility of a rogue biological entity replicating without the assistance of genes.

The prion story has only recently been worked out, and some details of how replication in proteins happens are still not understood. But how proteins are formed turns out to be crucial to the story.

Proteins are made of building blocks known as amino acids that are arranged like beads on a string. After the cell's machinery has put it together, one amino acid at a time, these linear chains fold up on themselves into unique three-dimensional structures—typically the shape that makes it most stable. This optimal arrangement is called the protein's "native state." Given suitable conditions, most small proteins will spontaneously adopt their native states. Since form and function are tightly coupled in proteins, this shape allows them to carry out intricate biological functions. For instance, the protein's shape might be a knob that fits into the cranny on some section of DNA, meaning it can bind there and thereafter inhibit that gene's expression.

Since it is difficult to observe the folding process accurately (it happens *very* rapidly and at a small scale), it would be nice to be able to infer a protein's shape from its amino acid sequence. But no one has yet been able to predict how a two-dimensional sequence will contort itself into a three-dimensional object. Protein folding is governed by the formation of electrochemical bonds between different parts of the growing molecular chain. Biochemists know in principle how these rules work, but they can't yet deduce how they are applied in any particular case, because the number of options is too vast. A protein just 100 amino acids long could theoretically adopt around 10^{90} different shapes—a truly vast number. The problem thus appears, from most perspectives, to be devilishly hard. The computer company IBM announced in 1999 that it would spend $100 million to build the world's most powerful supercomputer—one 500 times faster than any available today—specifically to deal with the protein-folding problem. Even with this petaflop machine (one capable of executing 1 million billion operations per second), it will take nearly a year to simulate the folding process of an average protein.

But proteins are able to regularly fold inside your body almost instantaneously, which suggests something is amiss: Is the human organism really computing so much information so fast, or have the folks at IBM and elsewhere incorrectly specified the problem? Some recent results suggest that protein-folding may be simpler than it looks. It can't be the case that proteins are doing the same calculations as an IBM supercomputer in fractions of a second. They must be relying upon natural constraints to help perform the job of acquiring a shape.

Considerable success in modeling the folding of proteins has recently been achieved by ignoring what the IBM computational effort will focus on: the complex details by which each atom interacts with each other atom to construct the protein molecule. Instead the newer approaches focus on the coarse-grained features of relationships between the amino acid sequence in DNA and known rules by which molecules generally structure themselves. In fact, fundamental physical laws underlie the protein-folding process. These laws significantly constrain what can happen, which eases the modeling problem.

It turns out that molecular chains, as they grow, explore a far smaller range of different shapes than is theoretically available to them. In particular, the topology of a protein—the three-dimensional shape it winds up in—appears to determine major features of how it folds. Thus both protein structures and protein-folding mechanisms can be predicted, to some extent, using models based on simplified representations of the polypeptide chain—at least for small proteins. This means modelers can take a relatively large-scale view of the folding process, rather than worrying about the details of how this or that atom links up with others in its neighborhood. So new methods based on the end-state of the process— the protein's overall shape—have shown great promise in predicting protein-folding mechanisms.

Sometimes, however, the process goes awry, and the protein doesn't wind up with the shape it was supposed to have. In some cases, this misfolding turns the protein into a prion. This may not be a problem for the prion, but for an organism housing such proteins it can be disastrous. Perhaps the most interesting example of a protein-folding disorder is the class of diseases called "transmissible spongiform encephalopathies" (TSEs). The three primary TSEs are scrapie in sheep and goats, bovine spongiform encephalopathy (BSE) or "mad cow" disease in cattle, and Creutzfeldt-Jakob disease (CJD) in humans. Testimony to the significance of prions comes from the current public-health crisis concerning BSE and CJD, which are closely related diseases. TSEs have a long incubation period in which victims show no symptoms, followed by a relatively short clinical phase, ending typically in death. They are all characterized by the accumulation of a specific form of prion protein in the central nervous system.

The body is constantly producing the proteins associated with TSEs like CJD. Normally the protein folds properly, is chemically active (remains soluble by a variety of agents), and is disposed of without problem by the body's natural cleansing powers. But prions can adopt a second configuration that is inactive. In this shape, it becomes resistant to solution by enzymes and therefore aggregates in body tissues. That a protein can have both active and inactive states is not unusual. What is surprising about prions is that once a single protein molecule in a cell takes on this inactive state, it can influence other nearby molecules of the same type to adopt a similar state. Suppose, for example, that a small amount of the protein misfolds so as to become a CJD–related prion. If this prion bumps into a normally folded protein, it causes a refolding process in that protein so that it ends up as another infectious prion, despite its perfectly normal amino acid sequence. Significantly, this influence can continue even after the cell undergoes division and new protein molecules are made. So long as the body keeps producing the normal protein, each CJD prion can keep on creating more and more of its brethren out of these new recruits. The result is insoluble masses of prions (called plaques) that continually grow larger and larger, showing up as pockets in the network of neurons in the brain. In effect, the prion is replicating itself even though it includes no nucleic acid of its own.

Prion replication shows all the standard features of an evolutionary process. A single type of normal prion protein, inserted into the brains of living laboratory mice, for example, can be converted by the mouse into two different forms, depending on the type of abnormal human prion that initiates the conversion process. These "strains" of prions consist of the same protein misfolded in different ways. Even though all the different strains of a particular prion contain a common genetic sequence, each protein strain has a distinct shape, and each strain of prion can impose its own brand of misfolding on its normal cousin by simple contact. It thereby transmits the disease with the unique characteristics of that strain to other protein molecules. Like a disease, the specific manifestations of the molecule in the individual are determined, then, not by the abnormal gene or the sequence of amino acids in the protein, but purely by the abnormal shape taken on by the protein. It is this shape that is infectious. And the different strains of these prion diseases "breed true" as they are transmitted from one host to another. Odd as it may seem, inheritance without DNA *is* possible in the weird world of prions. A single nervous system can even harbor multiple prion strains with specific and independent consequences.

These strain differences are also associated with slight differences in the protein deposits that cause the disease. Scientists have recently used these strain differences to show that a few British people truly have "mad cow" disease, the form seen in cattle, rather than the usual human form of Creutzfeldt-Jakob disease. What is now called "new variant CJD" is a consequence of the transmission of BSE prions from cattle to humans; people are now dying of "mad cow" disease under a new name. So the ability of prions to cross host species boundaries is clear and appears to be easier than for DNA- or RNA-based viruses. Even when individuals don't exhibit the symptoms of a prion disease, they can nevertheless be "carriers" of that prion and then transmit it to others of their own, or a different, species. The "species barrier" that limits infection of one host species by prions from another type of host now appears to be highly porous.

Are prions likely to be widely found in biology? Leading prion researchers think so. Although nucleic acids will likely remain the main agent of biological heredity, it is possible that since we now know how to look for them, we may begin to find prions all over the place as explanations for diseases previously attributed to other evolutionary agents, like viruses.

Certainly the prion protein corresponds to a gene naturally found in the genome of all vertebrates (as well as yeast), which is expressed in most tissues of their bodies. And more and more types of prions are constantly being found; entirely different classes of proteins can be prions.

The truly ancient origin of many of these proteins suggests they have positive functions. However, the prions in older families of organisms (like yeast) are not as dangerous to their hosts as those in later-arriving species like mammals. The infectious proteins in yeast are prions in the sense that they have the ability to change configuration, but they do not cause disease or kill their hosts, nor even significantly disrupt normal biological functions. What they do is introduce new functions, which can nevertheless be quite significant. One yeast prion, for example, Sup35, can cause new sections of the yeast genome to be transcribed, leading to the expression of genes that would otherwise remain silent. Certainly if prions are found to be ubiquitous, they could only proliferate because most of them have some kind of value to the organism that has been selected for.

SURVIVAL OF THE PRIONS

The principles of replication discussed in Chapter 3, if they are general, should be applicable to these new replicators. We can begin by examining how the replication reaction works in prions.

It's not yet entirely clear, but experts believe that during prion replication, an as yet unidentified factor, called protein X, first binds to a normal protein molecule. Protein X is believed to act as what biochemists call a molecular chaperone—essentially a catalyst that increases the rate of conversion of reaction materials into the final product by moving the normal protein to the place where the replication reaction itself can take place. Once in place, this complex (composed of the protein together with the catalytic chaperone) then binds to a third molecule: a nearby prion. Then, by an unknown process, the normal molecule is transformed into a second molecule of prionic form. Finally the original prion and protein X come unhooked, leaving the new prion to float free, ready to infect other "innocent" molecules. The step that limits the rate at which the abnormal forms can be produced in the body has little to do with the availability of source prions; it has more to do with the first stage in replication, when the normal protein binds to the chaperone, protein X.

The need to get other players, specifically the chaperone molecule, involved during the conversion of protein molecules provides evidence that the replication process has (at least) two steps and involves an intermediate complex composed of multiple players. Further, "normal" prion protein molecules are converted into "infectives" at different rates depending on the type of chaperone used. Prion "birth" rates are determined, then, by the kind of chaperone protein available and by how fast it works.

The net rate at which prion-filled plaques accumulate, and hence the time-to-symptoms for a particular strain of prionic disease, is a function not just of the prion's birth rate but also its death rate, because it turns out that prions are not completely resistant to dissolution by enzymes. The body can work to clear these globs of protein from the brain. It's just that some strains of prion are more resistant to dissolution than others. So it is the birth rate (a function of chaperone type) minus the death rate (susceptibility to dissolution) that explains the overall speed with which prions accumulate in the host.

With this knowledge of the replication reaction, we can begin to think about the kind of evolutionary dynamics this sort of reaction suggests. It appears that prions can produce a copy of themselves if only one other one is in the vicinity. If one becomes two, then the number of prions will increase at a linear rate. Replicator Theory tells us this means these replicators will experience a selection regime in which only the fittest survive, just as in the case of genes. This means that types of prions that are slow to replicate (or quick to degenerate) can be outcompeted by new variants and can be replaced over time in the population of hosts. It pays in this case to pick the right chaperone to bring to the replication dance.

What else have we learned about prions as replicators that might be useful to us in our definition of a meme? Prions are a type of protein with a three-dimensional configuration that allows it to replicate through the conversion of an existing physical substrate into a different form. Like viruses, prions are parasitic replicators because the substrate they depend on is the end result of gene expression. In this case, DNA makes the protein molecules for the prions to infect. But the ability of prions to replicate is independent of genes, thanks to the specifics of their replication mechanism.

Prions, furthermore, do not appear to combine into larger structures; they don't react with larger *functional* conglomerations of material, although huge masses of insoluble prion protein can accumulate through sheer repetition of the replication reaction. Thus replication through conformational change is the only trick prions have learned. They are limited replicators; any prion can adopt only a few different shapes or strains. However, prions are modular: Prion molecules contain particular sections, or domains, that can be attached to other proteins that transform them into prions. These domains are modular in the sense that they are add-ons and are transferable between widely varying types of protein from evolutionarily distant species. For example, a novel prion was created by recombinant methods in which the crucial sections of a yeast prion were fused onto a protein from a rat. Further, this new rat prion (in a type of protein not previously infective) could be reinserted into the yeast and retain its infectivity. This modularity suggests that, as species evolved, other proteins might have picked up these prion domains and also become elements of epigenetic inheritance. If they did, this kind of protein inheritance pattern could have had an important effect on the process of species evolution.

As suggested above, prions exhibit some truly peculiar characteristics compared to genes. For example, they are reversible. Because prions can change a protein's shape and function without affecting the genes that code for it, the protein can switch back and forth between its normal and infectious forms—either spontaneously, with the help of molecular chaperones, or through simple biochemical manipulations by people in laboratories. Prions enable an organism to respond to environmental changes and to pass these changes on to its progeny. This ability to respond to the here and now endows their inheritance with a certain Lamarckian flavor.

Prions can also arise in many ways. The initial folding error that initiates a chain of replication reactions may be induced by a mutation in the gene sequence coding for that protein. This represents a kind of genetic predilection to become a prion. Alternatively the misfolding may just occur spontaneously, as a result of some purely local incongruity at the time the folding of a molecule occurs; perhaps the presence of a rogue molecule somehow influences the outcome. Shape can also be phenotypically inherited through the contact of two molecules within a host, as described above. Finally, as a variant of the contact scenario, there are cases of interpersonal transmission in which a prion molecule can migrate from one host to another if the tissue in which it resides is consumed by another host, and there continue its business. (This is what happens when people eat infected beef.) These are four different routes to the production of new prions, only one of which (spontaneous mutation) is found in nuclear genes (viruses can also perform inter-host transmission). Since they can arise in these new ways, prions will exhibit profoundly different ways of spreading through populations than genes are capable of. Prions are real eye-openers for those who come to memes thinking only of the example of genes as replicators. Because prions work quite differently from genes, the possibilities for a cultural replication mechanism are expanded considerably by taking this comparative viewpoint. As we will see, the next class of replicator we will investigate—computer viruses—shares a number of the prions' unusual qualities.

ATTACK OF THE BINARIES

The electric things have their lives, too. Paltry as those lives are.—Philip K. Dick

First law of computer security: don't buy a computer. Second law: if you ever buy a computer, don't turn it on.—Virus writer "Dark Avenger"

Computers are confusing things. We recognize them as machines, but they also appear to be intelligent. It has been shown that we tend to treat eyes blinking and mouths moving on television and computer screens as fellow flesh and blood. At a visceral level, humans don't know the difference between the real thing and a simulation. We can even put the calculating hardware inside a metal-and-plastic body that moves around, which makes a computerized robot even more confusing. We also treat computers as agents because they are becoming more and more integral to modern human society, doing many things we used to do for ourselves, like make cars, answer the phone, balance our bank accounts, and post our communications. As a result, we talk to them on the phone and get angry when they don't talk back.

Given this near-humanity, can computers take the next step, become even more powerful? Can the ultimate power—the ability to replicate—reside inside the brains of these creatures, just as it does (perhaps) inside ours?

Computer programmers often claim that relatively simple algorithms are capable of replication. Here's one version of this idea, a program written in a high-level language that prints out a copy of its own instruction set in quasi-English on paper:

```
main(){char q=34,n=10,*a="main(){char
q=34,n=10,*a=%c%s%c;printf(a,q,a,q,n);}%c";printf(a,q,a,q,n);}
```

But this tiny, one-line program, written in the popular language C, is not a replicator, at least not as defined here. The problem, of course, is that the source program consists of magnetic states on a rotating disk (the computer's hard disk), while the "copy" is ink on paper. But a set of logical states in computer memory and a bunch of ink marks on paper are not the same thing. Neither can the program produce a copy of the original memory states in another computer (at least, not without reprogramming).

This confusion about replication is probably due to a surfeit of "symbolic" thinking. Computer scientists tend to slip conceptually between programs written by the end user and the states of memory managed by those programs, considering them as synonymous. They are not. Combining talk of abstract information with the tendency to anthropomorphize intelligent machines can be deadly to clear thinking. When we talk of computers, it's perhaps better to avoid the whole world of symbols and simply talk of code—some bit of structure that can be translated into another kind of structure. So bid good-bye to bits and bytes; let's speak instead of binary states in electrical circuits.

My denial of the claims by eminent computer scientists about replication in computer programs is not meant to be taken in blanket terms. Replication in computers *can* and *does* occur on a daily basis. Further, it can occur both *within* and *between* computers. We call the programs that accomplish this feat "computer viruses" (or what I will call comp-viruses, for short). But before we proceed any further, let's attempt to be clear about what we're talking about. What is a comp-virus?

Frederick Cohen, currently a researcher at Sandia National Laboratories and an independent computer system security consultant, is the acknowledged intellectual "guru" of the comp-virus community. He single-handedly made the study of comp-viruses academically respectable. Cohen defines a comp-virus as a sequence of symbols in machine memory that, when interpreted by the machine, causes another sequence to appear somewhere else in memory. Cohen's shorthand definition: A computer virus is "a program that can 'infect' other programs by modifying them to include a, possibly evolved, version of itself." A comp-virus, then, is a program that reproduces its own code by attaching itself to other executable files in such a way that the virus code is executed when the infected file is called up.

What makes a comp-virus interesting, according to Cohen, is "its ability to attach itself to other programs and cause them to become viruses as well." All viruses work by linking themselves to the existing programming on a host machine in some fashion. For example, "shell" viruses completely enclose the original file or executable program, making the original code an internal subroutine of what is, in effect, a whole new program. "Intrusive" viruses overwrite some or all of the original host programming, perhaps to replace a subroutine or insert a new interrupt sequence that they can make use of for replication. Most comp-viruses, however, are "add-ons," because they are easier to write. They merely insert additional lines of programming somewhere into a sequence of code already in memory, making the original program grow larger. In any case, when one of these host files is activated, the viral code executes and links copies of itself to even more hosts.

Viruses are thus programs just like any other on your PC. They consist of instructions that your computer executes. What makes viruses special is that when they execute, their primary purpose is to place their self-replicating code in other programs, so that when those other programs are executed, even more programs wind up being infected with the replicating code. The infected program becomes, in effect, the replicator's interactor or vehicle for further replication. Since this is a replication process dependent on host resources, it is parasitic in nature. The virus is also detrimental in the sense that it co-opts its host's resources and can order up aberrant behavior ranging from the posting of an innocuous message to the screen (thus announcing its presence) to wiping out hard drives and clogging networked computer systems with superfluous data and messages. This last kind of activity, although distinctive, is not a necessary component of a comp-virus, however.

Interest in comp-viruses has been almost exclusively practical: how to stop the next comp-virus epidemic from bringing down the world's computer network. But what can comp-viruses tell us about how replication can be achieved in yet another kind of physical substrate? The similarities between comp-viruses and memes, it will turn out, are significant—especially in the way their replication reactions work and the sequence of stages through which each of these replicators has evolved. Getting the most out of this case study requires beginning with a look at how comp-viruses reproduce.

A DAY IN THE LIFE OF A COMP-VIRUS

Every viable computer virus must have at least two basic parts, or subroutines. As these sections of code are executed, these subroutines translate into stages in the virus's life-cycle. First, the virus must contain a search routine, which locates new, uninfected hosts that are worthwhile targets for replication. This routine will determine how fast the virus can reproduce, which is a function of how broad the search criteria are. If the virus can infect multiple kinds of media and a wide number of file types, then it will find fresh victims faster. But, as with all programs, a trade-off between size and functionality pops up here. The more sophisticated the search routine is, the more coding it will require. So although an efficient search routine may help a virus to spread faster, it will make the virus bigger, and so easier to detect.

Second, every computer virus must contain a routine to copy itself into the area that the search routine locates. This is the part of the life cycle during which the replication reaction must take place. The copy routine will only be sophisticated enough to do its job without getting caught. Again, the smaller it is, the better. How small it can be will depend on how complex a virus it must copy. For example, a virus that infects only COM files (part of the PC operating system) can get by with a much smaller copy routine than a virus that infects EXE (or software program) files. This is because the EXE file structure is much more complex, so the virus needs to do more to attach itself to an EXE file.

When the virus executes, it usually intends to infect other programs. What can vary is *when* it will infect them. Some viruses infect other programs each time they are executed; other viruses infect only when a certain triggering condition occurs. This trigger could be a day or time, a sequence of keystrokes by the computer user, the presence or absence of a suite of files on hard disk, certain file attributes, a counter within the virus, or some random event. If the comp-virus is resident in computer memory, it can silently lurk there waiting for someone to access a diskette, copy a file, or execute a program before it infects anything. This trigger can be a rare event, so you can never be sure that a virus isn't already present on your computer. Sophisticated comp-viruses are very selective about when they infect programs because this is vital to the virus's survival. If the virus infects too often, it is more likely to be discovered before it can spread very far.

In addition to the two basic stages—which enable the virus to locate suitable hosts and attach itself to them, respectively—many computer viruses have other routines appended to their "genomes." First, it is usually helpful to incorporate some additional features into the virus to avoid detection, either by the computer user or by commercial virus-detection software. For example, a virus can overwrite some part of system software and essentially co-opt the system (this is called using "stealth"). It then becomes immune from inspection because it is now part of the computer's operating system—in effect, it *becomes* the virus inspector and so can ignore itself.

As a second kind of supplementary section, a comp-virus may include program code that, when executed, stops normal computer operation, causes the destruction of data in computer memory, or plays practical jokes on the user, such as flashing silly messages on the screen. Such routines may give the virus a certain "character," but they are not essential to its basic biological functions. In fact, such routines—which add what we can call an "attack" phase to the viral life cycle—are usually detrimental to the virus's goal of survival and self-reproduction, because they make the fact of the virus's existence known to the user.

As might be expected, this secondary functionality is often omitted altogether. The more appropriate image for comp-viruses is therefore one of "search and copy" rather than "search and destroy." It is only due to the fact that those individuals doing the programming have malicious motivations that these often puerile sideshows get added on to a comp-virus's code. How else can we explain a virus that causes a quote from the television show *The Simpsons* to pop up on the screen whenever the time of day matches the date (such at 4:10 on April 10, for example), which the Melissa virus did?

Computer viruses, just like their biological cousins, are not inherently destructive. The essential feature of a computer program that causes it to be classified as a virus is not its ability to destroy data, but its ability to gain control of a computer and make a fully functional copy of itself. Being destructive slows down a comp-virus's replication rate and injects new selection pressures into the evolution equation, neither of which are desirable from the virus's own point of view. For similar reasons, a bio-virus typically evolves fewer destructive capabilities—that is, becomes less virulent—over time because harming hosts is detrimental to the survival of the virus.

The usually malevolent effects of these viruses can either be designed or arise inadvertently. Often the damage to a computer is not intended, but because many virus writers are not first-rate programmers, destruction arises from the fact that the virus contains poor-quality code. One of the most common viruses in the early days—the Stoned virus, which flashed the message "Your PC is now stoned"—was not intentionally harmful. Unfortunately the virus's author did not anticipate anything beyond 360K floppy disks, with the result that the virus will try to hide its own code in an area on larger diskettes that causes corruption of the entire disk. Comp-virus designers still often don't seem to mean to cause the harm they do. Even the worst comp-virus thus far—the Love Bug virus—was apparently released mistakenly.

The attack phase of a comp-virus life cycle is optional because an independent sequence of code is responsible for it. This is entirely different from the case of DNA or prions, where all such effects are a necessary consequence of the replicator's physical structure. The protein coded for by a gene is the only possible result of the nucleotide sequence at that point in the DNA molecule. Similarly, a prion's disease-producing ability is entirely due to its strain, which is in turn an automatic reflection of its conformation, or shape. As the physical nature of these replicators changes, so too do their effects on individual hosts.

The physical footprint of comp-viruses also influences the computers they infect. The section of their code for the replication function gives them a certain length, trigger, and location in memory. This means the virus takes up a certain amount of real estate that other programs cannot occupy and requires a certain amount of resources to maintain and produce copies of itself. These can be considered the automatic, or primary, consequences of its presence.

What is new with comp-viruses—what makes them very different from the biological replicators we have examined—is the ability of programmers to tack on extra bits to the end of the creature, the code producing the intentionally damaging effects, which can be called their "*secondary* phenotype." Having phenotypic effects isn't novel for a replicator, but the fact that they have to be separately coded for is; in this case, the "function" doesn't automatically follow from the "form" of the virus, as is true of biological replicators. There is a real physical division within the replicator between the sections devoted to these different functions: You can cut off the "tail" of a comp-virus—the extra bit that says, "Your computer is stoned"—and thereby remove its ability to inflict damage on a computer, without harming its ability to replicate itself. This division between primary and secondary phenotypic effects is only possible because comp-viruses are super-molecular in scale.

So how does this division of code and activity in comp-viruses work in practice? The particulars of a life cycle can vary from virus to virus, but for our purposes we can just consider how the most basic infectors are spread. This typical comp-virus life cycle begins with a single infected computer. The virus in that computer searches for a new host file by looking for files—either elsewhere in the same storage unit or on other computers—without the trademark fragment of code known as a virus's signature. A signature is a marker of identity that a comp-virus appends to an infected file to say, "I'm already here, don't bother!" because continuous reinfection of the same file can lead to a significant depletion of memory (in the form of very large files with multiple copies of the virus tacked onto it) and the possibility of compromised func-tionality—both of which increase the probability a virus will be detected by the user.

However, once an uncontacted executable file is found, the virus is free to attach itself, turning it into an infected file. When the user at-tempts to execute this file, it first performs the vital biological function: It infects another file. Then it executes the normal instructions of that file as if nothing had happened. With the exception of a slight delay (for execut-ing the viral search and copy routines), the infected file appears to still be the normal one—that is, until the triggering condition causes damage somewhere. This is because the last thing the infected file does is to check the secondary viral functions: Is it time to inflict damage? If so, and only then, the comp-virus does whatever injury it has been programmed to do.

CLASSIFYING VIRUSES

Not all comp-viruses get through this generic life cycle in exactly the same way. The mechanisms they use to accomplish their search and copy functions can vary. We therefore need to be clear about what *kinds* of comp-viruses there are. Although many typologies of computer viruses exist, few of these classifications are theoretically sound. Some identify viruses according to the type of file they can infect: boot sectors, the oper-ating system, software programs, or macro files. Other classifications are based on how the virus operates; yet others distinguish whether the virus remains resident in memory or is merely a transient presence there. Clas-sifications can even be based on what kind of defenses the virus uses to avoid detection. Often these different traits are combined in the same listing, confounding categories. Some principle must be introduced to eliminate the chaos. Replicator Theory provides a clear rationale for developing a classification of comp-viruses, but we still need to do a little investigation to determine just how the application of these principles will work out in practice, given current computer technology.

The architecture for nearly all present-day computers was first devised more than 50 years ago by the brilliant Hungarian mathematician John von Neumann, a codeveloper of the most popular formal theory in the social sciences today (game theory), as well as a founding father of modern computational science. His basic idea was to make computers more powerful by dividing the information-processing task from the memory task. In this way, "programs"—a set of built-in rules for processing information—could be distinguished from "data," which was specific to particular jobs. This division is manifest in contemporary computers in their very hardware. The central processing unit (CPU), usually found on a large, flat circuit card (called the motherboard), is there to manipulate data that has been acquired from various devices, either for storing data or communicating with the outside world, to which it is connected via a hodgepodge of wires, cables, cards, and jumpers. The CPU uses a solid-state network of wires and transistors that function by transferring millions of electrical pulses per second through a complex maze of circuits. Other microprocessors take software instructions from long-term data storage sources, such as read-only memory (ROM) or a hard drive, and place them in the computer's main memory, or random-access memory (RAM), where they await access to its CPU.

Data may be stored in a computer either permanently or temporarily. (Cognitive science later instituted this same division when it distinguished between short-term and long-term memory.) RAM is the area in a computer where data is held temporarily while an application or the operating system is using it. In RAM, signals are established by an electronic charge communicated to RAM by some other part of the computer. In contrast, in ROM, the memory chip has built-in, permanent electrical states. ROM can retain data after the computer has been turned off, while RAM cannot. When RAM loses power, the memory states that represent data there are lost.

———

Maintaining the accuracy of memory depends on control over any changes in the electrical values of computer registers. Such changes must occur only in response to a legitimate command from some other part of the system, which reach that register in the form of an electronic signal. For example, holding down the Shift key and then pressing the *m* key on the computer keyboard sends a signal to the computer's processor that is translated internally from "M" into the binary equivalent of the number 77 (the ASCII code for "M"), which happens to be 01001101. In those terms, the letter is usable as pulses of electricity racing through a microprocessor's various parts. That's why a computer needs to be plugged in (or have a source of battery power) to work. The various parts of the computer thus communicate information to one another in the form of signals: One circuit activates another, changing its state. This is accomplished through the exchange of electrons. So computation in machines—these days, at least—is all about electrons moving from place to place, changing the charge of some substrate.

In general, then, computers register information as the absence or presence of an electric charge. Since there are only two options—on or off, positive or negative—the computer can only distinguish two different numbers: 1 and 0. Sending electronic pulses through circuits that either pass through a transistor to indicate "yes," or "1," or are impeded to indicate "no," or "0," creates this binary code. In any of the computer's memory devices, one electrical state thus represents a single bit of data— that is, a single binary decision-unit. However, computer programmers, being human, often prefer to write down their ideas in a form closer to human than computer thought and so use software that allows them to produce a series of language-like instructions called "source code." Another piece of software called a compiler turns the source code into object code, a language intermediate between the programmer's language and the computer's. Then yet another piece of software, a translator, turns the object code into machine code, the binary language computers understand. A comp-virus can thus be written in the machine's binary code or in a higher-level language with a large "vocabulary" of commands, such as Pascal or C or Lisp, but to achieve the necessarily exacting control over function that parasites require, most comp-virus writers code directly in machine code. In any case, the virus always resides in the computer, stored in memory, as states of electrical or magnetic charge in physical registers. This is important to remember when it comes to thinking about how information can be replicated inside computers.

The only reasonable classification of replicators from the perspective of Replicator Theory is according to substrate and replication reaction type. As just described, comp-viruses replicate by modifying computer memory. Since only RAM and disk-based memory can be changed by the computer's own activity (ROM is permanently set by the computer's manufacturer), these are the two kinds of locations where computer viruses can reside. If we refer back to our first criterion of distinguishing replicators by their substrate, this means we can distinguish, at the most fundamental level, between two basic kinds of comp-viruses: RAM–based and disk-based viruses. I will call these, respectively, "short-term" versus "long-term" comp-viruses, because any RAM–based virus must die when RAM is inactivated, and that happens every time a computer is turned off.

Because replication reactions tend to be so specific, we should expect that a particular comp-virus would replicate in one or the other substrate, but not both. However, long-term memory can be stored either as optical polarizations or as magnetic charges on the surface of a disk of some sort. So comp-viruses have, in effect, three potential substrates: electronic short-term memory (RAM), and magnetic or optical long-term memory. This identifies, then—within the second major category of disk viruses—a distinction between magnetic and optical long-term comp-viruses. However, these two kinds of media work in roughly analogous ways, so this distinction is not very important. Other substrates may be developed as new technologies for storing information come on-line in the future. However, I expect the basic principles of replication we have developed will hold for them as well.

There is yet a third level in our classification scheme below the basic and secondary substrate distinctions: the type of copy mechanism a comp-virus uses to replicate. The copy stage of a virus's life history defines the replicator reaction and so is another consideration, after substrate, in a proper hierarchy of replicator types.

All comp-virus life cycles begin on a long-term memory device with a program being read off by the operating system. But all computer programs, including viruses, must utilize a computer's RAM to store data or instructions as they wait for the CPU. So the basic cycle is one that moves from long-term to short-term memory and back again. The differences in viral life cycles come in how the search and copy routines are implemented. As we will see, comp-viruses have gone through three major developments over time. Each represents a new complication to this basic viral life cycle of transfer between disk-based storage and RAM.

The simplest type of viruses, and the first to arrive, were confined to RAM in a single computer. This early age, which I will call Phase I, began in the 1960s with programs called Darwin and Core War and their descendants. In the simplest life cycle, the program on long-term memory is used as a source for instructions submitted to RAM. These instructions create a "virtual" operating environment within RAM that allows instructions to be copied inside that environment. These are the simplest comp-viruses, since they don't need to carry a lot of baggage for conducting complex search or copy routines. Their cycles are somewhat unusual in that it is a one-way trip from disk to RAM, where they replicate and then die. There may be information about what happened inside RAM written back to the disk after a run of the virtual program, but this does not typically modify the source program itself, being rather a kind of log file.

It is important to distinguish between the software program itself, the virtual operating system that results from the execution of that program, and the small, executable programs that are written by the virtual operating system into RAM. The software programs themselves, residing on hard disk (in magnetic memory), are not simple, since they have the job of creating the virtual environment and interacting with the computer's own operating system. To duplicate a piece of software, you must execute a file copy procedure, which is accomplished by the computer's operating system at the push of a button. The program itself doesn't instigate this copying; the computer user does. So the program itself does not replicate. In such cases, it's only the little programs, residing in RAM and created indirectly by the software, that qualify as comp-viruses.

The second type of comp-virus arrived in the early 1980s and introduced somewhat greater complexity into the viral life cycle. These viruses used residence in RAM as a springboard for the infection of new computer hosts. They spread by first infecting long-term memory devices (like floppy disks) inserted into the host machine. These "alien" disks, serving as intermediary hosts, would then be physically inserted into another potential host, a second computer, where the virus could go into action again by transferring to *that* machine's RAM. The object was to replicate the source program itself, residing on a long-term memory device such as a hard disk, rather than a RAM-based instruction, as was the case in Phase I.

In the third phase of development, beginning in the late 1980s, true network viruses appeared, almost as soon as significant numbers of computers became directly linked to one another through electrical connections. In Phase III viruses, the search is broadened to include any other computer that a signal can reach through the network. Further, the duplicate is now only linked to the source comp-virus through this signal, which has effectively mediated the replication process. In some cases, the replication reaction itself will be split *between* computers. The reproduction cycle now moves from one disk and its RAM to another computer's long-term storage through *its* RAM. This involves yet other complications than were necessary for Phase II—in particular, the use of specialized signals for transmitting information over an electronic network. I will call this kind of inter-computer replicator a *network* comp-virus. This is a bit of code that gets into a machine in the form of an electronic message dispatched by another computer, rather than through direct physical contact. It requires a host machine capable of producing signals that transmit viral information directly through a channel to a second definitive host.

In order to outline this history in more detail, we need to first discuss what has happened within the "brain" of a single machine and then talk about the evolution of information transfer in networked artificial intelligences.

EVOLUTION INSIDE A COMPUTER

Replication within computers can take simple forms, even simpler than the example provided at the beginning of this chapter. In fact, such replication is the goal of a major research initiative, begun about twenty years ago, which has been christened Artificial Life (or ALife, for short).

It isn't standard nomenclature to call programs from this burgeoning field comp-viruses, because they don't fit the standard image. First, they don't replicate between computers, but rather within them. Second, they don't engage in destructive activity, being devoted entirely to replication within the RAM environment. But I have already argued that destructiveness is not an intrinsic characteristic of a comp-virus. Third, they also don't append themselves to preexisting files. Instead their hosts are preexisting memory registers. Even so, they are programs that are able to duplicate themselves elsewhere in memory and so fulfill my basic definition of a comp-virus.

Recognition that ALife programs deal in comp-viruses also has the benefit of removing the blinder—which comes from concentrating only on network comp-viruses—of believing that there is no such thing as a "good" comp-virus. After all, bio-viruses are used nowadays as delivery vehicles for genes in gene therapy or for other molecules that serve medical purposes, so why shouldn't comp-viruses be used to "cure" sick computers of some of their ills? Already, comp-viruses are used by the ALife community to investigate scientific questions, and the potential of this approach for science and technology is exciting. Real benefits are being achieved through the novel use of this kind of replicator.

TIERRA, originally programmed by Thomas Ray (a wildlife biologist turned hacker) in 1990, is now seen as a classic experiment in ALife. TIERRA is a real place; it just happens to exist inside computers. It's an environment, an "alternative Earth" (*tierra* is Spanish for "earth") composed of a completely featureless terrain—space in RAM—that basks in the flow of electricity. Comp-viruses roam though this environment as strings of machine code that compete with each other for two resources: computer cycles ("energy") and memory space ("territory"). Further, every now and again, as a viral program is copied between memory cells, one of its infobits is randomly flipped by the operating system as a way to mimic the genetic mutations caused by cosmic radiation and errors of replication in real organisms. The result of this reproduction is a slightly different program that might exhibit useful new functions or contain deficits that make it unable to further replicate. Old viruses and mutants that crash are eliminated in each iteration of the program by a death algorithm (called The Reaper) as either decrepit or unfit, to keep population numbers down.

TIERRA comes complete with its own operating system, which functions only within the conscripted area of RAM set aside for the program. This program (resident elsewhere in RAM) determines how the creatures interact with each other, introduces the mutation events into their reproduction routine, allocates resources, keeps records of events, and provides graphical feedback to the "outside world" of the computer user. This TIERRAN operating system is part of the viruses' artificial biosphere, the laws of which are to be exploited in any way possible.

Ray's comp-viruses are self-replicating in the sense that they can directly call instructions from the computer's central information processor, rather than requiring any keystrokes from the computer user. Replication is essentially the ability of a program to copy itself elsewhere in memory. The virtual operating system executes the instructions it finds in memory in sequence according to address. This means the execution pointer moves along the genome of each virus in the population, doing what the instructions found there tell it to do at each point. Thus Ray calls each virus a "virtual CPU." Because the set of 32 instructions in the TIERRAN language is quite robust, variations introduced into a viral sequence can remain meaningful. The virus is also a replicator-cum-interactor, because its structure (or code) instigates the replication process, while also serving as the agent competing with other comp-viruses for space in the TIERRAN ecology. The virus, like its biological counterpart, has no other "body." It is directly selected by virtue of its functions; no layer of protection sits between the code and The Reaper.

In his now-legendary first experiment with TIERRA, Ray seeded his virtual environment with a single example of what he believed to be the simplest comp-virus possible. This was a piece of code he designed simply to copy itself elsewhere in memory; this copy could then become active and do the same. Its "genome" was 80 bytes long. He thought at the time that it would likely take years of additional programming before a process resembling real evolution would emerge. But, as he later recounted, "I never had to write another creature."

For the first few thousand generations, it's true, nothing much happened. But then a number of 81s—mutant viruses with an extra byte of program code—showed up. A little later, a 79 appeared. Because the 79 had one less byte of code than its "parents," it took less time to reproduce and hence was able to outcompete the 80s and 81s because it had shorter generations (the length of time between replications). It began to take over the TIERRA ecosystem.

Next something quite unexpected happened. A 45 emerged. A 79 seemed reasonable, but how could a 45-byte-long virus reproduce using only just over half the instructions of the original 80? When Ray examined 45's code, he found the logical answer: 45 was a parasite. Instead of reproducing itself, it hunted for the reproductive code of an 80, then called on that code to help it out. This was analogous to the way a biological virus functions. "Real" viruses reproduce by inserting their DNA into a host cell and then using the cell's reproductive apparatus to build copies of themselves. This new digital replicator was a type of parasite that had evolved the ability to borrow parts of the necessary code for replication from the more complex replicators in its environment. But as the "lean, mean" 45s grew in numbers, they began to crowd out the host 80s on which their own reproduction depended, and so they began to die off as well. This decline in the numbers of parasites led in turn to an increase in the population of host 80 codes, causing an oscillatory cycle in the numbers of the two types of replicators. TIERRA had now become a real ecosystem with multiple types of digital viruses competing for resources but maintaining a presence in the RAM soup. This pattern mimicked the cycles of growth and decline between hosts and parasites observed outside computers. What is more, the comp-hosts began to evolve characteristics that would make them resistant to the comp-parasites, and the comp-parasites then found ways of circumventing these new comp-host defense systems. An ever-escalating evolutionary arms race was in progress, the very phenomenon that is believed to have provided the springboard for the increasing adaptive complexity of living organisms over evolutionary time.

But this was not the end of the story. The 79s were immune to attack by the 45 parasite, but eventually a new 51 parasite appeared that *could* use 79s to reproduce. Even later came another surprise: the arrival of a program only 22 bytes long. What was more, it was completely self-contained—not a parasite. TIERRA had evolved a functioning program using less code than any human programmer was likely to imagine.

Further tinkering with the initial conditions of a run of TIERRA produced additional examples of processes analogous to those in biological creatures. Ray found he could select for longer replicators as well as shorter ones or evolve forms of cooperation among highly related replicators, cheating, and even sexual reproduction. For example, sometimes two comp-viruses attempting to reproduce would choose the same location for their offspring. Each of the competing viruses might then overwrite different bits of this same section of memory. The result would be a hybrid, with part of its code originating from one "parent" and part from the other.

Ray's TIERRA thus provided a powerful demonstration that evolution worked in a binary environment just as it did in nature. Although he had designed both the environment and its first inhabitant, no new code was written at any stage in the runs exhibiting the evolution of parasites. Reproduction, death, natural selection, and mutation seemed to be enough to cause parasites to appear from nowhere. Ray didn't need to—and had not—built them into the system.

Reading Ray's report convinced John Maynard Smith (one of the most eminent of contemporary evolutionary thinkers) to announce (in one of the world's most prominent scientific journals) that "artificial evolution" in computers was possible. However, due to the hazards of over-anthropomorphizing computer intelligence that I noted earlier, we need to be very careful about how we interpret the events recounted above. The question is: Are these viruses real replicators? After all, it might be "virtual" replication only, because TIERRA is running in a "virtual machine." Replication might be simulated rather than physical because the von Neumann architecture, which dominates contemporary computing, separates processing functions (by the CPU) from memory storage (on disk). We need to know where the processes relevant to replication are happening to know if they conform to our expectations from Replicator Theory. Real binary sequence duplications—copies of memory states—appear in the RAM of Ray's computer when he runs TIERRA. But did they get there through a process of true replication?

This question can be answered by determining whether these viruses exhibit the four characteristics, discussed in Chapter 3, that identify true replicators: causation, similarity, information transfer, and duplication.

Let's use a program from Core War—perhaps the quintessential ALife program and one of the first computer programs designed to replicate information inside a single computer—as an example.

This classic Core War program, called Imp, is simplicity itself. If replication holds in this minimal case, the more complex TIERRA viruses should perforce pass the replicator test as well.

Imp consists of a single instruction: "MOV $0 $1," or move from reference address 0 to address 1. This instruction, when executed by the operating system, causes the move instruction to be copied to the next location in memory (addresses are relative to the current instruction). Then, on the next clock cycle, the system pointer will move ahead one position and execute the instruction found there, which is, of course, the move instruction, which tells the program to move forward again. In this way Imp marches through memory until something stops it. If Imp overwrites a location in memory and an enemy program tries to execute that instruction, the enemy will be turned into an Imp. With Imp, the search routine is totally automated or removed; it is immediately copied into the next slot without the need for additional instruction. Copy *is* the attack stage of Imp's life cycle as well. However, this stage is not destructive, being instead competitive in nature. So Imp bundles all three stages of the life cycle—search, copy, and attack—into one move.

Now let's run through the replication criteria to see if Imp satisfies them. Imp's "move" instruction being read by the virtual operating system at one address causes the system to write data to the next address in memory. This satisfies the causation criterion. Both of these two addresses now have the same instruction in them (MOV), making them identical. Further, the operating system installs the MOV command in the second register by referring back to what was in the previous address, so information has been transferred from the source to the copy. Once the transfer is complete, there are two MOV instructions where before there was only one: duplication. This satisfaction of all four criteria means that Imp, despite being incredibly simple (only one instruction long), manages to be a replicator. Since the TIERRA viruses operate in an analogous manner, they too constitute digital replicators within RAM.

It seems on the face of it somewhat preposterous to call virtual bits of code replicators. After all, the viruses in these ALife programs are not active agents themselves, being just sets of electrical charges in computer memory. How can they cause duplication to happen? During an execution of TIERRA (or Core War), a "virtual" environment is set up by the operating system in a portion of RAM designated by the software program. In this environment, virus code is "planted" onto memory registers by the operating system, which flips their bits in accord with instructions found in other registers. The key to replication in these digital ecologies is that the information in the viral "genome" is data (a set of binary numbers) but simultaneously serves as a set of instructions intelligible to the operating system. When read off, these numbers cause the operating system to take particular actions on that data or to perform manipulations on other locations in memory. Through this causal feedback between the replicator and the operating system, the conditions for initiating a replication reaction are satisfied. This duality of function, as information store and interpretable instruction, is crucial to the nature of replication generally.

So Maynard Smith's apparent jubilation over a new form of evolution was well founded. This is real evolution. It even involves replication. Artificial life programs are bona fide evolutionary environments. That's why they follow evolutionary principles and exhibit standard evolutionary patterns over time. For example, it is only through such a feedback process between the memory and information processor that replication can be differential. The shorter viruses in TIERRA were able to reproduce more rapidly than the longer ones and so gained an advantage in time. This is the very stuff of evolution: There was selection for shorter genomes in TIERRA.

TIERRA too is a quarantined world, cut off from the rest of the computer—and the world at large—by a wall of self-referential code. The interplay between a localized operating system and "data" (the viruses) is what keeps TIERRA isolated. Is the need for such isolation real? Ray certainly believed so and indicates that he was told it was vital to install a virtual operating theater for his creation by major figures in the ALife world. The fear is that these infectious replicators might get out into the "wild" of the Internet if the computer running the program was physically linked to the network. But since TIERRA viruses can't replicate except in the context of the TIERRA operating system, in fact they are safe; their "genomic" code is nonsense outside this "virtual" world, just as DNA is useless outside the nucleus of a cell. Similarly, the computer simulations that operate well within the confines of the holodeck on *Star Trek* are unable to step out of the room where the conditions for their existence are maintained (a fact played upon in several episodes of the television program). Like any other replicator system, they depend on finding themselves in just the right niche to survive and replicate. Imp, the Code War program that passed the replicator test, is considerably simpler than the one-line "self-replicating" printout program reproduced at the beginning of this chapter because it lives in a special world and can rely on being coddled by it. Outside that protected environment, of course, it becomes just another bit of code.

Many features of biological systems have now been duplicated *in silico* by a variety of descendants of TIERRA. It is evident from these simulations that digital analogues to DNA- or RNA-based viruses exist: These are indeed viral agents that exhibit a wide variety of phenomena that parallel those of their biological cousins. This parallel suggests that evolutionary processes really are "universal" in replicators of whatever stripe and color.

FROM FLOPPY TO HARD

As these developments in Artificial Life were unfolding, parallel evolution was beginning in a related domain: the ability of code to replicate *between* computers. By being able to construct a proxy—a different form of the virus "living" in a computer's equivalent of short-term memory, ready to go on the attack—viruses were able to escape the confines of their original host and spread out into the world.

Computer-borne pathogens are considered to be in circulation, or "in the wild," once they've been found on two or more workaday computers. This first happened in 1981 in an incident involving Apple II computers, although it wasn't spoken of as a viral attack at the time. Since then, the number of comp-viruses in the wild has continuously escalated with time, doubling every year or two. The number roaming through computers in 1990, for example, was a few hundred. By 1994, about 5,000 viruses were circulating. With the turn of the millennium, the total topped 50,000. At this rate, there will be nearly 500,000 comp-viruses in the wild by the year 2010.

Fred Cohen had mooted the idea that the replication of programs across a network of linked computers represented a threat to their security and stability in 1984. In that year, he published an important paper including both the results of several experiments and the theoretical foundation for later work on inter-computer viruses. In 1983, Cohen had written one of the first such viruses and then performed the first experiments in network security by releasing it onto a local suite of connected computers. The virus traveled farther and faster than anyone had expected in this limited environment. Another early experiment at the University of Texas at El Paso showed that in a standard IBM PC network, a virus could spread to 60 computers in 30 seconds.

Although these controlled experiments indicated how quickly viruses might be expected to spread, real-world infection rates could not be determined accurately until attacks took place on computer networks outside laboratories. The real world obliged shortly after Cohen turned in his dissertation on comp-viruses.

In 1986, two brothers, Amjad and Basit Alvi, working at Brain Computer Services in Lahore, Pakistan, noted the fact that the boot sector of a floppy diskette contained executable code. This code was the first area of memory activated when a 1980s-era computer was turned on because it contained a small program that instructs the computer how to load the operating system and activate the screen, disks, and keyboard. The brothers realized that they could replace this code with their own program, guaranteeing that a virus hidden there would gain control over the computer's activity as soon as the computer was turned on or rebooted and prior to the execution of any anti-viral software. The virus would then be free to lie in wait in computer memory and copy itself onto each floppy diskette that is accessed in any drive. The user would then carry this newly infected disk off to other computers, where it could then transfer itself onto another floppy's boot sector and repeat the process again. In this way, the program could spread itself widely. Indeed the brainchild of these brothers proved to be one of the most successful viruses of all time, becoming known as the Brain virus.

After boot-sector viruses came the more versatile file viruses. A file virus attaches itself to an executable file (such as those that contain software) and is activated when a computer user runs an infected application or opens a document created by the infected program. These viruses, which can be either RAM–resident or not, activate behind the scenes without the user's knowledge, performing their intended operations before permitting the actual program to run. Such viruses, spreading from disk to disk, started to appear in larger numbers in the late 1980s, causing thousands of computer systems to become unusable for periods of time and hundreds of thousands of users of several international networks to experience denial of services.

The early history of comp-viruses was largely a kind of pendulum swing between file viruses and boot-sector viruses. File viruses were "popular" in the late 1980s simply because there are more ways to infect files than boot sectors. However, by the mid-1990s boot-sector viruses again accounted for around 90 percent of all virus incidents because file viruses tended to disappear as the use of Windows as an operating system increased (Windows is fragile in the presence of file viruses). Although boot- and file-virus infections were common in the early 1990s, sightings of such viruses have waned since operating systems started including built-in safeguards against their spread. So major shifts in the kinds of viruses circulating in the wild can be due to revisions in or replacements of operating system software. In this case, the increased use of OS/2 and Windows operating systems caused viruses written for the older DOS to basically disappear by 1995—but not because the newer operating systems were designed to resist viruses. Quite the contrary, viruses have been even more successful in this new environment because new kinds of viruses could be written for and spread by these operating systems. Rather the inability of DOS viruses to survive in the revised environment was due to their reliance on program features missing in the newer operating systems.

But, generally speaking, virus incidents were widely considered urban myths (like rumors of alligators in the sewers of New York) at this time because they were often touted by the media and anti-viral software vendors but failed to materialize except on a very few computers. Even for successful viruses, the rise in prevalence was typically quite slow, often taking months or even years to become relatively common around the world. The equilibrium level of prevalence was also quite low. Well-prepared organizations experienced about four virus incidents per year for every 1,000 PCs they had, and this rate of incidence did not change substantially through the early 1990s. This relatively slow rate of spread was due to the fact that infection by these kinds of viruses required the physical transfer of disks between computers. But this restriction was about to be eliminated.

LEARNING TO NETWORK

The exponential increase in computer use over recent decades not only has resulted in the growth of information distribution and a shift in the way people communicate but also has provided an environment in which a new form of replicator has arisen and evolved: the network comp-virus. The crucial development was the standardization around a single protocol for the communication of information over such networks. Then came the development of hypertext: a way of easily accessing the information stored on linked computers through specialized software (called browsers), which allowed even everyday users to navigate the network from computer to computer in search of information. The widespread adoption of the Internet and the World Wide Web as a means for the communication of information naturally followed.

The Internet and the Web are actually an amazing collaboration of thousands of networks around the world that can share information using a single protocol called TCP/IP. The key to downloading and sharing data across the Internet is a format that may be read and deciphered by any computer on the network, no matter what operating system it's running. A sophisticated wiring and switching system also helps route data from one computer to the next.

As networking caught on, the population of computers through which a virus could spread increased dramatically. By the mid-1990s, "local area networks" (relatively small sets of computers connected within organizations) became "wide area networks" (like the Internet, stretching around the world). This enabled some viruses to be successful on a scale not seen before. The linking of computers, together with the nearly universal use of the same software programs (presently we live in Microsoft-World), led to dramatic increases in the population of hosts comp-viruses could potentially reach. The kinds of uses to which computers were put (and so types of software) also widened, with people using e-mail, personal messaging, Web radio and television broadcasts, and other forms of information-exchange via networks. This provided comp-viruses with a wider range of opportunities as well. The result is that these days, viruses can sweep around the world in hours or even minutes, thanks to electronic transfer via the Internet. Thus the chance of your computer being infected by some kind of virus has risen from about 1 in 10,000 per annum in 1990, to 1 in 1,000 by 1995, to 1 in 10 as this is being written.

The first denizens of these new populations of potential hosts were file viruses, which tended to spread primarily as "hitchhikers" on documents attached to e-mail messages. But in 1995, Microsoft introduced WordBasic, a text-based programming language for writing macro commands that vastly simplified the writing of viruses. Macros are essentially small programs that add specialized functions to the operation of large software programs.

For example, on May 3, 2000, the infamous Love Bug virus appeared. It spread (as an attachment to an e-mail entitled "I love you") from its origin in Manila, the Philippines, through the rest of the world *within hours,* striking 55 million computers, infecting about 3 million of them (you had to open the attached document to be infected), and causing billions of dollars in damage worldwide. The Love Bug outdid the Melissa virus, the previous record-holder in terms of replication, by many orders of magnitude. It managed this degree of success by sending itself to everyone in each infected user's e-mail address book.

So by the millennium, only five years since the beginning of the previous era, there had been another reversal in fortunes, with file viruses being back on top of the frequency charts. Seventy-five percent of all viruses in the wild at the end of the century were macro file viruses. The flip in prevalence away from boot-sector viruses was due not to a change in operating systems this time, but rather to the global networking of computers. This tended to increase the spread of file viruses because disk-based transmission became less effective in a world where information could be exchanged electronically with the click of a mouse.

Also important was the development of WordBasic, the easy-to-code language for writing macro viruses. Previously the stereotypical virus writer had been a pimply faced, asocial teenage boy who could write machine code, seeking vicarious pleasure through the fame of his progeny. Today the virus writer profile is much broader, including just about anyone, male or female, of any age. WordBasic represented a ready-made virus-writing kit accessible to anyone because all you had to do was put together a couple of commands, embed these in an e-mail attachment, and send it off. Point-and-click tools for producing complex comp-viruses have also been made readily available on the Net by virus writers. As a result, virtual communities of virus writers have sprung up that are willing to share code through chat groups, and virus-writing tutorials have appeared on-line that ease in the creation of this new type of virus.

One manifestation of this "progress" was Nimda—"admin" (a shortened form of "system administrator") spelled backward—a worm that started spreading in mid-September 2001 and quickly infected PCs and servers across the Internet. It was the first worm to use four different methods to infect PCs, making it the Swiss Army knife of comp-viruses. First, it spread by sending e-mail messages with infected attachments; second, it could scan for and infect vulnerable Web servers; third, it copied itself to shared disk drives on business networks; and fourth, it appended JavaScript code to Web pages that downloaded the worm to surfers' PCs when they viewed the modified page. This multipronged attack made it quite successful.

What's next? We already have viruses that preferentially infect popular personal data assistants, a new form of hardware that is often plugged into personal computers and shares operating system instructions with them. Further, communication protocols have now been established for mobile phones, and their use has increased dramatically around the world, so we can expect these to soon become victims of viruses tailored to their needs. The comp-virus story is just beginning.

ACCOUNTING FOR HISTORY

To better understand the history of comp-viruses, we need to think about computers in a new way. A computer is like an organism in many respects. The CPU and its attendants (RAM, ROM, hard disks) are the computer's brain, the many wires its circulatory system, the keyboard its sensory system, and its screen the face it presents to the world. Computers even have a primitive metabolism: the ability to suck energy from the cord plugged into the wall and, with this energy, to manipulate their own internal components. This makes them more like an organism than most artifacts. Unlike its biological counterparts, however, a computer can evolve new organs, or a more complex internal structure, with time. These are the new kinds of files it can accommodate. These new kinds of files then get targeted, often as soon as those new kinds of host "tissue" arise, by viruses. Coevolution thus occurs between an expanding number of host organs—new niches to fill—and the parasites to fill them.

Another difference from biological organisms is that computers can simply swap organs. Different versions of the same organ exist: For the word-processing function, a computer might use Microsoft Word or one of its competitors. Comp-viruses are typically organ-specific parasites, so those viruses that can attach themselves to the most prevalent organs spread best. On the other hand, comp-viruses that live in unsupported software are likely to die out.

Since each virus specializes in particular computer organs, there can be a microecology of parasitic programs within a single host computer, each resident in a different section of its "body." These viruses can even interact or come into conflict with one another, just as biological viruses do. The virulence of each virus is then affected by this interaction among them. One virus trying to cause damage to some host tissue may tread on the territory of another comp-virus, which blocks or accentuates this damage. Comp-viruses must therefore learn to coevolve in a situation of software conflict. For example, the DenZuk virus will overwrite the Brain virus if it finds a Brain already resident on a computer it is attacking.

Like biological parasites, comp-viruses are tied to a particular host species. In the case of computers, species are defined by operating systems, which are entirely different kinds of organisms for comp-viruses to live in. Thus a PC virus typically can't replicate in Apple Macintoshes because it can't depend on the same programming "calls" working in the microenvironment it lives in. Other participants in the replication reaction simply aren't present. Historically most viruses have been linked to the numerically dominant species, IBM PCs, and the dominant maker of software for that platform, Microsoft.

Continuing with this ecological analysis, a network of PCs can be seen as a social group of computers, because they're in communication with one another, sharing a common language and set of "customs" for the exchange of information. This group may have a number of parasitic viruses feeding off of its members, and each virus can be considered to have a certain prevalence in this population. The Internet then becomes an ecosystem of these strange, silicon-based organisms, composed of multiple computer species that have found a common language through which they can exchange information. This common communication protocol doesn't necessarily enable comp-viruses to switch host species, however, as their replication still depends on a replication reaction that expects local resources to be just right.

With these ecological relationships established, we can now analyze how comp-viruses have evolved over the course of their brief history. Certainly, their history suggests intensive selection of comp-viruses by the ecological conditions just outlined.

Comp-viruses also exhibit one of the other main components of evolution: mutation, and lots of it. However, this all takes place when they are being programmed by people. Computer transmission through networks is almost flawless, so computers hardly ever introduce variation into data themselves. Comp-viruses therefore can't rely on random flips of bits to lead to new wrinkles in their offspring, as TIERRA did. Further, almost any variation introduced through flaws in transmission is likely to cause the resulting program to become inoperative. So how does variation arise? Effectively all mutations are intentional, in the sense that humans must take a copy of the virus "off-line" from the course of its spread through a network, reprogram it, and then send it out into the wild again with a new form. This makes evolution in comp-viruses very limited. They are quite fragile replicators dependent on highly sophisticated assistance for any real evolution.

On the other hand, comp-viruses engage in simple manipulations of their environment to make it more amenable to their continued presence and replication. They alter memory states, switch program pointers to hide themselves, or mark their domain to keep other programs of the same type from intruding into their territory. So comp-viruses are able to engage in a wide variety of responses to environmental conditions. They just can't *evolve by introducing variations into their code*—at least for themselves.

This makes the network comp-viruses quite different from the RAM-resident ALife viruses we looked at earlier, whose forte seemed to be the development of truly novel abilities to respond to changes in ecological circumstances (like the sudden appearance of a predator). This might be expected, because ALife viruses are tiny programs living in a

protected environment, whereas network viruses are much larger programs living in an environment bent on their destruction. Network
viruses have to be much more clever than their ALife counterparts to
survive. And their environment keeps changing in ways that few biological viruses have to deal with: There are complete changeovers in
operating conditions and host species types every few years. Only
"assisted" evolution (through human design) would seem able to keep up
with such drastic fluctuations in selection pressures.

SURVIVAL OF THE COMP-VIRUSES

We have thus seen that comp-viruses face difficult selection pressures in
the wild. From the perspective of Replicator Theory, I now want to ask
whether it is likely that only the fittest comp-viruses survive (as was the
case with our other comparative case of a replicator, prions). For this to
hold true, the replication reaction for comp-viruses must exhibit a one-
for-one dynamic. Is it true that the replication reaction for comp-viruses
exhibits a one source–one copy pattern, as expected by Malthusian
growth?

Well, what about the macro virus Melissa, which lands on your computer, only to turn around and fire itself off to the first 50 names it finds
on your e-mail list? Isn't that a much higher rate of reproduction at
work—50 copies at one go? In fact, it is not. Fifty copies of Melissa don't
reside on the source computer, each one being fired off to a different destination. Instead the computer's operating system is having to do a lot of
work; it repeats 50 times the "now send this" command, each time to a
different location. That is 50 events, constituting in each case the beginnings of a replication reaction. And it is only a series of signals that
Melissa produces. Each e-mail simply includes instructions the destination computer needs to create a copy of the Melissa file on its own disk.
Before any copy of the virus is actually made, the destination computer
must be an IBM PC with the right e-mail software package on it. And
users must execute the file they receive with their e-mail. This doesn't
always happen either. These filters on success can significantly reduce the
effective distribution of the comp-virus.

But the most salient point is that, from the point of view of the host
computer, a 1-to-50 rule is not at work: In fact, hidden underneath, the

operating system is engaging in many independent activities, each one of which qualifies as a reproductive act. The point can be seen more clearly in the ALife case, which doesn't involve signaling in the same way. Only one memory cell is being operated on at any given moment; there is no parallel processing in von Neumann computers. So the operating system plods from one register to the next, seeing what it is supposed to do with the information there. Comp-viruses are "made" one at a time, both within and between computers.

So the answer to our question is clearly yes: Each comp-virus could inspire the production of a copy of itself. A linear rate of growth thus appears to characterize the behavior of comp-viruses. This trait makes comp-viruses like DNA and prions, and is consistent with an active selection process in a rapidly changing environment of new kinds of computers and software. Only adaptive comp-viruses should survive, as suggested by their history, in which the dominant forms are rapidly replaced.

Can we now characterize the nature of comp-virus replication reactions more clearly? Where do they occur? The first place to look is RAM, since all instructions go through there. Still, no virus is likely to find everything it needs already present in RAM at the time it wishes to replicate. Many of the calls required by a virus in RAM reside on the hard disk, where the software programs that assist in replication are stored, or in ROM, where many basic computational functions permanently reside. So code stored either in ROM or on the hard disk (as well as RAM itself, of course) can assist viral replication. But when these calls are made, the instructions are "loaded into" RAM, where they become actively present. It is as if RNA had migrated from the cytoplasm of a cell into the cell nucleus to help with DNA duplication. We can therefore think of RAM as being the enclosed domain within which the replication reaction happens, where all the relevant parties come to circulate in an electronic soup, waiting for their cues. Variation will arise in just how dependent a given virus is on making calls to ROM or a disk, so dependence on "external" resources will always be a matter of degree. Even in cases of disk-to-disk replication, RAM is still the place where the intermediate steps occur, although the actual duplication of information takes place only on the disk, as a result of signals sent by the information processors down the cable attaching the disk to the motherboard, where RAM resides.

How do these considerations play themselves out during the history recounted above? In Phase I, infection was controlled, localized to the RAM disk within individual computers, and short-term. Replication reactions, thanks to the involvement of virtual operating systems, were almost entirely restricted to RAM, approaching the ideal of true isolation.

In Phase II, replication still took place within the confines of one computer. Typically the spread of a virus was dependent on instructions that remained resident in RAM. Once activated, such RAM–resident programs used operating system commands to remain active in memory. This gave them the chance to infect the many other bits of information that go in and out of RAM during normal computer operation. And sometimes this information would be copied to "hard" media inserted into the machine, such as floppy disks. Then Phase II viruses could spread through what has been called "sneaker net"—human vectors padding in their tennis shoes, handing these physical media (floppy disks) from person to person. In effect, the virus copy was being *physically carried* in some long-term storage medium to the next host. Because of this contact between computers, "epidemics" were not restricted to just one computer. But this method of transmission worked slowly and laboriously, since human vectors don't move as quickly or as far as the electronic ones that came later.

This *modus operandi* involved a reversal in roles from that characteristic of the players in Phase I, where the program in RAM was the replicator. The ALife program itself, sitting on the hard disk, was relevant only to the extent that it produced the virtual operating system that catalyzed the RAM–based replication of data structures. Now, with Phase II, the program in long-term memory has become the replicator, and any manifestation in RAM is the catalyst for the duplication of long-term memory structures. In Phase II, the data structures to be duplicated exist on hard disks, which are accessible by the general operating system on the boot disk. This is what Thomas Ray and other ALifers were worried about: Use of the computer's "real" (as opposed to "virtual") operating system as a catalyst would lead to the potential destruction of real data in long-term memory. (Of course, what the ALifers most fear is just what network comp-virus writers desperately want: their progeny replicating out in the world.)

Phase II involved normal copying from RAM to disk using local operating system commands. No communications protocol was required. Infections are now long-term, but with a relatively slow rate of spread. With boot-sector viruses like Brain, two copies of the same virus exist simultaneously on the same machine: one copy in the boot sector of the boot disk, another copy in a sector of a floppy disk inserted into another drive, together with the operating instructions in RAM derived from the viral program, which functions as a catalyst for replication (assisted by the operating system).

Phase III, in contrast, is defined by the electronic replication of network comp-viruses. These viruses have a vastly increased rate of spread, leading to real viral epidemics that have crippled sections of the Web for short periods. Thus, as time has gone on, the different kinds of vectors underlying the epidemiology of computer viruses have had a major impact on the speed with which viruses can disperse, and hence on the "reality" of viral epidemics. The rate at which sneaker net can connect computers is vastly slower than that of electronic transmission. This has meant a considerable reduction both in the time it takes for an epidemic to occur and in the size of the potential population of susceptible computers. Electronic transmission is largely responsible for comp-viruses becoming a real threat to the security and functioning of the Net.

The new reality of epidemics has in turn led to the need for anti-viral software distributed to the users of computers tied into the Web. But this has itself prompted the development of techniques among virus writers to increase the likelihood their viruses can avoid detection. So an arms race of offensive and counteroffensive, each time with more sophisticated weapons, has begun between viruses and anti-viral software. This race, which began during Phase II, continues today.

What makes Phase III possible is a change in the nature of the physical links between computers: As they get wired together, the flow of electrons is smoothly extended from inside one machine to exchanges between them. Signals can now escape the narrow confines of their source host to wander the network. But the involvement of two physically separated sets of electronic memory (in the RAM of two different computers) means that the RAM in the source computer is responsible only for producing signals, while the receiving computer's RAM must complete the replication reaction based only on information provided by those signals. In effect, the *signal-assisted* replication of Phase II (which brought all the information together in RAM) becomes *signal-mediated* replication in Phase III.

Taken together, these considerations mean that we can have RAM–RAM or disk-disk methods of replication, but not RAM–disk replication. Some viruses replicate within RAM, others within a single disk, others between disks. But in each case, the same replicator does not replicate in *both* electronic and magnetic media: The surmise (derived from Replicator Theory) that replication is typically specific to one set of circumstances is upheld. All these routes lead to Rome, or the Holy Grail of replication, but the roads differ in the kinds of terrain they pass through to get there.

It is important to recognize that the signal sent through a channel between computers is not a comp-virus itself. The signals traveling between computers are different than the sequence of information that initiates the process and that gets copied onto the destination computer. Networked computers must *translate* the information in the virus into a form amenable for transmission through a particular kind of channel to another computer (which may be running a different operating system). So a coding procedure falls between the instruction to send a message and the message actually being sent. In the case of the Internet, the communication protocol involves adding on bits to the signal to make it capable of reaching its destination in a form interpretable by the destination computer. In particular, when one computer is told to send data across the Internet, the data is broken into sections called packets. Special identifiers are added to each packet to help it reach its destination. It's like placing each packet inside an envelope with the destination address printed on the outside. The difference is that many packets are typically required to deliver a message, and each packet flies to its destination by a potentially different route. The packets thus need to be decoded and reassembled by the receiving computer to reconstitute the message sent by the source computer. This should make it clear that these signals are not themselves a virus-in-transit.

Of course, replication reactions taking place within the confines of a single device may use signals to communicate with the operating system or long-term memory, but they won't have to switch coding schemes. At the same time, the instructions sent to the operating system that result in writing a sequence of magnetic states to memory are not the same sequence as the comp-virus either. It's always the case that messages are not the same as "meanings" in computer communication (nor its human equivalent). This is a fundamental point of profound importance in the hunt for the memes.

CONVERT THY NEIGHBOR

We can conclude that Replicator Theory applies to comp-viruses as well as prions. Genes and prions are molecular, but comp-viruses are super-molecular, being composed of states in multiple molecules (actually sequences of electrically charged molecules). But these differences in scale don't seem to matter. The principles we identified with replicators earlier—of sticking to one substrate and of information transfer (or con-straint transference)—both hold true at each scale. This is not that sur-prising, since we already noted that Replicator Theory has been applied to a variety of problems in evolutionary theory besides the description of DNA duplication. Since Replicator Theory applies to all the replicators we know about, I will assume it can be usefully applied to memes as well.

The ways in which comp-viruses replicate parallel in many ways those of their biological counterparts. The basic process is essentially equivalent in feel to what I called conversion, as in prions. Comp-viruses work by converting existing memory states to states that can produce copies of themselves. This is just what prions do. Such a method relies on the existence of the right conditions. Luckily for virus writers, nearly everyone uses the same software for email, word processing, and other basic tasks, so the number of potential hosts runs into the millions.

Comp-viruses are also like prions in using a co-opting strategy: They convert an existing physical substrate. All comp-viruses share a funda-mental strategy: to flip the electrical state of some unit of memory from one kind of charge to another. This means the use of the word "virus" is entirely appropriate: They are dependent replicators.

On the other hand, comp-viruses are very robust replicators and so don't exhibit the same limitations as prions in this respect. Fred Cohen proved that viruses could evolve into any result that a universal computing machine could compute, thus introducing a severe problem in detection and correction. In effect, the fact that computer languages are like real languages means that programs can express—or compute—anything. The problem is that they typically have to be designed by human hands to do those things. So the fact that they are *potentially* robust does not necessarily lead to their full-throated evolutionary potential being realized. They are limited by human imagination, which in many cases is less creative than the natural selection process.

However, it is important to note here that the comp-virus case brings up one important qualification of the argument I have emphasized thus far, that replicators always remain tied to only one physical substrate. The qualification is that *information transfer can be mediated.* That is, the transfer of information from one place to another, within the context of a replication event, can be conducted by an accomplice, rather than by the replicator itself. Signals generated by replicators can serve this functional role of message mediator. This is a new feature of comp-viruses not seen in genes or prions.

Further, this mediation tends to loosen the bonds to individual substrates. Comp-viruses, as we have seen, can jump from magnetic to optical substrates of memory, with the chance of flipping back and forth between magnetic and optical forms in the future as well. Admittedly, both long-term memory substrates work with electrical charges. Nevertheless it is the first indicator we have of any flexibility on this point. The question is whether it is the super-molecular nature of comp-viral replication, its mediation by signals, or the digital nature of computer code that allows this degree of slop in the lineage. This is an important question from an evolutionary point of view.

Of course, Phase I and II comp-viruses don't exhibit this feature, so it appears not to be strictly due to digital coding. Since these early forms are also super-molecular viruses, this cannot be the whole story either. What is new to Phase III viral life cycles is signal mediation, so it appears this is what is crucial. It is the easy, reliable code-decode protocols that can be arranged for binary data transfers introduced by such transmission that make it possible to begin life on one substrate and end it on another.

The essential point about this is that the messenger need not be of the same substrate as the replicator. The source and copy replicators still need to be made of the same material; it's only the signal in between that introduces a degree of play in the system. This point will be important when we come to discussing how memes propagate themselves.

If we now look at both of the replicators we have investigated, we can see a similarity between them. Both prions and comp-viruses act like religious fundamentalists. It's as if they lie in wait for you at airports or come directly to your door and try to convert you into something like themselves. Their mission is to spread their version of what's valuable: They want to share their state of being with you. In short, they seek to *convert* you. Persuasion is their *modus operandi.* They can't make whole new people from scratch, being quite willing to rely on other replicators for that. But once you're around, they go to work on you, trying to refold you into something more congenial, which they can trust to go off and do the same job on yet other innocents. In effect, comp-viruses are manipulative memories, and prions proselytizing proteins. As we shall see, memes use these same methods of persuasion in their attempts to dominate *their* world.

Chapter Five

THE DATA ON INFORMATION

Where is the wisdom we have lost in knowledge?
Where is the knowledge we have lost in information?

— *T.S. Eliot*

Memetics is very far from the "mimetic" judgment Plato passed on
materiality — that it consists of poor imitations of "True Forms." Memetics
is the antithesis of metaphysics, insisting that matter matters.

— *B.C. Crandall*

We have learned a lot about the nature of replication with our excursion through Replicator Theory and Universal Darwinism, as well as from our foray into "Comparative Replicatorology" in the last chapter. But some edges of the replicator notion still remain fuzzy.

In particular, one criterion for replication is that information be transferred from the parent replicator to its offspring. But is information something "hard" or "soft," physical or abstract? The nature of information is a much-vexed issue in the philosophy of science, and it will require an entire chapter to deal with. I can't boast of finding The Final Solution to the Information Problem, the one true point of view that all philosophers henceforth must accept. After all, they've been arguing about information since the ancient Greeks. But now enough real headway can be made on this issue to firm up our ideas about the nature of replication. In fact, we will be able to derive a couple of corollaries to replication in the process. This clarification will then prove to have useful spin-offs when it comes to catching a meme.

INFORMATION IS PHYSICAL

There is a widespread sense abroad—in both academia and the world at large—that information is an abstraction. But I argued in a previous chapter that information can be transferred from one replicator to another. So how exactly can you inherit an abstraction? Steven Pinker says that "information can be shared at negligible cost: if I give you a fish, I no longer possess the fish, but if I give you information on how to fish, I still possess the information myself." Pinker is suggesting that the "inheritance" or passing along of information is rather special because it comes with no price tag. How is this possible? Because information is not a thing of this world. It is immaterial—after all, anything you can give away and still have must be something magical and ineffable, right?

Pinker's "free gift paradox"—that you can pass information to someone else without losing possession of it yourself—springs from a common tenet of cognitive science: the equating of ideas with information. This allows Pinker to say that "mental states are invisible and weightless. . . . The content of a belief lives in a different realm from the facts of the world." If this is the case, then beliefs can't exist in the brain, which is composed of a messy network of cells no one knows what to do with. We must therefore separate the mind from the brain, introducing a dualism that has been prominent in intellectual circles ever since René Descartes, the French philosopher, introduced the distinction between mind and brain in the seventeenth century.

In this view, an idea, and perforce any other form of information, doesn't have mass or charge or length. Likewise, matter doesn't have bytes. You can't measure so much gold in so many bytes. Matter doesn't have redundancy, or fidelity, or any of the other descriptors we apply to information, which make it possible for information to be a replicator. This dearth of shared descriptors means matter and information have to be discussed separately, in their own terms. This leads directly to the claim that replicators like genes—or memes—and interactors like organisms belong to not only separate conceptual worlds but different realms of existence as well. This kind of dualism had its forebear even earlier than Descartes, in the philosophy of Plato. For Plato, ideas existed in an abstract heaven, which he called the realm of Forms. Although this realm was independent of everyday physical reality, it could be accessed by the mind through deductive thought. Platonism implies that replicators reside in the domain of Forms (where information resides), while interactors live in the domain of material things, with us and our minds.

The replicator, then, isn't a material object. The physical gene—that is, the string of atoms comprising the DNA molecule—is not what evolution conserves and passes on. Only the information embodied in the nucleotide sequence gets from one generation to another; the atoms actually bearing the message get swapped over each time transmission occurs. A gene is just a kind of "cybernetic abstraction," a message that is transmitted in a kind of magical way from generation to generation.

An informational replicator, then, may exist on different substrates, using different codes. This is what makes the replicator notion abstract. It defines a replicator simply by its function. Such an idealistic interpretation may have some analytical interest, but it can't be used to understand evolutionary lineages, which are distinct historical entities limited to one time and place.

Nevertheless, this "info-mystical" viewpoint is common, both among our better thinkers and in the popular imagination. George Johnson, the distinguished science writer, for example, recently claimed that

> these [computers]—the most complex things produced by the human mind—can be made indefinitely small because of a crucial distinction. While ordinary machines work by manipulating stuff, computers manipulate information, symbols which are essentially weightless. A bit of information, a one or a zero, can be indicated by a pencil mark in a checkbox, by a microscopic spot on a magnetic disk, or by the briefest pulse of electricity or scintilla of light. The special nature of information confers another advantage. The power of computation can be leveraged and leveraged again. Design a computer and then use it to help you design a better computer, ad infinitum.

The rise of computers has had a lot to do with the popularity of ideas like these. The constant process of transferring information from one physical medium to another and then being able to recover that same information in the original medium brings home the supposed separability of information and matter. In biology, when you're talking about things like genes and genotypes and gene pools, you're talking about information, not physical objective reality. They're patterns. Even if computation is tied to a physical substrate, there's a difference between computation and information. Computation is just a way to juggle numbers with machines.

But we can see now that Johnson's argument—this time taken from computer science rather than biology or psychology—is fundamentally mistaken. The exact *opposite* of what Johnson says is happening. As our computer chips become ever smaller, they are beginning to brush up against constraints set them by fundamental physical laws—even quantum confusion about just what state they are meant to remember. So as we head toward quantum-scale information processing and begin to compute with individual molecules or photons, it will quickly become evident just how "entangled" the fundamental units of information are with their physical substrates.

It's true that information doesn't have a number of physical properties like mass or charge or length. For instance, you can't know how heavy an idea is because when ideas change, the brain substrate still weighs the same—it's just that the bits of material have been rearranged. In this respect, ideas are a bit like consciousness, or the soul. When the soul (supposedly) departs the body at death, no change in the body's weight can be observed.

But ideas are not immaterial. Information doesn't exist independently of the material through which it is made manifest. Even our thoughts and ideas are in the structure of gray matter and the form of electrical fluctuations in our brains. Changing ideas can require the expenditure of energy needed to rearrange the bits of matter. Information is a measure associated with a *quality* of matter. It may not be matter itself, but information is still a physical quantity.

The message of this chapter is simple: It is a fundamental fact that "information is physical." This phrase has become the mantra of a group of scientists called information physicists. It was first adopted by one of their most august members, the late Rolf Landauer, of IBM's Thomas J. Watson Research Center. The mantra doesn't mean that information is a separate kind of entity whose bits get totted up alongside the number of molecules and joules when accounting for all the material in the universe. Nevertheless the presence of information can change the course of events, and it is in this sense that information is physical. Any complete description of the universe must include information. Physical reality must be considered to include not just matter and energy, but information too. As Norbert Weiner, one of the founders of cybernetics in the 1940s, put it: "Information is information, not matter or energy. No materialism which does not admit this can survive at the present day." Why does information matter? Because the *structure* of the universe— that is, of matter and energy—is a function of the distribution and quantity of information. If matter and energy are randomly distributed through space, there is no structure to the universe; if there is useful information in the universe, then there are pockets of organized matter or energy within it, capable of doing work, such as being converted from one to the other. Information is a *quality* or property of bundles of matter or energy.

Information has not always been an important concept in physics. In earlier centuries, theoretical attention was focused on apparently more fundamental concepts like mass, momentum, and energy—on the task of finding the universal properties of nature that could be described using simple, elegant equations. Nevertheless information has come to the fore in recent years. In fact, it is now believed that information may be the most basic concept in physics—potentially even more important than matter or energy as an organizing principle. Just as Theodosius Dobzhanzky famously said, "Nothing makes sense in biology except in the light of evolution," so too one might argue now that nothing in physics makes sense except in the light of information—in particular, the generalized information theory that is now emerging. Indeed the two domains of physics and biology appear to be intimately related, since it is possible to describe evolution itself as a process of computation.

Rolf Landauer's work, beginning in the early 1960s, makes clear why physicists are interested in information. Before then, it was widely believed that processing a single bit of information—each 1 or 0 of binary code—inevitably consumed some energy, which placed a fundamental constraint on computer power. But Landauer showed, to the surprise of many people, that this was not true. As computer technology becomes ever more efficient, each calculation can be performed with less energy expenditure, because it occurs on smaller and smaller scales. But there is still a cost. At some point, the bits must be flushed from the computer's memory so that the machine can be reset for another computation. You go through a history of calculations to get a result that contains less information than the process itself. To take a simple example: 2 + 2 = 4. But once you've gotten to 4, you can't recover how you got there: was it 2 + 2, or 213 – 209? The sequence of steps that got you to the end of the calculation is forgotten. What Landauer demonstrated was that it is only at that point—when information is erased—that computation dissipates a small amount of energy as heat. This result, now called Landauer's Principle, has been described as the "thermodynamic cost of forgetting." So energy consumption is not an intrinsic quality of computing; it is *erasing* data that consumes energy.

It seemed to be a physical necessity that all computers must perform this erasure of information and thus incur the energetic cost. But Landauer also postulated that this inefficiency in contemporary computers resulted more from a coincidence of design than from any physical law: Computers were built with irreversible logic because it was easier to design a circuit that trashed unwanted data than it was to design a closed system that would reuse it. Landauer's colleague at IBM, Charles Bennett, soon convinced himself that a reversible computer could be designed that would circumvent Landauer's principle. Computation need not be a process in which information is discarded. When such a computer completed its task, one would essentially run it backward, returning it to its initial state without tossing away any information. Such a computer could operate with unprecedented efficiency, because it would not use any energy. The work of these physicists at IBM eventually led to the development of reversible computing—a system that recycles leftover bits and their energies instead of dissipating them as heat. Reversible computing has remained mostly theoretical thus far but may yet revolutionize the way computations are done. At any rate, the fundamental idea that computation can occur, at least in principle, without an expenditure of energy—or a loss of information—is now well established.

Information's physical nature becomes particularly obvious when it starts obeying counterintuitive laws, such as those of quantum mechanics. For example, in a phenomenon called "quantum teleportation," objects or information can be transported from any point A to any point B (theoretically across a galaxy) without traversing the intermediate space. I don't use the word teleportation casually here; I mean to invoke the fantastical machines called "transporters" in the television series *Star Trek,* which can immediately make a person or object disappear from one place and reconstitute a perfect replica in some other place, regardless of distance. Exactly how this piece of legerdemain is supposed to be accomplished is not explained in the television program, but the general idea is that the original object (such as Captain Kirk's body) is first scanned, extracting all the information from it. Then this information is transmitted to some receiving location and used to construct the replica of Kirk's body from atoms of the same kinds, arranged in exactly the same pattern as the original. Captain Kirk disappears with a spacey kind of sound effect from inside the spaceship Enterprise and reappears a moment later on some planet's surface. The news is that a way to achieve this bit of science fiction in real life was recently suggested by IBM's Charles Bennett, and has even been demonstrated by transporting a quantum state of light from one side of a lab table to the other without it traversing any physical medium in between (believe it or not!).

Quantum mechanics is non-local, meaning that distant objects may become "entangled" together physically. In this peculiar quantum state, particles of matter may have effects on one another even though they are not close together—what Einstein called "spooky action at a distance." The entanglement of two particles is achieved by first bringing them into contact and then separating them. These two particles thus become twinned in a way idiosyncratic to the world of quantum mechanics. They share information that has no independent existence in either particle alone, such as the fact that they are in states that are opposite to one another. Once this pair of particles is entangled, they can constitute the communication "channel" through which a message can be passed, despite being taken any chosen remove from one another. This is because altering one particle of this entangled pair causes the other to be affected in a highly correlated way without any material link between the two.

The first step in quantum communication (which involves the teleportation of information) is to give one of this pair of particles to "Alice" (the physicists' traditional name for the message sender) and the other to

"Bob" (the guy who receives the message). Bob may be very far away from Alice "as the crow flies." For communication of the message to happen, the obvious difficulty is getting the particle to Bob without disrupting its state. So once Alice and Bob have received their entangled particles, Alice brings another particle—the one whose state she wants to teleport to Bob as a "message"—into the vicinity of her entangled particle. At this point, Alice scans the message particle together with her member of the entangled pair of particles. This yields some information about the state of the two together but disrupts both their states. This is because Alice here comes up against the constraints of Heisenberg's Uncertainty Principle. At very small scales, the act of observing a system—for example, by shining a light on some molecules in a tiny container—can inject enough energy into that system to disturb its state. So the very act of attempting to measure something introduces into the equation a fundamental and unavoidable uncertainty about the state of what was to be measured. In effect, you can never measure both the position and momentum of a tiny particle at the same time with infinite precision because observation itself disrupts what you're looking at. This means that, in quantum communication, the measurement of a particle does not reveal to Alice anything about the separate particles, but only about the two of them with respect to each other. But even this kind of measurement destroys the state she teleports.

To reconstruct Alice's message requires two bits of information: First, are the states of both the message particle and the entangled particle presently the same (yielding a binary yes/no answer); and second (regardless of the previous answer), what is that state (another binary up/down answer)? Quantum teleportation permits the instantaneous transmission of one of these answers via the original entangled pair of particles, while a "classical," less-than-light-speed connection such as a telephone line must be responsible for communicating the other answer. Since both answers are required before either is useful, the duplication of the original state at Bob's location takes time to produce.

Still, the answer to one of the questions is already at Bob's place, in the state of his entangled particle. Because Alice measured the particle to teleport and the particle entangled with Bob's together in the same special space, they too will become correlated in the peculiar quantum mechanical way—that is, entangled with each other. Although measurement destroys the original state of the message particle, Alice sends her measurement results to Bob by calling him. After hanging up the phone, Bob may have to twiddle his entangled particle to make its state agree with the one at Alice's location and thus retrieve the original, pre-measurement state of the message particle. Bob uses the information Alice sent him through a classical communication channel to determine which transformation of his particle is required. Once this is done, he has resurrected at a distance the original quantum state of the message particle (which was unknown to either Alice or Bob) and so duplicated the message.

Quantum teleportation is thus a process in which the complete information in an unknown quantum state is decomposed into classical information and quantum information (the latter called an Einstein–Podolsky–Rosen Correlation). These are sent through two separate channels and later reassembled at a new, potentially very distant location to produce an exact replica of the original quantum state that was destroyed through the message-sending process. Because you can determine the state of a classical physical system without disturbing it, there is no problem making a copy of information at macro scales. But at the quantum scale, this destruction is a necessary feature of the process, due to the Heisenberg Uncertainty Principle. Alice's determination of one property of the particle (such as its rate of spin) to a certain precision reduces the precision with which she can know about a complementary property of the particle (such as its direction of spin). Therefore Alice cannot retrieve all of the information that is stored in a particle. That's why it's impossible to copy quantum information perfectly. While you can move quantum information around, you can't copy it—you can't xerox a particle. Nevertheless, the message has been transferred; Bob has consumed the signal. It is therefore an instance of communication.

Note that the communication channel has no length in this case, even if it joins widely separated points in space. This is quite different from electrons proceeding down a telephone wire—the phenomenon Claude Shannon was thinking of when he developed the mathematical theory of communication. In a telephone wire, the electrons must interact with the atoms between points A and B, potentially leading to disruption of their state and hence the message. But with quantum communication, there is no interaction with the world, and hence no possibility of disrupting the message. Quantum communication thus makes it clear that Shannon's theory of information is not universal. It applies only to cases where the transmission goes through space. So, theoretically at least, you can have perfect communication, without information loss, using quantum methods: Bob can be absolutely sure of obtaining the quantum part of the message exactly as intended by Alice, regardless of how far apart they are.

But true teleportation is not in prospect because reconstituting a duplicate from local materials would require some way of getting round the physical barrier of instantaneously transmitting *all* the relevant information about an object, which has been proven impossible. Still, "beaming down" to a planet's surface some seconds after being in a circling spaceship—a staple plot device in the *Star Trek* series—is possible. Realization of this truly amazing prospect awaits an engineering solution to the considerable problem of constructing a state of entanglement that can encompass simultaneously the very large numbers of atoms making up objects like Captain Kirk.

The fundamental lesson of this counterintuitive phenomenon—the point I want to draw from it—is that you can think of transmitting information in just the same way as transporting bodies. That is, you can send a message or move Captain Kirk from one place to another *using the same equipment.* In both cases, it is just a question of rearranging atoms into the desired configuration. How much more physical can information get?

At any rate, it should now be obvious that the dominant conception of information in science is a down-to-earth, rock-bottom physical one. Information may take a variety of forms, but it always manifests itself in some physical structure. This justifies the cry of the information physicist: "No information without physical representation!"

THE NATURE OF BIOLOGICAL INFORMATION

Information may be physical, but what does this mean for biology? Remember, our interest in information comes only in the context of figuring out what the implications are for replication and evolution. Physicists treat *any* kind of structure as information. But *biological* information isn't just about structure. In evolutionary theory, more subtle discriminations are required. For example, in DNA we have to distinguish the special structure exhibited by the sequence of linked nucleotides from the physical structure of the double helix itself. The positioning of atoms in that special staircase arrangement has many qualities that make DNA a superb choice for storing information, such as redundancy (each strand is complementary to the other), closure (it wraps up on itself like a cocoon into something called chromatin, protecting the messages within), built-in error-correction (the right base has to match up or the double helix goes out of kilter), and so on. Double-stranded DNA probably evolved from a single-stranded, RNA–like precursor for this reason. It is a superior molecule for information storage.

The problem is that all of these features of DNA molecules count as structure but not as biological information. DNA molecules are adapted to replicate. The features of DNA molecules that allow them to replicate were selected for, beginning with their precursors at the origin of life. But the global aspects of the molecule's structure just mentioned aren't what get involved in its major biological function: the transmission of information across time. That role goes to the ordering of the base pairs themselves. The genetic "message" is made up of these units, strung one after the other along the helical backbone. That's why, for example, the bonds that connect the two bases that make up each of the rungs of the DNA ladder are easier to sever than those that connect successive nucleotides together. This makes it is easier to chop up the helix into bits than to cause it to split down the middle (although it makes this latter move when replicating). This means you can insert a new bit into the middle of a genetic "word" but not so easily tear an existing word in half.

Unfortunately, in spite of the massive amount of work done by a variety of scholars to explain the notion of information, none of the suggestions made thus far is adequate to distinguish information as it functions in evolution from other sorts of structure. The physical information contained in the arrangement of atoms in a double spiral staircase in your living room is no different in kind from that exhibited in the linear sequence of base pairs that actually produce the proteins that evolution works on. Structures are not physically separable from concrete systems: They may be distinguished only analytically.

We also can't just say that it's the sequence of base pairs that gets passed on. It isn't a particular sequence that transfers when DNA replicates; instead the *complement* of each base does (that is, for every cytosine that comes along, a thymine gets linked to its opposite strand, and for every adenine, a guanine is affixed to the other side). That is, the original ATGC does not equal its transcript, GCAT; they may match one another, but they are opposites. While the transcript may later be used to reconstruct the original sequence because they *are* complementary, they are still not the *same* sequence.

Information thus can't be equated with structure or order, pure and simple. Biological information is not equivalent to physical information; it is some kind of subset. Only the aspects of structure that matter to evolution actually count to a biologist. The question, then, is how to distinguish the special *sort* of structure exhibited by sequences of base pairs in molecules of DNA from the physical structure of the double helix itself—that is, from the layout of atoms in space. Perhaps we can find inspiration from an earlier generation of thinking about genes. What did genes "do" before they transmitted information?

Inheritance was not always considered a problem of information transmission. In fact, until around 1950, molecular biologists described genetic mechanisms without ever using the term "information." The prevailing view in the early part of the twentieth century, championed particularly by Linus Pauling, was that genes transferred (not "transmitted") chemical *specificity* through time. By the 1930s, work in biology had determined that interactions between molecules in living organisms are highly specific—that is, a given molecule would typically react with only one reagent. In effect, enzymes act only on particular substrates. This was encapsulated in the catchphrase of the time: "one gene–one enzyme." As more was learned, this organizing principle became codified in the idea that the behavior of large organic molecules was determined by their conformation, or three-dimensional shape. Pauling, the great pioneer in determining the structure of large molecules like proteins, was particularly fond of the image of biological interactions taking place as a kind of precise fit between a "lock" and "key" as the surfaces of these molecules came into contact and intermediary aggregations were formed, then broken apart.

This lock-and-key arrangement is typified by a general phenomenon called "molecular recognition," which results from an exact fit between the surfaces of two molecules. The random bumping together of large macromolecules in solution is the first step toward molecular recognition. When the surfaces of two molecules match well, enough weak bonds can form between their surfaces to temporarily withstand the thermal motions that tend to break them apart. These chemical bonds lend relatively high stability to their relationship. It is these physical requirements for matching that naturally account for the specificity of biological recognition on the molecular level. For example, enzymes can typically only catalyze very specific kinds of chemical reactions due to their ability to latch physically onto just one or a few types of molecules.

Pauling thought this kind of process would extend to gene-based inheritance as well. And it does. For example, a particular protein can attach itself to a stretch of DNA by "recognizing" this specific sequence because its physical shape is extensively complementary to special surface features of the double helix in that region. In most cases, the protein makes a large number of contacts with the DNA. Although each individual contact is weak, the numerous contacts that are typically formed at the protein–DNA interface add up to ensure that the interaction is both highly specific and very strong. Once in place, the protein can prohibit that gene from being "expressed" (or read off by RNA) for the manufacture of another protein that is the gene's "natural" product. By the 1960s, the idea that "structure determines function" had become the dominant principle in molecular biology.

The "problem" of biological information, then, is this: The information-transfer criterion that is an essential feature of replication isn't just an abstract similarity relationship between source and copy. That relationship is already covered by the similarity condition, another of the formal criteria for replication. Informational inheritance involves a real transfer. But of what? Abstractions reside in a metaphysical domain like Plato's Forms, never moving. They don't have to move because they are immaterial; we simply refer to them when we need to. So what *moves* when replication happens?

The answer to our problem—why biological, as opposed to physical information, has a more precise nature—becomes obvious, I suggest, if we return to the old idea that information is tied to the specific structure of matter. What I want to suggest is that certain kinds of matter have a special *ability* that makes them biological.

Certain classes of proteins can adopt, and then retain, two or more stable atomic configurations. They are called *allosteric;* in other words, they are flexible. Different allosteric proteins can interact, much as keys fit into locks, with the result that information is transferred from one molecule to the other. How does this occur? The source molecule makes the atoms of the receiver molecule deploy themselves in a spatial pattern analogous to its own. In essence, there is a change in the receiving molecule's shape, in the orientation of the atoms that make it up. It flips into the configuration of the other molecule (or at least something closer to it, if it can adopt multiple states). There is no "communication at a distance" at the molecular scale of organization. This requires very close interaction, where the relatively weak powers of electromagnetic attraction can work to align the two molecules properly, making them ready for the configurational change to occur. Provided with a sufficient store of energy, such molecules can transfer information serially: One molecule passes information to a neighbor, which in turn passes that information to another neighbor, and so on, in a chain. Biological information thus moves and *flows*; it can be communicated. The flow of information generates an ordered sequence of conformational changes, propagating through the chemical mixture.

This capability, I suggest, defines *biological,* as opposed to physical, information. Biological information is capable of *transferring* constraints or specificities of structure. It arises when one bit of matter can help another bit to become like itself. Linus Pauling thought large molecules in general were able to do this. Biological information is structure that can, through lock-and-key manipulations, change the conformation of other matter into its own shape. Only those kinds of matter that can be flexible, like allosteric proteins, are going to be flexible in this fashion, and hence biological. Certainly the split-and-reassemble quality of DNA replication has this special trait.

In an evolutionary context, it is this *aspect* of structure that counts as information. Other aspects of structure don't matter in biology—only the component of structure that has been passed to an object by another one is important. Of course, you can't "see" which aspect of structure this is: All structure still looks like atoms arranged in space. You can only know which aspects of the current configuration have been "acquired" through transmission by observing the history of that bit of matter.

So here's how the new view sees information transfer. Information is *communicated* between molecules when the lock and key are differently shaped: The lock fits into the key, and both are changed in the process. The message is the *change* in shape, not the new shape itself. But the new shape is what matters in the next go-round, because that is what the new interactant sees. The key—now in the form of a lock, for the roles are reversible—runs off with a new conformation, so that a new kind of key is required to fit it the next time around. There is no constancy of players in such chains, no lineage of similar forms, but rather a constant back and forth between locks and keys as they flex from one shape to another.

Information *replication* is a special case of this communication process. It occurs when the transfer of structural constraints from one molecule to another results in the duplication of the original configuration: The lock fits into the key; the key turns into a lock, but the original lock doesn't change conformation, so the two locks now coincide. Of course, this kind of complementarity between lock and key is how DNA works. This is essentially how prions work too: Some unfolding and refolding goes on in both of the molecules involved, but both also wind up in a similar state to one another.

In the physicalist tradition of Pauling, information is determined by both a structure *and* a substrate. It is not disembodied, but always a function of some ordering of physical matter. By contrast, in Shannon's mathematical theory of communication, information transmission is modeled as a reduction in uncertainty about a message's content. Reducing uncertainty sounds a lot like restricting possibilities, which is one way of interpreting the transfer of specificity. But Shannon's theory is neutral about physicality. It was specifically designed to be "general," to apply regardless of how the message is embodied. In particular, it makes no reference to the meaning or content of the message. From Shannon's standpoint, a nonsensical message and a meaningful message can be equivalent. So the message "man" and "mna" are equally good in the sense that they contain three valid letters; it's just that only one of these makes up a recognizable word. But clearly, in a biological context, a message should have a meaning, or function. This is accounted for by a structuralist approach. "Mna" doesn't have the same physical structure as "man"—and only one of these two sequences of letters "unlocks" the mental imagery associated with the male of the species. The more abstract ("informational") and more physical ("structural") approaches to information thus differ in how they can be interpreted.

There is no need to speak of information in the abstract; it is structure as well as order. The anthropologist Gregory Bateson famously defined information as the "difference that makes a difference." That difference is the constraints information introduces—in our case, to the outcome of replication. Replicators are entities that can *transfer structural constraints* on what is possible from one location and time to another. This is what Pauling had in mind with his notion of "molecular specificity." But more particularly, replicators physically limit the kinds of reactions that can occur to those which result in duplication. What makes replication replication, in this view, is the fact that the constraint set is highly specific: The similarity between the duplicate and the original is very good. In replication, rather a lot of information is transferred, and consequently the constraints on what is produced are tight. This suggests that replication is but one end of a spectrum of reactions—the end where the outcome is relatively guaranteed, and in which the produced configuration is largely determined by inputs from a single source: the original replicator. Various mechanisms are used to ensure that this similarity between source and copy recurs over and over, buffered as much as possible from environmental disturbances.

THE STICKY REPLICATOR PRINCIPLE

Transfers of information are tied to particular material substrates, due to the transfer mechanism's reliance on physical "background conditions." This is why genetic base pairs depend on molecular helpers like various kinds of RNA to work their biological magic. From this basis, it follows that replication will typically be specific to one kind of physical substrate. If you try to copy DNA using different assistants or materials, you don't get the right kind of material to make new organisms at the end of the line.

The restriction of a given replication reaction to one suite of conditions and materials is also implicit in the basic replicator equation from Replicator Theory. To exhibit the same kinetics in interactions with other molecules, the source and copy replicators must presumably share many physical characteristics, such as being made of the same substance.

So a crucial upshot of our investigations into the nature of replicators is the suggestion that replicators pick one kind of material to live on, and then stick to it. This can be elevated to the Sticky Replicator Principle, for those who like fancy phrases to remember.

Just to recap this section, a replication process, which involves true information transfer, is likely to be tied to a particular physical substrate. An important consequence of the Sticky Replicator Principle is that abstract conceptions of information simply won't wash when thinking about replication issues. This means that any theory involving replication must specify some physical, rather than merely functional, claim about the mechanism that a replicator uses to achieve replication. This is because replicators are not only picky about what kinds of things they will stick to, but also about what kinds of contexts they are willing to be found in. Since traditional memetics doesn't make such a claim, it is inadequate as a theory.

THE SAME INFLUENCE RULE

We can now couple the Sticky Replicator Principle with the knowledge gained earlier from our examination of the replication reaction and equation to go back and say something else about the similarity condition for replicators. Taken together, these two constraints imply that the source and copy in a replication relationship are not just similar in some vague sense, but similar in terms of their material substrate.

This conclusion flies in the face of standard memetic theory, of course, which suggests that substrate specificity is not implied by the fact that memes are replicators. The common belief that a memetic lineage can consist of a meme passing from brain to computer to book and back to brain without compromising the notion of replication has been cast into considerable doubt by our investigation.

Where did this idea come from anyway? This traditional view of replication suggests that if two things can do the same job, then they are the same kind of thing. More technically, we can call this idea "functional equivalence." For example, it is possible that brains and computers can serve as equivalent storehouses for memes because both can produce the same behaviors. This kind of view has been legitimized by reference to the famous Turing Test, dreamed up by the brilliant British mathematician Alan Turing, in the first half of the twentieth century. Turing suggested that a way to determine if a computer can think (like a human) is if you can't tell its answers to a series of questions from those of a person—that is, if you can have an intelligent discussion with the machine. Obviously the "programs" at work in the human and the computer will bear little resemblance to one another. Nevertheless they produce the same kinds of outputs, so we can call them functionally equivalent agents, at least for the job of answering questions. Functional equivalence is thus the idea that the mechanisms "inside" don't matter, as long as a process can exhibit stable input-output relationships: Ask a given question, expect a particular answer, from both the computer and the functionally equivalent human being. Different mechanisms for generating the output, then, are for all intents and purposes the same.

When this principle is applied to replication, it makes memetics substrate-neutral. In essence, what kind of "machine" grinds through the replication process isn't supposed to matter. So the same meme can exist in brains or on paper, no problem.

Functional equivalence does have its good qualities. It is consistent with heredity: It obviously provides a good correlation between parental "inputs" and offspring "outputs." And this, as I have asserted, is all that evolution requires. But it means that those who adopt the functionalist stance in memetics—and the list includes most of the major players, such as Dennett, Blackmore, and Dawkins—cannot hold the memetic line against encroachments from the Gadjusekian "let's make do with genes" view of cultural evolution. This is because functional equivalence can achieve heredity without replication—without memes.

Functional equivalence is also the source of another problem I identified earlier: substrate neutrality, which is tied to the explanatory futility of memetics thus far. Functional equivalence means memes are not tied to any one kind of code. Source and copy can be on different substrates because function is potentially transferable between contexts. The Little Red Riding Hood story is essentially the same, whether told in pictures or a song or written words, because you can still get the "moral" from it in any case. But functional equivalence is not, so to speak, what I have in mind.

Functional equivalence is fine as a principle for computational science. We already know that machines using different substrates—silicon chips, DNA, or quantum states—can perform computations. They can also rely on different mechanisms, such as operating systems and software programs, and nevertheless accomplish the same thing. But replication is a more precise notion than computation, so functional equivalence is too loose a criterion for evolutionary purposes. We want to exclude this kind of equivalence, which leads us to call genes on paper and genes in cells the same thing, because lineages involving brains and those involving computers are not the same—their evolutionary dynamics are bound to differ. Replicators are not Turing machines, capable of generating any kind of solution. In fact, the whole idea of replication is just the opposite: to make sure only one kind of product comes out at the end.

Replicators, then, to be defined as similar, cannot just do the same kind of job. They have to do *the same kind of job in the same kind of context.* If replication is always specific to one substrate, then information transfer can take place only within certain restricted kinds of circumstances. This condition suggests that true replication involves what I will call "structural equivalence" between the source and the copy.

The notion of structural equivalence is taken from social network theory. This theory attempts to describe, in a general way, how the structure of the network—the particular links that exist between people—hamper or facilitate the flow of goods or information through that group of individuals. The relationships in one network may be defined by kinship; in another, by positions within a company.

This idea obviously resonates with the technique of viral marketing mentioned at the beginning of this book, which is based on role-playing in networks. For example, there may be an old lady in the Bronx who is treasurer of her local Senior Bingo Players Association and so passes money on to the society's president. Similarly, an old lady in Brooklyn does the same thing, thanks to being the treasurer of the Bingo Players in this other New York City borough. The two old ladies are similar to each other, but not the same person, and neither are the two Bingo Player presidents. Nevertheless the same consequences result from the relationship between treasurer and president in these two cases. Both old ladies are "wired up" in the same way to similar "downstream" players in their respective bingo associations. They are what I have called "structurally equivalent"; they have the same kind of influence on the overall situation, even though they occupy different places in the overall network.

Whereas functional equivalence can be summarized as requiring "same input/same output," structural equivalence necessitates "same input/same *influence.*" The same suite of effects must ripple through the larger context in which the evolutionary agent (such as a replicator) finds itself (like our example of a social network). This means that what must be similar is not just the immediate, proximate output from the agent's own action, but also the downstream consequences, due to the agent's similar connections into a larger context. So the similarity criterion we have discussed as crucial in replication implies that source and copy must have the same causal influences on the world at large. This insight— recognition of the need for structural, not just functional, equivalence in replicators—can be enshrined as the Same Influence Rule. It essentially argues that function follows (physical) form.

This rule must be applied to the set of replicators defining a lineage; it is comparative quality (like the qualities of our two Bingo Players, above). It is a quality that replicators in a lineage share, not just the source and its immediate descendant, produced by a particular replication reaction. In effect, if the whole sequence of replicators identified by a chain of replication reactions is considered together as members of a group, they must not only share a single substrate to be similar, but also play similar evolutionary roles. All these copies of a replicator will use the same set of skills and tactics to get ahead in a game in which all of them face similar selective forces. This will allow selection pressures on this group of replicators to be consistent over time and for the adaptations associated with this lineage to accumulate. And the accumulation of adaptations is the *sine qua non* of evolution.

AT DETOUR'S END

If we now stand back for a moment to look at what we have learned through this foray into how various sources of information can be put to work on culture, we can see that we have gained considerable insight into how memes will have to work if they are to be true replicators. The notion of a replicator is itself a fundamental generalization of the Darwinian idea of a biological unit of inheritance. Crucial features of any replication event—causation, similarity, information transfer, and duplication—have been identified. Replicator Theory taught us that entirely different evolutionary regimes are possible. This theory complements nicely our earlier suppositions about the central phenomenon of replication. In fact, all of the criteria we originally specified for replication are explicit in the replicator equation and replication reaction concept. Information transfer is present in the notion of the replicator serving as its own catalyst. Causation is also obviously present if the replicator catalyzes itself. Duplication, as noted above, is the place where Replicator Theory begins, and similarity of instigator and product is the obvious goal of the whole process. The description of inheritance as *transferred specificity,* inspired by Linus Pauling, is obviously also an image congenial to Replicator Theory as developed by Manfred Eigen and his compatriots. So the replication reaction is a very good representation of the phenomenon we are so vitally interested in here.

We have also learned that any replicator must be a physical thing. This runs contrary to a lot of contemporary thinking. Memes, after all, *appear* not to have a single, archival kind of medium. Consider Miguel Cervantes's book *Don Quixote:* a stack of paper with ink marks on the pages. But you could transfer the book to CD or a tape and turn it into sound waves for blind people. No matter what medium it's in, it's always the same book, the same information. So it seems that it is information content that defines a replicator, not its material embodiment.

Genes are not—nowadays at least—any different from memes in this respect: They too have a variety of media in which they can be "archived." A gene can be lifted out of its home inside the nucleus of a cell and its code deciphered by modern gene-sequencing machines. The code can then be written down and stored in a library or in computer memory. Later, potentially many years after its original extraction, the gene can then be reconstituted in a cell, and the cell can in turn be injected into a living organism to go back to work normally. All this through the wonders of genetic engineering. Does this mean that genes are like memes, that they don't have a single medium? Are they protean replicators, capable of taking on a variety of different forms to suit the occasion, like the eponymous figure from Greek mythology who had that ability?

In fact, neither memes nor genes can exist in all those media. If this proposition were right—if replicators were merely abstract—this would be a disastrous state of affairs. It would lead us straight back into the morass of moribund memes, memes that are everywhere and yet nowhere because they can be anything and nothing: memes in your head, memes in your movements, memes in your toaster. In short, replicators that aren't tied down to any particular *thing*. Memes or genes, then, like information itself, would have to exist in some Platonic never-never land. At bottom, this kind of abstraction just can't handle the fact that the evolutionary process is not simply one of information transmission.

The idea that replicators can migrate from one form to another without consequence simply isn't true. Just look at how crucial evolutionary parameters have changed as genetic engineering technologies have improved. Technicians have been able to write down gene sequences for quite a while, but it used to be the case that you did it by hand, after making a judgment about how far the gene had traveled down a special gel. Then, on transferring the data to a computer, lots of errors crept in because you were pecking at a keyboard with 26 letters on it where only four (A, C, G, and T) were required. Today the transcription machinery has improved significantly, as specialized computer software has been developed to take information directly from an equally automatic gene sequencer and plug it into a computerized database. Many fewer errors crop up as a result—mutation is constrained by technology.

Prior to the invention of sequencing technologies, genes were stuck inside organisms and couldn't get out. Gene lineages followed their normal course, mutating occasionally, to produce the standard bushy family tree of life. But now genes can be inserted into just about anything and made to persist in that new genome, whether it belongs to the species of organism that donated it or not. So the "horizontal" transfer of genes between species is becoming much more characteristic of gene lineages, thanks to technological developments, even in the higher animals, where it was relatively absent as a "natural" phenomenon. (Viruses have long performed some introductions of "alien" genetic material into the genomes of complex animals from time to time.) The storage of genetic information in "artificial" media will have a tremendous impact on the dynamics of evolution in the "real" world. That is what genetic engineering is all about: changing the way things normally work when left in nature's hands. So you can, in the abstract, represent a gene in many different ways these days. But how you do so influences the bio-evolutionary dynamics that result from that choice. The point is that several aspects of a gene's life are dependent on how it can manifest itself in the material world.

I conclude that information is not *meta*physical; it's merely physical. This conclusion should, I hope, prove to be the death knell for the protean replicator, the shape-changer always popping up around every corner, often where you least expect it.

Further, the principles we have just derived—the Sticky Replicator Principle and the Same Influence Rule—are important deductions about replicators that will color all later argument in this book and help narrow our search for memes significantly. We have gained all these insights only by examining the general theory of replication—something that has not previously been done by memeticists, but that provides us with the possibility of making real progress toward a science of memetics.

With these many and lengthy preliminaries now behind us—and keeping our new, hard-won wisdom about the nature of replication in mind—we are now ready to begin our quest for memes in earnest. The question we now need to answer is: Where are memes hiding?

Chapter Six

STALKING THE WILD MEME

If the event o' the journey
Prove as successful to the queen, — O, be't so! —
As it hath been to us rare, pleasant, speedy,
The time is worth the use on't.

—William Shakespeare

Originally memes were defined as shared elements of culture learned through imitation from others, with culture being defined rather broadly to include ideas, behaviors, *and* physical objects. It was thought that memes might be present in any of these various forms. In this context, Susan Blackmore invites us to consider a bowl of soup. Does the wet stuff itself count as a meme? Or the instructions about how to make soup stored in a person's brain? Or the kitchen-bound behavior of making the soup? Or perhaps the written-down recipe? Blackmore has recently argued that "memes are not magical entities or free-floating Platonic ideals but information lodged in specific human memories, actions and artifacts." So for Blackmore, memes are in all three kinds of things I have identified: the brain, behavior, and the products of behavior. Dennett too suggests that memes are both in the head and out in the world, as part of a complex life cycle. Imitation, taken in the "broad sense," is consistent with memes being in any of these kinds of things.

The primary problem with this state of affairs is that it doesn't tell you where to look for a meme, or how to tell when you've actually located one. Just look at what we still don't know about memes: Where are memes to be found? Are memes in behavior, in brains, in artifacts, or some combination of the three? Or perhaps something in the brain might be a "memotype," with some behavior as its memetic phenotype. Or the opposite. (All of these positions have adherents.) How many different ways can you get infected with one? From how many different "cultural parents" might you acquire memes? How long can a meme live? How do memetic alternatives compete with each other? What do they compete for? How many variant memes might there be? What different kinds of forces select among alternative memes? What forces cause them to mutate? How fast do they mutate? Are memes translated from one form to another? If so, how? *All* of these questions are open. At present, memeticists can't even act like the drunk searching for his keys under the lamppost because the light is better there. We don't know where the lamppost is or where the light is coming from.

In the last chapter, I concluded that to truly identify a replicator we have to specify how its replication reaction works. Now we have to get our hands dirty and determine the physical mechanics of memetic replication.

Why be concerned with the mechanistic details? No one in memetics has previously done this; no one has even felt it was necessary. The consensus seems to be that we can ignore implementation issues altogether. Memeticists commonly note that a lot was accomplished by Darwin prior to the identification of genes. They often remark that the study of cultural variation is presently at the same state of development as the study of biological variation was in Darwin's time: It seems reasonable to suppose that an evolutionary process underlies cultural elaboration, but the mechanisms of inheritance—the cultural analogue of Mendelian rules—remain unknown. But, the argument goes, this is okay because we can still use basic Darwinian thinking about selection, variation, and heredity to get us some way toward a memetic science.

Well, quite a bit *was* accomplished in genetics before 1953, when Watson and Crick accurately modeled the genetic structure, including Mendel's ingenious experiments with peas, which suggested there were two copies of each gene in an organism. But since Watson and Crick broke the genetic code, the entire field of molecular biology has developed. In essence, heredity, a biological phenomenon, was shown to have a chemical basis, and the laws of chemistry were demonstrated to apply to the fundamental problem of biology. This kind of reduction has proven to be extremely powerful in the case of biology and has led directly to an explosion of knowledge about the actual workings of genes, from their physiological activities to their different kinds of roles.

We can also take the example of other bio-replicators to heart. Thanks to the identification of infectious agents such as prions, which lack genomes altogether, and subviral genomic particles like viroids and virusoids, many diseases previously believed to be genetic in origin are now being characterized as transmissible. Some people also think that cancer—another large category of illnesses—is just a lineage of rogue cells setting out on their own reproductive course independent of the host body, much to the chagrin of the cell lines that are "behaving." This list may only get longer as we find that viruses are behind many diseases of strange etiology. Viruses are being found to have more strategies up their sleeve than previously thought, like long incubation times. *Kuru* and "mad cow" disease were only the beginning. Prions are now being touted as the agents responsible for a variety of diseases previously thought to be innate, like Alzheimer's and Parkinson's disease. These developments have come from a growing recognition that parasites have evolved many different and quite complex life cycles that do not conform to the standard etiological pathway of direct exposure and quick development of symptoms, the easiest causal relationship to establish purely by coincidence and correlation. Diseases we have long ascribed to genetic or environmental factors—including some forms of heart disease, cancer, and mental illness—may in many cases actually be found to be infectious. Some think even obesity is caused by viruses! And this process of reclassification of diseases can only be expected to continue as our understanding of the evolution of bioparasites increases.

So the move to look for a physical correlate of the replicator concept in culture is a natural progression. The history of genetics and epidemiology provide clear parallels. My hope is that a similar theoretical reduction will occur in memetics: that some aspects of memetics can be reduced to the principles of a highly progressive science and so give memetics the chance of becoming full of portent for the future. Once we recognize what memes are, we can begin to trace their pathways through society. Just as in the case of diseases, so too, I predict, will our apparently innate knowledge be found to have been transmitted to us through epigenetic means, either recently or in the past. Memes will acquire greater significance as we turn away from the standard dichotomy of nature or nurture. It's therefore likely that tremendous advances await us if only we can "get physical" about memes. Once we nail down the replication mechanism, then all kinds of new hypotheses about memes will suggest themselves and whole new fields of empirical endeavor will open up.

There's another reason this argument in favor of continued ignorance about mechanisms simply won't work. The analogy to Darwin's time is faulty. Things are *not* the same now as then. It's rather like arguing that currently "underdeveloped" countries should go through the same process of development as occurred in the West the first time around, when the world was younger. In fact, such countries cannot rewind history back to the time of the Industrial Revolution, even if they wanted to. This is because they are the poor relations, with little power, in a world now dominated by global financial markets, nuclear weapons, international media conglomerates, and mobile labor forces. The weight of the First World is on their shoulders, but unlike Atlas, they don't have the gigantic powers to keep it aloft, to pretend as if it didn't exist and freely exercise their own will. The developing countries also cannot tap into the manpower pool on which the original revolution depended because they *are* that reservoir of labor.

In much the same way, memes live in a world of genes. Memes began, at least, as the genes' poorer cousins, with relatively underdeveloped abilities for replication and control over their tendency to mutate. So, in some sense, a straightforward analogy to genes is bound to fail, simply because genes got there first. This changes the evolutionary context utterly and completely for any replicators that follow. Generally speaking, memes are necessarily dealt the weaker hand by history. Memes can't simply replay the videotape of evolutionary history again to get a different ending to the plot, with genes parasitized to do their bidding instead, because there is only a tiny chance memes would get the scenario they desire at the end. Evolution never works out in just the same way twice because each sequence of descent with modification is a long history of coincidences.

A major error of previous stories about memes has been to forget this general context. As soon as the exciting prospect of a cultural replicator was recognized, theorizing generally went about as if memes existed in splendid isolation, rather than being "Third World" replicators. Memetics has forgotten that memes are parasites, dependent on hosts for a living, and that they necessarily play second fiddle to genes. So memes must deal with the fact that they live in a world full of genes. And just as champions of memes argue that a "genes-only" view is wrong, so too is a "memes-only" perspective insufficient. It's a genes-plus-memes world we all live in. Each replicator must coevolve with the other and handle those activities of the other that impinge on its own goals and well-being.

There is a yet a third reason the "but Darwin did okay" argument doesn't hold water. As I've stated before, it's possible that memes aren't necessary at all. Genes might be sufficient to explain cultural inheritance. It might, in the end, be a "genes-only" world after all. In effect, the only way we're going to be able to tell whether cultural selectionists are right about the nature of cultural evolution is to actually put our hands on a meme. We simply must determine where memes live and identify how they satisfy the criteria for being replicators for memetics to survive.

The challenge, then, is to provide some empirical support for the meme hypothesis. Memes cannot be left as intangible, ineffable, unknown "stuff" that is somehow created, transferred, transformed, and preserved to account for changes in cultural traits. We need to find out: What's the *matter* with memes?

THE QUEST BEGINS

Taking the need for a physical model of memetic replication seriously means we need to establish in which kind of substrate memes can be found. Although Dawkins's original catalogue of memes suggested he believed they could be found in beliefs, behavior, or bottle-tops, he later refined his definition of memes, primarily to restrict them to one substrate—the mental—due to what he perceived as confusion about the memetic "genotype" versus "phenotype." He believed, as do most Western intellectuals, that beliefs causally precede behavior, so due to their priority, ideas are more important than their products. Since replicators are also causally prior to, and more important than, interactors in evolutionary terms, we should therefore identify memes as ideas. "A meme should be regarded as a unit of information residing in a brain," he declared.

But both Dawkins and Dennett continue to assert that memes can exist in artifacts. For example, even after his switch to brain-based memes, Dawkins again declared that memes can "propagate themselves from brain to brain, from brain to book, from book to brain, from brain to computer, [and] from computer to computer." Dennett says, "A wagon with spoked wheels carries not only grain or freight from place to place; it carries the brilliant idea of a wagon with spoked wheels from mind to mind." Since Dawkins explicitly argues that a vehicle "houses a collection of replicators" and "works as a unit for the preservation and propagation of those replicators," memes must exist *inside* artifacts like Dennett's wagons, using them like Trojan horses to infiltrate the defenses of innocent minds. Brains, behavior, and artifacts are equally good homes for memes. They exist as information, regardless of where or how this information is stored, which makes memes "substrate-neutral," Dennett says.

We can use the Sticky Replicator Principle, which holds that memes can exist only on one physical substrate, to rid ourselves of this tendency to what is more properly called "substrate profligacy." But is it ideas, actions, or artifacts that memes are hiding in? I'll let the cat out of the bag by saying that memes will be found only in brains. However, it's important to note that Dawkins did not have principled reasons to think ideas are more likely to be replicators than behaviors; he did not couch his argument in terms of Universal Darwinism. His reasoning for preferring the mental identification was philosophical, rather than biological, in nature. A justification based on *functional* criteria for preferring a restriction of memes to ideas remains to be produced, and I will produce it. So we can't simply take Dawkins as the authority defining the new scientific field of mentalistic memetics. Rather we must provide a more substantial defense of our choice of a substrate for memes.

Let me begin the process of elimination by asserting that in light of our definition of replication, artifacts don't have the requisite qualities of causation, similarity, information transfer, and duplication. For proof of this statement, let's take the quintessential example of artifact production, the automobile assembly line.

Cars are made inside factories—an enclosed, protective environment, full of raw materials and machines for converting those materials into the desired final form. This is all well and good, and in accord with Replicator Theory.

But cars aren't born; they are assembled, these days mostly by robots. That is, they do not develop from a chunk of raw material but are constructed by having various bits added onto them. And, in fact, the materials are not so "raw" in this case—many of the bits used are preprocessed, ready for attaching to the developing chassis. So construction is not entirely encapsulated in the local site but involves many prior activities, such as the making of spark plugs (a product of some other factory), which get plugged into the engine block at the assembly plant.

Of course, it can also be said that proteins are cobbled together in a similar kind of sequential process. In particular, ribosomes, which construct DNA in the cytoplasm of cells, are themselves made through a complex process as part of the growth of a cell (which begins when instructions are read from previous bits of DNA). So it isn't the "assembly line" nature of the process that distinguishes replication from production. Replicators can be assembled, rather than grown.

What _is_ the difference, then? The crucial point is where the directives controlling the robotic movements come from. In the car factory, the assembly robots don't have the information "on board." Their movements respond to instructions from a computer program somewhere. In this program, there are no instructions that say, "If the previous car was painted blue, then paint this one blue," or, "If there was a hatch-back on the previous car, then add one to this car as well." Rather, the instructions that direct the movements of the robotic arms may be stored in a database, perhaps even more remote from the factory floor than the computer itself, reflecting orders for new cars from customers around the country. Even before robots took over the factory floor, the automobile workers would work from written instructions that told them what bit to weld onto the next car. What happened with the previous car on the line had no bearing on how they behaved with respect to the next hulking mass of metal headed their way.

So multiple cars come out the factory door (fulfilling the duplication criterion of replication), and these cars may look enough like one another to be called similar (they all have the same brand of name plate slapped on them, at least). But one car does not play any role in generating the next; computer programs direct the machines that perform that job. So it seems that artifacts fail our replicator test because two of the fundamental qualities of replication are missing. First, one car plays little or no causal role in making the next. (At most, the fact that one car can be ticked off as completed means that the pointer on the computer program moves to the next line down the order list, where the specifications for the next car can be found.) Second, there is no information transfer from car to car, even indirectly. No quality-control machine reads out the condition of one car, and then feeds those criteria to the machines controlling the construction of the next—duplicating an accidental dent in the hood down the line, for instance. Instead the specifications are pre-set and come from a location remote from the assembly site. In short, automobiles are produced, not reproduced.

The reader may recall that I earlier allowed that artifacts *can* be replicators. Indeed there are whole classes of them, ranging from Xeroxed patterns of ink on paper to computer viruses. So why am I now saying that memes can't be considered artifacts that replicate? Because the whole point of invoking memes is to explain culture. We commonly attribute the special features of how particular groups of people act and think to their culture. But the specialized, highly technological categories of artifactual replicators we've discussed cannot explain the phenomenon of culture, which is both much bigger and older than these types of artifacts. Xerox machines and computers very recently arrived on the scene, while culture is at least as old as the so-called "cultural revolution" of 40,000 years ago, during the Upper Paleolithic, when people first began leaving permanent records of their cultural activities in the form of cave paintings, sculptured rock, and decorated bone fragments. Even today the existence of computers themselves or jet planes or, more prosaically, plastic lawn furniture can hardly be explained as the concerted activity of replicating artifacts. Not all kinds of artifacts can be replicators. Of course, not all proteins are replicators like prions, either, nor are all computer registers occupied by viruses. Further, it's not likely that any protein could be converted into a prion, nor that any artifact could be turned into a replicator. The idea of replicating cars is even a bit frightening.

But any register in computer memory could *potentially* be converted to a viral state. This is one way of seeing the difference between limited and unlimited replication, of course: While prions are limited replicators, comp-viruses are not. And what we are looking for is an unlimited replicator capable of accounting for the complex phenomenon of human culture. So we need to find a physical substrate that is robust enough to play a causal role in *all* of culture's myriad manifestations, not just a few exotic ones. Our search for a cultural replicator must continue.

EINSTEIN'S TEA PARTY

This elimination of artifacts from consideration still leaves two major traditions of thinking about the nature of memes as contenders for the crown of True Replicator. These are also the traditions that are presently competing for numerical dominance in memetic circles: behaviorism and mentalism. Each is essentially the inverse of the other. Behaviorism sees the cultural replicator as a behavior. Think of a chimpanzee mom demonstrating her nut-cracking skills in exaggerated, slow-motion style, as a way of teaching the same task to her young son, who is watching intently. For behaviorists, the signal transmitted to the immature chimp by this behavior is the replicator. This signal must first live in the macroenvironment of the physical world, and then in the microenvironment of the young chimp's mind, before generating a new replicator in the form of another demonstration of the skill by the young skill-learner. Memetic mentalism, on the other hand, argues the reverse: The knowledge underlying the mom's skill-production and other cultural behaviors are the replicators, while the generated behaviors are the interactors of culture. Both positions have some intuitive appeal and have attracted adherents.

A little vignette involving a very obvious kind of behavior—speech—is perhaps the simplest way of examining the possible nature of behavioral replication:

Two men meet in the street. "Hello," the first man says. "Hello," the second replies.

The first man's friendly "hello" goes into the second man's brain as waves of compressed air tickling his cochlea; another "hello" comes out the second man's mouth moments later. What happens to the spoken phrase in between seems to be replication. Let's go through the criteria we've identified to check whether this impression is justified.

First, it seems pretty clear that the similarity condition is satisfied: One "hello" equals another. (Think of looking at the sonograms—records of the amplitude and frequency of the sound waves—for the two spoken phrases; they will look alike.)

What about causation? Presumably the second man wouldn't have said anything unless the first man spurred him to do so by issuing *his* greeting. So the generation of the first signal seems to have initiated a process that resulted in a second "hello" being produced. This looks like causation all right.

Had the first man said, "*Guten Tag,*" the reply might have been quite different: "Say what?" for instance, or, "Sorry, no *sprechen* the *Deutsch!*" This indicates that the choice of language used in the greeting had an impact on the nature of the reply as well. This is an instance of information transfer in action: the information transmitted being that the first man is an English-speaker wanting to be friendly, which influenced the nature of the speech made by the second man. Wishing to reciprocate the good intention, he too produced a copy-cat "hello" instead of a different English phrase or something in another language.

What about the final condition, duplication? Do we go from one "hello" to two "hello"s with this exchange? Well, my ability to describe the interaction as an "exchange" already suggests the answer is no. There are two ways of looking at what has happened in our little scenario. The first is that the "hello" signal dies when it reaches the second man's ears. In that case, where did the second "hello" come from? The second man must have generated it separately using some internal system independent of the stimulus from the first man. One signal is produced and dies; then another signal is produced somewhere else at a later moment in time. There is a break in the existence of the behavioral pattern. This doesn't suggest continuity, much less duplication.

Alternatively, the signal can be thought to transform itself into another kind of signal, in a new code suitable for living inside the brain. In particular, the spoken phrase must progress from air-wave code to brain code and then back again as the second "hello" is emitted into the air. Just as parasites like the wily *Trypanosomea brucei* (a fluke that causes sleeping sickness) can take on wildly different morphologies as they wend their way from pigs to humans to snails to pigs and so on, a signal can go from compressed waves of air to a set of action potentials in the brain and back again. But in this case, the exchange of greetings represents the complex life cycle of a *single* signal, passing through multiple brains. No replication takes place; it is just different phases of the one signal's life. Even if we grant continuity to the interchange, then, it is not clear that anything was duplicated—one thing, a signal, might just persist through changes in form.

So the "hello" can be considered either to undergo a single complex life cycle or a sequence of independent productions. In either case, it seems the duplication condition of true replication is not fulfilled by behavior: There is no point in time at which two copies of the signal coexist. Just as the two instances of "hello" do not get printed on top of one another in the narrative representation on the page of this book, neither do the actual conversational "hellos" overlap in time. Because behavior doesn't duplicate, it fails our test for replication.

What this example shows is that you need to open up the black box of the brain and look inside to see whether what is happening in there is proper replication, transformation, or something else again. The advocates of memes-as-behaviors might disagree with the necessity of the duplication criterion, but for each type of replicator, temporal overlap between the source and copy is a characteristic feature of the process. Going from one to two occurs in every known example of replication (just remember the cases we looked at earlier: prions and comp-viruses).

There is another argument that memes can't be in behavior. It begins with another tale. I call it "Einstein's Tea Party."

In 1904, Albert Einstein could frequently be found muttering "$E=mc^2$" as he tramped around his apartment in Bern, Switzerland. After a few days of this, his parrot, Jolly, began squawking "$E=mc^2$," copying the German physicist's tone and inflection exactly. Although Einstein loved the bird's companionship, he found this mimicry often upset his train of thought. He determined to give Jolly a companion, hoping the parrot would pay more attention to a new lady-friend than the scientific discoveries of his human benefactor. So the next day, after finishing at the patent office, he purchased a second parrot, which he called Polly.

By this time, Einstein had stopped muttering "$E=mc^2$." But the amorous Jolly hadn't. Polly soon learned to imitate Jolly's vocalization perfectly. Einstein hadn't counted on that and was quite annoyed, since now Polly was also interrupting his meditations.

Max Planck visited Einstein's apartment for tea a week later. Einstein was reluctant to discuss his recent work in front of another physicist until it was published. However, as Einstein's maid, Gertrude, came into the room with tea, Planck heard the new parrot saying something. He instantly recognized it as the solution to a problem he had been trying to solve for years. "Yes," Planck exclaimed, "E=mc²!"

Gertrude, just then serving the guest, wondered why he was so excited about the parrot's nonsensical chatter as to nearly upset the tea service.

Find the meme in the story. Is it in the behavior (the spoken phrase) or in the idea (the mental representation)? As we will see, this simple story illuminates many of the issues separating the behaviorist and mentalist camps. I will argue that it provides another basis on which the mentalist position can be preferred.

Our basic goal in analyzing this story is to explain how the great German physicist, Max Planck, came to learn Einstein's novel idea, because that may involve a meme being transmitted. What are the basic data to help us decide the case? Well, notice that a bit of behavior—speaking a particular phrase—appears to go from Einstein (a human) to Jolly (a bird) to Polly (a second bird) and finally to Planck (a second human). Of course, the bit of speech ("E=mc²") is very similar in the humans and the parrots: sonograms of the vocalizations, which I take to be good empirical measures of the behavioral "meme," would be demonstrably similar in frequency and amplitude throughout the sequence. In particular, Planck's sonogram would be very similar to Einstein's (since Einstein and Planck would both be speaking native German), although less similar than in the Einstein and Jolly comparison. This is because even though Einstein and Jolly are different species, parrots can imitate the incidentals of human vocalization better than most humans, and Jolly learned the phrase directly from Einstein.

On the surface, at least, it is surely impressive that the same sound-stream is reproduced a number of times, in two different species. This reasoning suggests we should focus our attention on the behavior. In particular, it seems a robust replicator indeed that can skip over species boundaries in its ability to find hosts. Add to this the fact that the transmission chain was from a human to a bird and back again, with perfect repetition of the meme by the last link in the chain, a human who had never heard the meme before, and we have quite impressive fidelity of transmission. So should we therefore say that it is the behavior, rather than the idea, that is the meme, because, arguably, there appears to be a line of information transfer for the spoken phrase between the original mumblings of Einstein and the excited moment of recognition by Planck at the end of the story?

Before we record a victory for the behaviorists, what about the idea behind the phrase "$E=mc^2$" and all its connotations? Can they be said to have passed through two birds' minds on their way to Planck's? It seems unlikely that there can be very similar mental representations in species with different kinds of brains. No common mental structure should be able to make it all the way from one physicist's brain to the other's.

This reasoning would indicate that we should take behavior as what is replicated during social transmission, through imitation. Complex ideas are too individualized to be instances of a replicator lineage. The behavior is much more likely to jump the species barrier intact than mental representations, surely? So the incident at Einstein's tea party favors the behaviorist approach to memes, unlike our earlier story of the chance meeting in the street.

But the emphasis on comparing mental to behavioral "similarity" is misplaced. What matters more in determining where the replicator lies is evolutionary dynamics, as indicated by the discussion of the replication reaction in an earlier chapter.

Focus for a moment on Gertrude, Einstein's maid. Gertrude, on hearing Polly or Jolly, did not "understand" what the parrot was saying. This is true even though Gertrude speaks German, because the maid's scientific and mathematical education had not led her to the position where the sequence of vowels and consonants in the phrase "$E=mc^2$" could be applied to any context meaningful to her. Perhaps Gertrude didn't know that "c" is a mathematical symbol for the speed of light, for example. The problems would be greater, of course, if Gertrude had been, say, Portuguese instead. Then there would have been decoding difficulties as well as interpretational ones. On the other hand, if Polly the parrot only spoke Parrotese, the problems would be the same: how to translate the bird's language into one the human could understand. If the bird had said, "Energy equals mass times the speed of light squared," the same meme would no doubt have been formed in Planck's mind, but it might not have helped poor Gertrude, without the ability to do anything with even this version of the math.

Let's stretch this story a little further. Let's say that Gertrude was able to remember the phrase "$E=mc^2$" exactly. When her husband asked about her day's activities at home later that night, she could then tell him that Planck was excited about the parrot saying, "$E=mc^2$." Thus Gertrude could carry the mental memory, with few of its implications or meaning available to her, and even become a link in a transmission chain to her husband with high fidelity. From a mentalist perspective, Gertrude is much like Polly: an innocent bystander to meaning, but a good transmitter of the necessary message anyway.

But it is more likely that Gertrude would *not* remember the sequence of sounds correctly (unless she had heard it repeatedly from the parrots while moving around the apartment cleaning) *because* she lacked a good understanding of the phrase's meaning. She might tell her husband that Planck had said "$E=cm^2$," for example. Her husband, if equally untrained in mathematics, might then further butcher the phrase when recounting the story of Einstein's surprised visitor to his cronies in the pub the next evening. Transmission fidelity would obviously be extremely poor in such a case.

Meanwhile, the parrot lineage would exhibit no such variation, despite an equal lack of appreciation for mathematical niceties on the part of the birds, who were more likely to remember the phrase as a single chunk of information rather than a sequence of consonants and vowels, thanks to their more limited brains. The point is that the fidelity of replication in a purely human lineage (Einstein to Planck to you, the reader) is a function not just of the ability to imitate vocal behavior but also of the ability to correct errors. Had Polly the parrot slightly mangled the all-important phrase, Planck might have "heard" the right one anyway, had he been primed, perhaps by some earlier discussion over aperitifs with Einstein, to be thinking about physics at the time.

This is not what would have happened if the listener had been Dolly, another parrot. Evolution of the representation in any parrot lineage would be minimal. Variations are unlikely to be introduced through transmission of the phrase from Jolly to Polly to Dolly to Zolly and so on, even though the words are meaningless to the parrots, because the birds are highly adept vocal mimics. But should some mutation in the phrase arise in a lineage made up purely of birds, that changed phrase would be successfully passed from parrot to parrot thereafter *ad infinitum,* with little further modification. However, it would do so with never a chance to recover the "right" version, because parrots have poor comprehension of the theory of special relativity.

So the novel dynamics of bird lineages should be very different from those in human ones. In particular, a bird lineage will exhibit high fidelity with low mutation rates, but no error correction. Human transmission chains, on the other hand, will exhibit higher mutation rates (due to their innocence of mathematics), but reverse mutations back to the "correct" phrase will pop up, thanks to the ability of some people, at least, to recover meaning, due to their knowledge of mathematics or physics. This is one lesson we learn from Professor Einstein.

I have still further ammunition against the behaviorists: Transmission modes will be very different in the human and bird lineages. When Einstein publishes his result, thousands of people Einstein has never met will start talking about it among themselves. By now, millions more people in faraway lands have heard or seen the phrase in some context—as an advertising jingle, or on a T-shirt—which has caused the formula to enter their memory. In contrast, parrots in these times and places will not be picking up the latest issue of the *Physical Review* or observing bumper-stickers to catch up with Einstein's work. These are not transmission methods available to parrots. So humans can maintain a communication chain even when the acquisition of the message is from a machine. Parrot, book, e-mail, whatever—it doesn't matter to people. This is certainly not the case for any parrot lineage. They don't read or use computers; it matters a lot to them whether someone is speaking or reading the phrase silently to themselves. So again we have very different kinds of dynamics in the replication chains, due to the kinds of channels for information transmission available to humans as opposed to parrots.

In conclusion, these two stories—the meeting in the street and Einstein's tea party—have shown that behavior does not exhibit features that are indicative of replication. First, behaviors, like signals, are not duplicated. Second, tracing the path of a behavior does not identify distinct evolutionary lineages. The same spoken phrase went through a progression of people and birds in Einstein's apartment, but this conflated two different kinds of evolutionary dynamics. Inheritance does not work the same way in lineages involving different host species. I think these stories effectively kill the possibility that memes exist in behaviors.

The behaviorist "mistake," which it is all too easy to make, is to concentrate on the obvious, on what is observable. And what is observable, typically, is some relatively brief exchange of information during a social interaction, as in our example of men in the street above. But the fundamental question of evolution is how long-term dynamics play out, and the fate of replicator lineages. These processes are not so readily observable, but the explanation of such patterns is the basis of any evolutionary science, including memetics. Certainly the chain of very similar "$E=mc^2$"s in our second story is startling, and it would seem that explaining *that* should be the central focus of memetics. But it is not. The big picture is what happens to repetitions of the spoken phrase when the chain becomes longer, when it is iterated to 20 or 100 duplications, and perhaps just in a series of birds or people.

There are some remaining qualms I should deal with before we settle definitively on the position that memes must be something in the brain. Most importantly, the issue of what constitutes mental similarity remains. More particularly, the question is what kinds of copying events in the mind classify as proper replications of mental content. Behaviorists might insist that the measure of mental similarity must include all the intricacies of implication and inference that a person can draw from the signal "$E=mc^2$." This would mean links to *everything* the individual knows about Einstein, for instance—his great shock of tousled white hair and mustache, his eminent position at the Institute for Advanced Study in Princeton, his reluctance to acknowledge quantum mechanics, his protests about the Manhattan Project—and other famous phrases besides the formula, such as "Only two things are infinite, the universe and human stupidity, and I'm not sure about the former." Some people might not know that "$E=mc^2$" should be associated with Einstein's name, erroneously thinking that Max Planck was responsible for its original formulation. Should all these ramifications be counted in the tabulation of mental similarity?

They should not. In complex brains, the connotations of any idea may ramify indefinitely. But the complicated network of associations to "$E=mc^2$" will necessarily be individual. They cannot be part of a single, replicating meme. In fact, the phrase "E=mc²" may itself be several memes, cobbled together, that become associated only in particular contexts, as when someone thinks about Einstein for some reason ("Isn't he the bloke that came up with 'E=mc²'?"). But this is a natural restriction. Only small bits of information are likely to be repeatedly replicable through multiple generations. What happens to those bits in terms of mental structures, as they come together during use by host individuals, is a different question.

The behaviorist has one final refuge: the empirical impossibility of a mentalist memetics. It would be ironic indeed if we identify a physical model for memes but it turns out to be one that can never be tested! Behaviorists have argued that any memetics that begins in the brain is doomed to failure because memes are then defined as intrinsically unobservable entities: Who can see what happens inside brains? But this confutes the subjective with objective aspects of mental events. To observe the activity of living brains, we don't have to miniaturize ourselves into a small spaceship that gets injected into the cerebral fluid of some hapless victim, as in the 1970s science-fiction film *Fantastic Voyage*. Instead we can nowadays rely on various sensors and imaging devices, such as EEG recorders or brain scanners. In the age of brain-imaging technologies like MRI and PET, we can "see" memes in action, although brain imaging studies don't yet give us pictures of the memes themselves.

It should be remembered that genes were once unobservable as well, but this didn't stop them from being biological replicators. Genes have moved from being a hypothetical entity to an observed, experimental phenomenon since they were physicalized by Watson and Crick, who identified the mythical units of inheritance as a particular molecule inside the nucleus of each cell in the body. Basically the moral of this story is never say never to a clever scientist. This is, in fact, one of the "Laws of Technology" of Arthur C. Clarke, the science-fiction writer and inventor: "When a scientist says something is possible, he is almost certainly right. When he states that something is impossible, he is very probably wrong." So the apparent advantage of focusing on behavior, if indeed it exists, is only likely to remain an advantage temporarily. Convenience isn't reason enough to change our minds about such a fundamental issue as where to find a new kind of replicator.

A HOME FOR MEMES?

With behavior now out of the picture, and given our earlier elimination of artifacts, the only remaining candidate as the substrate for memes is ideas. Our conclusion by exclusion, then, is that the brain must be where memes reside, if they exist anywhere. But before we try to identify a mechanism by which ideas might be replicated, which means delving into the somewhat mysterious, highly intricate territory that is the human brain, I first want to convince you that the brain is actually a likely place to find replicators. My goal is to show you that the brain is a highly isolated, protective environment in which replicators could evolve.

In fact, the brain is just the kind of place Replicator Theory would suggest as the birthplace of a replicator. It is full of ambient energy waiting to be harnessed and provides lots of scaffolding for a replication reaction in the form of a network of cells with support structures on which to hang components, all in standardized configurations, and so on. Just as DNA is protected inside a double envelope of outer cell and nuclear walls, so too is the brain encased inside a system of doubled protection, this time made of bone, tissue, and chemicals. The hard braincase is an important first line of defense against environmental insults. It ensures that physical trauma won't deform the brain and so impede its work. The bony skull also protects it from radiation, which would introduce noise into the signal-transmission process. Inside the skull is cerebrospinal fluid, which fills the spaces between these closely packed cells, providing buoyancy and mechanical support for the brain, while helping stabilize the flow and chemical composition around the central nervous system and flushing away the metabolic products of neurons and glial cells.

The brain, as a signal processor, must also be protected from sources of noise and signals coming from the rest of the body that might be harmful. This includes what genes might try to "say" to the brain: Viruses are potential messengers knocking at the door with bad news. For this job, the brain has instituted the blood-brain barrier (BBB), which functions just like the software and hardware-based "fire walls" that companies anxious to preserve company secrets install around their intranets. The BBB is present in all vertebrate brains, so it is an ancient adaptation for this purpose, nearly as old as the cortex itself. It is also laid down within the first trimester of human fetal life, which implies it is an important early step of self-defense. Perhaps nowhere else in the body are cells attached as firmly and closely as they are in the multi-mile-long network of capillaries in the brain, forming a barrier of tight intercellular junctions.

Why do these capillaries form such an impenetrable barrier? This anatomical arrangement shields delicate brain tissue from toxins in the bloodstream and from biochemical fluctuations that could be overwhelming if the brain had to continually respond to them. By restricting the passage of molecules between the bloodstream and the brain, this microvascular system ensures the proper maintenance of the neuronal microenvironment. This defense works against most viruses, which are stopped by the BBB (although HIV, the virus that causes AIDS, manages to sneak through—how is still unknown). So the BBB is a second line of defense against illegal information and disruption of the computational functioning of the brain.

The brain is secure, stable real estate—a good place for a cultural replicator to set up shop.

Chapter Seven

MEMES AS A STATE OF MIND

If my opinions are the result of chemical processes going on in my brain, they are determined by the laws of chemistry, not logic.

—*J.B.S. Haldane*

Memes must have initially replicated within individual brains because that's the simplest context for a replication reaction. How did brains evolve to allow such an event? Perhaps surprisingly, it is possible to see quite clearly how and why a novel replicator such as a meme could evolve inside nervous systems. More importantly, it is now possible to define how a replication reaction would work inside a brain. There are even clear reasons for thinking such a reaction arose during the history of life on Earth, as we will see.

GETTING MORE NERVOUS WITH TIME

In the beginning, there were no replicators. Sometime after the Big Bang, starstuff congealed. Planets formed. But there was emptiness upon the face of the deep. A certain quiet prevailed.

Then the first primitive entity that could copy itself arose, its identity obscured by the mists of time. Although "born naked," these early replicators quickly learned to sheath themselves in a coat of protoplasm for protection from the elements. This was the humble beginning, the birth of life.

Next came single-celled organisms, the first specialized interactors. They possessed facilities both to receive information from their environment and to take action on behalf of the genes stored inside. Primitive bacteria, for example, could orient themselves toward particular chemicals; advanced protozoa could even habituate to stimuli or become sensitized to them.

However, some 680 million years ago, living creatures arose that were composed of multiple, cooperating cells. Biological processes originally collected into a single cell could now be divided among them. Information-processing abilities, in particular, became the specialization of particular types of cell within these more complex organisms. Such organisms harbored receptor cells that responded to particular chemicals in the environment, while other cells stimulated their primitive motor apparatus and so began doing things like vibrating. Some of these cells began to release hormones into the body that diffused through intracellular spaces and, later, blood vessels until they reached other cells able to read the message. Such primitive nervous systems consisted of "hard-wired," two-step links from sensory to motor cells, transmitting messages from place to place within the organism. This was the birth of intercellular communication.

As time passed, even more specialized cells—neurons—evolved that were able to produce electrical spikes rather than chemical signals. This allowed messages to be delivered from point to point, even across large bodies. Then these neurons began to aggregate in one portion of the organism to improve their ability to coordinate behaviors. In effect, a "central" nervous system developed: Neurons and sensory receptors were localized at the "head" of the organism, with sensory and motor neurons relegated to peripheral status, organized as nerve cords running along the length of the body. Ancestral annelids and other invertebrates then slipped a neuron into the circuit between receptor and motor cells. These "inter-neurons" created a three-layer network, freeing its members so that they were able to specialize. Certain neurons began firing only when specific kinds of inputs were fed to them by other neurons; each time around they would become more likely to fire in response to that same class of stimulus. They were, in effect, "tuning in" to certain features of the external environment. This was the birth of memory.

Other neurons could make use of the information stored in their colleagues when attempting to make their own classifications as well. This led to the formation of clusters of neurons collaborating to make finer and finer discriminations about the world around them. Increased numbers of neurons led to the need for some internal organization as well. These administrative tasks came to be fulfilled by yet other specialized neurons. This was the birth of (real) brains. The rest is prehistory.

This brief fable about the emergence of the brain ends before we come to the birth of memes. Let me move our timeframe forward even more, paying greater attention to detail as we go along.

Several major events can be discerned in the history of brain evolution. First, nervous systems become increasingly centralized, evolving from a loose network of nerve cells (as in the jellyfish) to a localized brain. This was a trend toward encephalization—that is, a concentration of neurons and sense organs at one end of the organism, which allows greater coordination and control by the decision-making center.

With the rise of vertebrates and the development of a spinal cord (in animals such as fish, amphibians, and reptiles), we also get a division of the brain itself into a series of three major swellings: the hindbrain, midbrain, and forebrain. In certain animal lineages, the brain becomes both much larger and still more complex. Eventually a gradual increase in the *relative* size of the brain compared to the body is also seen.

With the rise of mammals during the Mesozoic era, the continuing evolution of the brain becomes marked by changes in the ratio of midbrain and forebrain structures, and in the expansion of the forebrain itself. The mammalian brain keeps the major division into three sections but also adds two new structures: the neocerebellum ("new little brain"), looking much like a fungal growth at the base of the brain, and the neocortex ("new cortex"), which grows out of the front of the forebrain. In most mammals, these new additions are not particularly large relative to the primordial brain stem. But when we come to primates, they become much larger, and in the human they are so large that the original brain stem is almost completely hidden by this convoluted mass of gray matter.

With our arrival at human brains, we find about a hundred *billion* neurons have gathered together to form a truly complex communicative network. Since a typical neuron in the cerebral cortex receives input from a few thousand fellow neurons, this makes for about 100 *trillion* connections in the 1.5 kilograms of gray matter inside your head. (If you started counting synapses at the rate of one per second right now, you'd be finished in about 32 million years.) Each cubic millimeter of your cortical tissue—a volume about the size of an average fleck rubbed off the end of a pencil eraser, and roughly the same consistency—includes 100,000 neurons, 500 million dendrites, and close to 1 billion synapses wired together with 4 kilometers (about 2 1/2 miles) of axonal cabling. This is all supported in a mass of cells called glia, which together constitute neuropil, or gray matter.

Neurons are structurally and electrically polarized. One end of the cell receives chemical or electrical signals from sensory cells or other cortical neurons that destabilize its polarity (or electrical conductance). The opposite end of the cell transmits signals to other neurons or to muscle or gland cells. This receptive end is called a dendrite (from the Greek word meaning "tree-like") because it usually has numerous branches. Dendrites can be likened to a bushy antenna system that receives signals from other neurons. When a dendrite is stimulated in a certain way, the polarity of its cell body exceeds some threshold value, causing a specialized section of the neuron (called an axon) to generate a millisecond-long pulse. This action potential or spike sends a signal out along the axon to a synaptic terminal at the opposite end of the neuron, where it may be picked up by the dendrites of other neurons.

For one neuron to influence another, the two must be connected. This is accomplished by junctions called synapses, which abut the dendrites of other neurons. Each neuron has up to 10,000 synapses impinging on it from about 1,000 other neurons. This means that, on average, each neuron touches another neuron in about 10 places. This is a large number of ways in which one neuron can influence another and is sufficiently complex in itself to ensure that interesting evolutionary dynamics could take place even at this small, inter-neuronal scale.

Synapses work by transmitting a chemical signal (neurotransmitter) or electrical signal (current-carrying ions) across the narrow gap between its cell and the next. These messages activate receptors on the membrane of the target cell, which change *its* polarity, and thus make it possible for the message to be sent further through the network of interconnected cells that make up the brain.

Each neuron in this network is itself a small computer, capable of sophisticated non-linear operations, such as the multiplication, division, or amplification of signal strengths, or establishing a threshold for its own activity. If we put this knowledge together with our recent awareness of the role of other agents in information processing—such as support structures like glia or gases like nitrous oxide permeating the inter-neuronal space—it becomes clear that a neuron's decision to fire is influenced by a broad range of factors and is a truly complex decision-making process.

Just imagine what happens when you link millions of these into a communicating network! It obviously gives a big-brained organism tremendous computing power. But from an evolutionary point of view, the question is why an animal would need such power, especially given that it comes at a significant cost, that of maintaining such a large, energy-sucking organ in the first place. Why bigger and bigger brains have evolved remains a puzzle, especially as each neuron seems so complex. It is also an important question for us, for only in big brains do memes become possible.

THE PLASTIC BRAIN

Bigger brains must have evolved because they made the lives of the animals with bigger brains better. Primitive nervous systems that have been around more than a half-billion years are still fairly typical of the vast majority of such systems in the animal kingdom. These are essentially reactive systems: A sensor sees a condition, such as heat or light, and passes that information directly to a fixed decision-making process that converts this real-time information directly into a command to respond in some fashion. For example, organisms that rely on light for power can sense in which direction the most light can be found and then move toward it. Animals with these rather simple nervous systems exhibit what the scientists who study animal behavior call "fixed action patterns." Variety in these behaviors arises only through changes in the genetic code underlying brain development. If a situation presents itself that is unexpected by the gene-based algorithm, the animal acts inappropriately, and without remedy, time after time.

Complex responses can be genetic too. More complex behavioral sequences, with multiple decision points, can be coded as relatively simple rule systems into an animal's genome. A famous example in the animal behavior textbooks comes from bumblebees. In their hives, they can engage in the precise, multi-step removal of wax caps on the cells of dead pupae, even though they have no cortex at all to speak of.

What distinguishes learning from such programming is the ability to *change* responses because of new experiences. Human twins, for example, do not necessarily react in the same way to the same situation, despite their genetic identity. Take an extreme example: the case of Chang and Eng, the nineteenth-century Siamese twins who were joined at the chest. They had virtually identical genotypes, right down to the nucleic acids. And two people simply can't be raised in a more similar environment— there was no spatial separation between them, and no birth-order effects in the way their parents treated them. Their experiences throughout their entire lives were as similar as could be, given that they were different individuals, because wherever one went, by necessity the other did also. They even married sisters (with whom each fathered about 10 children). Yet Chang was the dominant one, more intellectual and volatile, while Eng remained submissive and sober. They even sometimes voted for opposite parties in American elections (the pair settled in Virginia as adults). Their example is wonderful evidence that genes don't control everything in big-brained creatures. Something made Chang different from Eng, and that thing was his brain.

So *any* big brain is unique. It even differs from itself over relatively short periods of time, because the state of its network is constantly in flux. Thus the outcome of growing bigger brains, with their increased numbers and layers of neurons, is greater plasticity of response. This plasticity produces true *behavior,* or selective movements and activities in answer to specific stimuli. Flexibility of response within the life span of the individual organism (rather than on a generational timescale) becomes possible with more complex nervous systems. A more contingent response, dependent on a larger number of cues from the environmental circumstance, can be created. The number of surprises such animals encounter over their lifetimes may not decrease, but the number of behavioral wrong-footings presumably does.

With the human brain's 1 trillion connections we get truly staggering possibilities for flexible responses. Could all this be coded into our genes? Or is there room for another replicator in there?

The human genome has between 40,000 and 60,000 genes, with a considerable fraction of these not coding for anything. This makes a difference several orders of magnitude in size between the number of genes and the number of synapses. Simple division suggests that each gene must therefore control at least 1 million of the brain's decision-making units—more, really, since fewer than half our genes code for features of the brain. This simple analysis makes it clear that each neuron cannot be directed to its final resting place, and all its connections dictated, by a specific genetic program. In effect, there is no gene specifying the state of the 45th synapse down the axon of the 2,649th neuron on the left, just past that third crease of the temporal sulcus. The numerical disparity between genes and synapses proves that genes cannot directly specify the exact topology of our neural networks.

It is possible, however—by setting the rules by which neurons find their places and roles—for genes to *indirectly* determine the final form of the brain. Many models have shown us that very complex objects can be created by a small number of heterogeneous objects interacting with one another. Evidence from brains themselves also show this ability. Transplantation experiments indicate that neuronal connectivity is largely determined by very general rules rather than any characteristics specific to individual neurons. Neurons transplanted into the fetal brain of another species wind up perfectly well connected. Thus, "when a pig neuron grows up in a rat brain environment, it integrates with other neurons according to rat rules." This is because the genetic program for neuronal development is able to compensate for whatever body it happens to find itself in and allows the neuron to wire itself up to that network appropriately.

So the simple arithmetic above is not enough to tell us that our behavior isn't genetically "determined." General principles of construction—the *contingent* expression of genes—could be at work, and the many possible interactions between genetic rules could do nicely to create the complex structures we have inside our heads.

Still, it is well known that genes don't act as blueprints. Organisms are not preformed in any way; there is no homunculus (or miniaturized person) inside egg or sperm cells that simply grows larger as cells are added to the developing body. Rather genes are like recipes. The recipe doesn't have to be as big as the cake it helps you to make—a whole cake shop can be filled with the products of a very slim recipe book. Similarly, a few genes can have major effects on the final shape and function of a body. Just look at the way the brain grows. A special class of genes, called Hox genes, are now known to control very general processes during development, such as the basic divisions in the body plans of vertebrates, including segmentation of the brain. So one gene can have massively ramified effects on a developing structure.

But if every last interconnection is not spelled out in the chromosomes, then how *do* neurons get wired up properly during development? If genes are not the sole source of information shaping the brain, what else is at work? Presumably an organism's experience also shapes its mental contents. This certainly seems to be the case.

We now know that there are at least five distinct phases during the life span of primates during which the number of synapses significantly changes. Three of these phases take place before an individual has even left the womb. The fourth occurs during the first couple of years of life and leads to truly remarkable changes in the number of synapses. In humans, the infant brain sprouts neurons and synaptic connections at a phenomenal rate through the first year of life, so that a toddler actually has almost twice the number of neurons as an adult. But this proliferation is rapidly followed by significant pruning of neurons and their connections to sculpt the brain's processing pathways. Throughout our early childhood we lose about 20 billion synaptic connections each day. Poorly placed or unneeded cells and connections die off as the brain tunes itself to the sensory world and the body in which it finds itself. The last, recently discovered phase of change in synaptic architecture takes place during adolescence and is characterized by a slower decline in the density of synaptic connections in the brain. Neuronal networks are thus trained by, and come to incorporate information from, the environment. This has been encapsulated as the "use it or lose it" hypothesis of brain development. Social upbringing—our culture, if you will—is thus important in framing the way our minds work.

In sum, the genome suggests the general structure of the central nervous system (the number of cells of various types, their relative placement, and so on), while nervous system activity and sensory stimulation provide the means by which the system is fine-tuned and made operational. And since new stimuli are constantly flooding sensory inputs, this means the brain is always in flux: Its contents are extremely volatile.

This conclusion is consistent with a theoretical reason to believe genes must have relinquished at least some control over the operation of big brains. There is a fundamental limit to the complexity of form that can be achieved by natural selection. Increases in the complexity of phenotypes is subject to constraints, as first recognized by Manfred Eigen. What he called an "error threshold" follows directly from Replicator Theory. Think of the likelihood of mutation as being constant for each gene in an animal's genome. Then begin to increase the total number of genes. At some point it becomes inevitable that the disruptive forces of mutation, acting on each gene, will dominate the directional force applied to the genome by natural selection: The mutations can be creating centrifugal force in a variety of directions, while selection pulls only one way. Combinations of genes required to produce the organism's adaptive features simply cannot be kept together in the face of assaults on each gene by mutation. As the pressures put upon natural selection increase, new mutations will no longer be efficiently filtered out and can begin to sneak into the larger genome. Then one or another component of the gene clusters within that genome will tend to be replaced by some variant that can't play its proper role. At this threshold in the size of the genome, no further accumulation of useful variants, and the adaptations they produce, will be possible. In effect, the maximum number of beneficial genes selection can preserve as a team will have been reached. For any additional accumulation of complexity, a new evolutionary force must arise (or the mutation rate, which we have assumed to be constant, must be reduced somehow). The evolution of larger nervous systems can therefore be seen as a means of increasing the complexity of body shapes and functions once increases to the complexity of the genome have reached their limit. The organism must then begin to store information in the convoluted tissues of the cortex rather than in the double helix in the center of each of its cells.

Anyway, it doesn't make much sense for genes to exert total control over our thinking. The whole point of growing a big brain, after all, is to introduce some flexibility into responses. In essence, the idea is for the brain to accumulate information not instructed by the genes, but selected from the environment. Learning is favored over genes in a range of circumstances. Increased brain size should be selected for in lineages of organisms when it is advantageous to acquire information that is difficult to code in the language of genes, such as remembering a migratory route or recognizing each member of one's social group as an individual.

There is another consequence of growing a big brain that is worth mentioning. With increased volume, the number of neurons "inside" the cranial sphere increases faster than the number of ones on the surface, those connected to the outside world (assuming sensory receptors are also limited to the surface of the body). It seems almost inevitable, then, that a big brain will turn inward on itself.

Still, the behaviorist credo that animals are devices for translating sensory input into appropriate responses dies hard. In fact, "the notion that behavior is always a reaction to a stimulus is so ingrained that the name applied by psychologists to an element of behavior is 'response.'" But contrary to this commonsensical notion that the brain is constantly trying to respond to sensory stimuli, much of the information a big brain is working on at any moment has been called up from memory, rather than having just come in as the latest update about environmental conditions. One reason that so much of the activity of big brains is internally generated is that the downside of considerable plasticity is the need for increased memory management. In big brains, neurons are busy trying to keep the "books" straight, or to supervise what is happening elsewhere in the cortex. The brain must try to keep what has already been learned stored somewhere as new impressions come on-line, with the potential to disrupt previously established links and associations. Big brains are thus largely talking to themselves, modulating what they think about things that happened long ago or that might happen in the future. This process even continues during sleep, when memory continues to be consolidated and events of the day are rehearsed to extract meaningful connections from experience.

As a result of this continual reassessment of the current situation, the human brain gains an extended temporal horizon about which it can "think." This makes possible our sense of the past. Despite its ever-changing nature, the brain includes an internal model of the world, which it compares with new experiences. It also has a model of itself. That is, the brain has a point of view. It represents its own place, both in the body and in the world at large, and uses this model to continuously test whether signals coming in from the environment can be ignored or must be incorporated into its worldview.

Thus, the brain never stops sucking information out of its sensors to update what it has in its database. Each new input reverberates throughout the brain, altering it at many levels. The brain also decides how to respond to external stimuli based on its own activity at the moment. The reaction to exactly the same impulse can be different each time the impulse arrives because a response has to do with attention, context, and the behavioral relevance of the stimulus. Every millisecond, the brain sees things differently.

In a sense, then, the brain becomes its own environment, to a significant degree, as its size increases. It becomes a world unto itself, largely responding to stimuli that it has *generated,* not received. This makes the brain a "solipsistic" organ—an entity living in its own house of mirrors, truly convinced of only one thing: its own existence. As a structure, the brain is like a maze in which most of the routes circle back to the starting point, rather than leading out.

This feature makes the big brain a natural medium in which replicators might bubble into existence, because self-referentiality is but one step away from self-replication. To point to yourself, for example, you must form a physical circle with your arm. Referring to oneself can thus be represented as a loop, like a snake coiled up with its head biting its tail. Replication too is a recursive process that iteratively returns to the same place it has visited before, hoping to begin again. It's just a temporal loop, rather than a physical one. The fact that a major proportion of the links in the brain feed back to areas that feed forward to them is therefore highly suggestive. The self-referential nature of the brain gives us every reason to expect that the replication of information is possible in such a place. The only remaining question is whether the replicator that benefits from all this looping of information is genes or a parasite on biological evolution arising in this new microenvironment: memes. The answer to this question, it turns out, depends on how fast you want to store new information—that is, on just *how* plastic the brain is.

THE MILLISECOND MEME

DNA is at work in the brain, and not just during the period when the brain is developing in the womb. When does the genetic programming of brain states become potentially memetic in nature? To answer this question, we can take the prion case as a source of insight: Prions are replicators because they can convert protein shapes (and hence functions) without the gene responsible for the protein being changed in any way. This is independent change, worthy of entitlement as a new replication process. Similarly, that other class of parasitic replicators we looked at earlier, computer viruses, are defined by their ability to duplicate memory states without recourse to computer operators, using only the operating system and software pre-existing on the computer (or network) to achieve this goal. True replication, after all, must exhibit a direct informational link from one replicator to the next. This is what is required by the information transfer condition for replication. So what we are looking for in the brain is a new causal pathway for information transmission. Memes, by analogy to these other replicators, can therefore be said to arise *when changes in brain states are effected without recourse to DNA*.

Luckily we don't have to look far for evidence that memes might play a role in brain functioning. Everything we know about learning mechanisms suggests that short-term changes in neuronal states are not accompanied by the expression of genes. In fact, the concept of "short-term" memory is *defined* by the lack of protein synthesis, which is the normal consequence of gene expression. The primary difference between short-term and long-term memory is therefore the direct involvement of genes. Since this condition is so important for our definition of memes, elaborating on just how these two mechanisms of memory work is worthwhile.

Memories are thought to be stored in the brain as changes in synaptic connections between interconnected neurons. Change in such networks can arise in three ways. First, new neurons can be recruited from generative areas of the brain to participate in memory storage (as added nodes in the network). We have recently learned that this process, which changes the physical wiring diagram of the brain, can occur in the neo-cortex even in adulthood. Second, new synaptic connections between *existing* neurons may be formed, which also modifies the topology of the network (as new links, but without new nodes). These connections between neurons are largely determined by the early experience of the individual, with particular links either being strengthened or withering away as a result of feedback from the environment. And third, there can be changes in the *strength* of existing synaptic connections linking neu-rons, with no new links or nodes involved. This strength of connection between two neurons can be expressed as an increased or decreased like-lihood of one neuron firing as a result of being fired at by the other.

Thus a variety of mechanisms for storing information are available in the complex brain. The first two methods are physiological, requiring the growth of either entirely new cells or parts of cells. They are there-fore restricted to relatively long-term changes in memory. On the other hand, the third mechanism—the ability of neurons to alter the strength of their synaptic connections with activity and experience—is a complex phenomenon involving a variety of physical and chemical factors. This *plasticity* in synapses is thought to play a critical role in memory storage. Existing connections can be strengthened, or weakened, within a time-frame that staggers the imagination: in milliseconds (thousandths of a second). Since it is only through such plasticity that short-term memory independent of new protein synthesis can be identified at all, we focus on this mechanism here.

First, however, it is important to show how this mechanism, synap-tic plasticity, contributes to *long*-term memory storage as well, because this will set the terms within which memes can be defined. A number of studies, involving animals as varied as the fruit fly (*Drosophila*), nematode worm (*Aplysia*), and mouse, have demonstrated the importance of a par-ticular protein, called CREB for short, which is involved in the process of consolidating "long-lasting" changes in synaptic plasticity (those that occur over a period of seconds to hours). This body of data supports the hypothesis that gene transcription is critical to long-term, learning-induced changes in neuronal networks. The transcription of specific DNA sequences in the cell, induced by CREB, initiates a cascade of activity in which other genes are also read off. This ultimately results in the production of proteins that are responsible for the structural changes underlying long-term changes in the responsiveness of neurons.

Further, regulation of gene expression by CREB can be localized to individual synapses, which are chemically "marked" by the local translation of a kind of RNA located at specific dendrites. This RNA, called "messenger" RNA (the same molecule that transports the genetic "signal" from nuclear DNA to the site of protein synthesis), is shipped out to these extreme locations in the neuron. There, it is translated to produce proteins under the control of local synapses. These proteins then hang around, regulating the sensitivity of that dendrite to incoming stimuli. Decentralized protein production, under the finely tuned control of synapses acting as local managers, thus contributes to the critical task of storing memories. The flexibility of neuronal responses to incoming stimuli is increased significantly thanks to the highly sensitive repertoires of proteins circulating around individual dendrites.

Essentially the cell is making use of messenger RNA's ability to produce proteins as a way to store information in these stable molecules. It's true that there is no change in the neuron's gene sequence, and none of this activity feeds directly back to the organism's gametes, to the cells that will be responsible for determining the genetic constitution of the next generation of brainy organisms. So this use of messenger RNA and protein may appear immaterial from the point of view of biological evolution, since the regulation of a neuron's propensity to fire occurs without immediate instruction from nuclear DNA. But there *is* an indirect link: The particular kind of messenger RNA sent out to a dendrite is determined by the transcription of a specific gene. For this reason, there will be selection on those aspects of the genetic system present in neurons for their ability to respond to the outside world appropriately, and for the ability of those proteins to stabilize memory. In this sense, there will be biological evolution in these memory-production mechanisms, and thence on the genes that produce them. In effect, these longer-term changes in brain states are just flexible genetic phenotypes, which could still preclude a role for memes.

To steer clear of this possibility, we need to identify a domain in which there is no gene expression at all. Such a domain does exist. No known mechanism of gene activity is reactive at the temporal and spatial scale necessary to explain short-term changes in synaptic plasticity—

that is, in the strength of connections between neurons in complex brains. To do so, the product of gene expression, a protein, would have to be produced and go to work *in milliseconds, and at individual synapses.* Regardless of what we learn about how brains work in the future, it is unlikely that gene expression will ever invade the territory framed by this temporal and spatial scale. This is change in neuronal states without recourse to DNA (even indirectly)—which satisfies our condition for independence.

Therefore some activity in the brain is not directed by genes. The basic physiology of neurons and their connections are essentially fixed at the timescale of a second or so. But the *state* of a node can be changed within this timeframe. This doesn't mean that replication of neuronal states occurs, but it does imply there is *room* for another replicator to be responsible for brain activity within the confined domain we have identified. However, this alternative replicator must work very fast indeed, and in a highly localized situation—in the context of a few neurons at most, working in concert.

Let me be clear about this crucial point. Even if they are quite distinct kinds of response—one electrical, the other structural—short-term and long-term memory *are* linked. Any short-term neuronal response probably triggers some cascade of long-term memory consolidation, which suggests that the brain's short- and long-term responses are integrated. Short-term changes trap the trace of memory, which the long-term mechanism then fixes into place. Short-term memory may not depend on genes during the instant of trapping, but it does depend on genes to produce its mechanisms. And then gene transcription is needed to consolidate its gains. So long-term memory formation is dependent on short-term changes in neuronal plasticity.

Nevertheless the *mechanisms* for these two kinds of memory are independent. If the mechanisms are distinct, then so too are the results. We can legitimately disentangle two kinds of information in the brain, each being the product of processes with different dynamics and causal structures. The distinction is not only analytic—it's not just a distinction born of my mind but not in my brain. It's a real, physical difference, because the mechanisms are physical things.

Here's the decisive point: Short-term mechanisms for changing neuronal states will be under natural selection because they reappear in each generation of brains. But the *products* of those mechanisms—memetic information, potentially—is not subject to the same selection effects as it scoots around the brain. Similarly, prions are produced using mechanisms put into place originally by genetic processes for the folding of proteins, but contain information acquired independently of genetic assignment (through inter-molecular contact) that then flits between molecules. The key factor isn't what generates the mechanism in the first place but what happens to the information generated by that mechanism. A lineage of memetic information can be created independently of gene-produced mechanisms for storing memories, even short-term ones.

In effect, what we have located is the possibility of a "millisecond meme." With the involvement of such a replicator, the state of a neuronal node would be determined by the sequence of events experienced by the host organism and endogenous memetic evolution. This state would thus be (to a greater or lesser degree) independent of genes in that its current configuration has been produced by a history of stimulation by previous meme products, as well as other stimuli. It would be in this domain that, as the eminent neuroscientist Jean-Pierre Changeux suggests, "the Darwinism of synapses replaces the Darwinism of genes."

THE NEUROMEME DEFINED

We now have a temporal and spatial framework within which to continue our search for memes. But what about getting even more physical? Can we define a substrate and replication mechanism for memes?

Most definitions of memes are abstract, couched in terms of information or the mental representation that results from imitation. But replicators exist as specific substrates, as physical complexes. So too must memes be, if they are replicators.

This isn't to say that physical models of memes don't already exist. In fact, a number of memeticists have suggested that memes should be associated with the physical network of neurons—the map that links each neuron, through synapses, to others. Dawkins, for example, argued that "if the brain stores information as a pattern of synaptic connections, a meme should in principle be visible under a microscope as a definite pattern of synaptic structure."

But it is clear, I hope, from the preceding discussion, that this argument cannot be true. It commits the same error as an earlier generation of neuroscientists, such as Karl Lashley and Wilder Penfield, fell prey to when they attempted to locate the traces of memory (which they called an "engram") in the brain. They were victimized by an overactive concern with tissue. Memories simply aren't warehoused in the miniature bodies of specific neurons, to be elicited by an electrical prod to the exposed gray matter or excised by surgically removing an area of the cortex. You can't separate some molecules from the brain and expect to hold a memory in your hand. No bit of information has meaning except through its relationships to other bits of information. Our model of memes must be consistent with what we presently know about the *distributed* and *contextual* nature of memory and learning in the brain. I therefore argue, contrary to Dawkins and others, that a meme must be an *aspect* of the neuronal network.

Memes must also be memories—although here I'm not using the word in its everyday sense, as something that can be consciously recalled about a specific event in an individual's life history. Instead let me use the word more generally to refer to any information that persists over time in the brain, even if it is the stored value of some variable internally generated by a solipsistic brain. Memes, then, are just a class of memories that can copy themselves.

What is the context in which such copying might occur? What neurons do is communicate with one another—pass information from one to the other in chains of exchange. That is their job. So the replication of memes must occur in the context of communication between neurons. However, it's also the case that for Darwinian evolution to occur in the brain, something must change along the way. But this change can't involve new tissues being formed and added to the network—that can't occur at the millisecond timescale we require. What *is* reproduced through millisecond generations, however, is *states* of neurons: Molecules like neurotransmitters, the electrical potential-producing elements in a neuron, can agglomerate or dissipate very rapidly indeed. This means the electrochemical states of neurons can be reconstructed on this timescale and so can serve as the foundation for what might be called neuromemetics.

These ideas demand further discussion. But given our current ignorance about how the brain stores and manages information, it isn't worth trying to be too specific about the physical scale at which replication occurs. I will simply use the word "node" to indicate where replication happens. A node is, in effect, some portion of a neuronal network. I will also take the most elementary case—limiting myself to only two nodes— and treat communication between these two nodes in a rather schematic way to keep my discussion simple. My reasons for being vague about all this are derived from our current ignorance about the details of neural functioning. Let me next outline the kinds of ways in which two nodes can interact in an attempt to identify if any of these possibilities qualifies as replication.

The most frequent kind of event between nodes is probably one in which a source node, 1, is in some state, A, prior to firing. As a result of being stimulated and emitting a spike, node 1's state changes to B. The spike then stimulates a second node, 2, converting it from state Y to state Z. Since the state of node 1 at the end of this process (B) is not the same as node 2's state at the end (Z), no state has been duplicated and hence there is no possibility that replication has occurred. I will call this simply a case of neuronal *communication:* Some information has passed "hands," but that's all.

A second possibility is that node 1, prior to firing, is in state A, and that after being stimulated by 1's action potential, node 2 winds up in the same state A. However, let's say the exertion of producing the spike converts node 1 to another state, B. So the state in question (A) has effectively moved from node 1 to node 2. For this reason, I will call this form of communication *state-switching.* Again, there is no duplication of states and hence no replication of state A.

A third possible kind of interaction between nodes can, I think, reasonably be called the true replication of neuronal states through a process of communication. This occurs when node 1 begins in state A, fires a spike at node 2, converting it to state A, and then itself returns to state A after firing. How does this satisfy our four conditions for replication, defined in Chapter 3? Node 1 has caused node 2 to acquire the same state as itself by stimulating it, and this state is similar to its own. Causation and similarity are thus fulfilled by this example. The two states are also related by descent, because at least part of the information that made node 2 take on state A was derived from the signal coming from node 1. And the state has been duplicated as well: The two nodes both exhibit this state simultaneously. Thus this process exhibits all four of the necessary characteristics that identify true replication. State A is now what philosophers call a "dispositional state," or what dynamical systems theorists call an "attractor." What these terms mean is that the node has an ability to return "home" after being jostled away from it. These are "prodigal son" nodes.

Yet another possibility is when two downstream nodes, 2 and 3, are both stimulated by node 1, and wind up in the same state, A, as node 1 began in. But in the process, node 1's state gets altered to B. Here, we go from one copy of state A to two copies in the next timeframe, but the location of the original copy has moved from node 1 to node 2, say. Is this still replication, good and true? Since we care about changes in the number of states, not where they are, I would argue that this qualifies as a case of memetic replication as well. The table below provides a summary of the various communication types I have just discussed.

Types of Neuronal Interaction

Type of Interaction	State Changes Rendered	Replication Criteria
Neuronal communication	Spike leaves source node in new state B, while changing receiving node to new state Z.	No duplication of states
Neuronal state-switching	Spike changes source node from state A to B but leaves receiving node in new state A.	Movement, not duplication
Neuronal state replication	Spike leaves source *and* receiving nodes in state A; alternatively, spike changes source node from state A but leaves two receiving nodes in new state A.	Causation, similarity, inheritance, and duplication fulfilled

What this analysis implies is that if some nodes in the neuronal network automatically return to their original state after firing, and at the same time (thanks to their connections with other nodes) simultaneously cause another node to settle in the same state, then that node has caused itself to be copied. Alternatively the node may simultaneously cause two related cells to adopt its original state. In either case, the "Midas touch" of such nodes has the defining quality of a replicator, making these new nodes "golden" from an evolutionary point of view. I therefore argue that the states of these nodes are memes. Like other replicators, memes are physical things. They are, in fact, *electrical* things—propensities to fire—

tied to the special kind of cells called neurons (but are not the neurons themselves).

Here, then, is my definition of a neuromeme:

A configuration in one node of a neuronal network that is able to induce the replication of its state in other nodes.

Let me explain what several of the terms in this definition mean. A "node" can be, at minimum, a single neuron or even a single synapse. However, a node may also be an ensemble of neurons in some cases: a set of linked cells that settles on the same state while causing another set to do likewise. Some clusters of strongly interconnected neurons may acquire a configuration in which they tend to return to the same state after excitation: that is, their collective state is an attractor. Such ensembles replicate themselves *in situ.* Alternatively a meme can move to another physical group if that group adopts (perhaps for the first time) the same potential to fire as the source ensemble after excitation.

Our definition therefore means that memetics—for the moment, at least—is agnostic about how the brain codes information and should therefore remain mute about issues of physical scale. A neuromeme is basically just a brain-based, super-molecular structure capable of replication. The analogous nature of the mechanism through which memes replicate and those used by prions and comp-viruses is clear. In all cases, it is the state of a preexisting substrate that is changed by the replication process. This neuroscientific perspective on memes also makes it clear that replication is not the primary function of information processing in the brain. Not every synapse or cell in the brain is going to be in a memetic state at any given moment.

What, then, do I mean by a node's "configuration"? A configuration is defined by those factors that determine a node's propensity to fire. These factors, although presently somewhat mysterious in nature, would be things like the quantity and type of neurotransmitters and other proteins present in the node, or the glia and nitrous oxide surrounding the node, which contribute to its propensity to fire. What is clear is that, taken together, these must be the factors that determine and predict the primary product of memetic activity—a spike that can assist in the conversion of other nodes into similar memes, either in the same or another brain. Depending on the scale of the node concerned, this could involve a matrix of values, with each cell of the matrix measuring the values of particular contributors (such as acetylcholine, a neurotransmitter) that determine the propensity of a particular synapse or neuron to fire. Let me emphasize that a node's configuration is not defined as a probability. Rather it is measured in terms of the physical complement of proteins and other molecules that modulate a node's responsiveness to stimuli.

A node's propensity to fire is also likely to be *stimulus-specific,* or to arise only in response to particular kinds of originating events. Signals are impoverished in content (only hinting at what needs to be said or done), so local conditions must be just right for replication to occur. Given that memetic information must be converted from signal, back to meme, and thence to signal again, complex regularities in information-processing at each stage of the information's progression through brains will likely have to evolve to support replication. A very precise relationship between memes and the signals they produce is surely required for a lineage of similar neuronal states to arise.

Further, because each neuron comes under the influence of many other neurons, it is likely to take many states as a result. Only specific kinds of inputs are apt to cause a neuron to adopt the particular infectious state of a given memetic lineage. Again, this gives us a physical reason to suspect that only certain kinds of stimuli will have the necessary causal relationship with neuronal states to result in replication.

I see this definition of a meme in terms of a state rather than a material substrate as the key conceptual move in neuromemetics. This move is crucial to achieving a legitimate perspective on brain-based cultural replicators. The implications are many. First, this switch in perspective preserves the basic distinction between replicators as informational templates and interactors as dynamic entities in the world. Genes, for example, aren't molecules; the molecules were there since early in the evolution of the universe and can be explained by chemistry. At the same time, genes aren't put together from nothing but take material already organized to a degree by other forces and add a twist: It is the *new state* of matter that, in appropriate circumstances, can replicate. What matters is the sequence of amino acids along the double helix of DNA. Similarly, in prions, another class of biological replicator, it is the conformational state of the infectious particle that is important in determining which "strain" of prion is passed on through contact. Its own three-dimensional shape determines how a local, non-infectious protein is converted into a new prion. In the same way, my distinction is between what a cultural replicator *is* (a state of matter) and what it *does* (like copy itself and have other effects in the world).

The second major consequence of the move from substrates to states is that memes are not bound to particular cells. The molecules in a neuron that define its electrochemical state, as a readiness-to-fire, come and go; it is the readiness-to-fire that endures. Memes are still physical; it's just that their substrate is energetic—"electric," if you will—rather than material.

If a meme does not depend on the identity of individual neurons, it also doesn't depend on the unique connectivity of some neuron to other neurons, nor its positional relationship to the cortex as a whole. This means the qualities of a meme don't depend on the unique features of the wiring diagram of an individual brain, which reflects the sequence of local environments it experienced during its development and learning activity. Thus the memetic state can be replicated, move from area to area, and presumably stimulate other nodes in the network to join it in the infectious state. The ability to replicate is, of course, the crucial feature of a replicator. But this freedom from materiality means that replicator information can travel, across nodes, into the environment. This transmissibility is crucial to the continued replication of a meme in different locations, and hence the persistence of memetic lineages.

The division of state or condition from physical material also implies that memes are separable. Although the brain is composed of a single neuronal network, such that any individual neuron can be connected to any other through a small number of links, this is not the case for the memes that "ride on the backs" of those neurons. Only a specific node is involved in any meme's manifestation. The fact that one meme's substrate is physically linked to another's does not mean the two memes are joined together. Rather they remain functionally distinct. This is evident in the fact that one meme can fire without the other doing so. With ensembles of neurons being tuned to several different types of stimuli (what neuroscientists call "overlapping coding"), multiple memes can even be layered in a single node, being manifest as firing patterns separable through time. So several memes can be latent in the same node, each lying in wait for the proper stimulus to come along and express it. This divisibility makes memes the "corpuscles of culture."

Finally, in my earlier discussion of replication in computer viruses, I argued that the information in a replicator must serve two functions: as a store of information and as an interpretable instruction, in order to be able to initiate the replication process when that information is "read off" by some external agent. How is this accomplished by the kind of replicator I have suggested in the brain? The neuronal state stores information about kinds of things "out there" (because it only responds to certain suites of stimuli), as well as about the recent history of the organism (or the brain itself). This makes it "about" something out in the world or something the brain cares about. At the same time, that state is capable, when induced by the input of energy (much as when an operating system "reads" a computer address), of firing a signal that instigates the process through which another similar meme is created elsewhere. In this way, a neuronal state can both be read or "interpreted" by a catalyst and implicitly *be* an interpretation or representation of other things.

La Même Chose

What makes two nodal states the "same" and therefore duplicates of one another? Being the same in evolutionary terms means being "identical by descent." This technical phrase implies that two replicators share some physical features and have the same "parents." So for two genes to be identical by descent, they must be copies that arose through replication from the same ancestral gene sequence. The copies of a particular gene in two siblings are identical by descent if they have been inherited from the same person, such as their paternal grandmother. By analogy, for two memes to be identical by descent, two nodes must have the same parental nodes and be similar in quality. What kind of quality is relevant?

Remember that our similarity criterion for memes requires only that they pass information through the neuronal network in the same way— this was the import of our Same Influence Rule, outlined earlier. This does not mean the nodes must be absolutely identical in terms of their physical layout but rather that, when stimulated, they perform in ways that have similar consequences to the larger network. This means, for neurons, that they must communicate roughly the same message to other nodes that themselves play equivalent roles in the network. The considerable redundancy in the brain gives memes ample opportunity to engage in parallel activities, despite their different locations.

Further, at the small scale at which memes are defined, there should be comparable connectivity matrices in different brains for small groups of neurons among the billions of connections made in each. Indeed there could be quite a few copies within a brain, allowing for movement of the memetic state to different locations or replication within a host to produce a (small) population of similar memes simultaneously. The combination of these two factors—the use of the Same Influence Rule to define similarity, and the fact that neuromemes are defined on a quite small scale—should guarantee that memetic states can be reproduced with some fidelity in different parts of the same brain (as well as between brains, as we shall see).

But what about the fact that two nodes releasing the same pattern of spikes at different places in the network seem to be saying something quite different by virtue of the fact that each drives the dynamics of the brain from a divergent angle? In this case, one state-A neuron in the primary visual cortex and a second state-A cell in some other region of cortex would *not* have an equivalent influence on brain activity. Doesn't this negate the possibility of replication within a brain—according to the criteria we have set up—because no two neurons will have exactly the same effect on the larger network of which each is a part?

This is where it becomes important to remember the crucial distinction between local, or "value-added," processing and global meaning, which reflects the cumulative decision-making by an information-processing stream. The node in the visual cortex from our example may be participating in exactly the same way in one decision-making circuit as the node somewhere else is doing in its own, independent circuit. In each case, the meme is telling its neighbor down the line something that has the same meaning for that neighbor *in its context*. The subject of conversation may be quite different between the two cortical domains, but that is irrelevant to our concerns. What each meme is doing is adding the same kind of value to its local information-processing stream; what the brain will later make of these decisions when bringing these parallel processes together is a separate matter.

There's also another way two nodes can be equivalent. Neuroscientists have told us that most of the thousands of input lines to the average brain cell are actually parts of feedback loops returning via neighboring neurons. Barely a tenth of a neuron's connections come from sense organs or levels lower in the information-processing hierarchy. In effect, every neuron is plugged into a multitude of mutual relationships with other neurons. If much of brain communication is a kind of back-and-forth message-passing to synchronize spike streams, then memes may mostly be talking to one another, with each keeping the other stimulated and in the right state. Neither is doing much except having the same kind of influence on the other. This makes them equivalent in the sense required for them to be considered replicas of each other.

Not only that, but in some cases the signal sent out by one neuron may not come directly back to it but instead may travel through some intermediary, forming a more complicated circuit. In effect, memes may replicate by forming what Manfred Eigen calls hypercycles. A memetic hypercycle would be a set of neurons in which each member catalyzes the production of the appropriate memetic state in the next member by firing a spike at it. This process of jolting the next guy into action goes round until it comes back to the first neuron, which gets returned to its original state, completing the causal circle. The whole neuronal circuit then constititutes a single replicator. If the circuit is small enough, it might be able to duplicate itself in another circuit by sending a spike to the neuron that originates the round of mutual stimulation in that memetic hypercycle. Such cyclic replicators are known to exist in sets of biomolecules and may have been the first replicators on Earth. But this story remains speculative with respect to memes because no one is currently looking for such circuits in the brain.

STATIONARY MEMES

I suggested above that memes can replicate in the brain through two different routes. The first method requires the source node to return to its original infectious state for duplication to occur, since only one downstream node also adopts the same state. I will call this the Stationary Meme Model. In the second method, the Mobile Meme Model of replication, the source node loses its original state but simultaneously induces two downstream nodes to become infectious. I now want to elaborate on how each of these models might work, beginning with the stationary one.

The first way in which a meme could remain in place is through a process operating within the cell to ensure it maintains the right state despite disturbances. This would most likely be some kind of electro-chemical process that kicks in automatically after the cell fires.

Another way in which the source node could be returned to its original state is through feedback from outside the cell. This could occur through either a short loop or a long loop. Both kinds of loop rely on the spike train produced by the memetic state eventually producing another signal that returns to stimulate the node which initiated the loop, returning it to its original infectious state. The short loop is distinguished from the long one by staying within the brain, taking advantage of reciprocal connections—which form a large proportion of all connections throughout the brain—between different areas of cortex.

Signals sent out into the macroenvironment can also be picked up by the host organism again. For example, what went out through the mouth can come back in through the ear, to reincite the stimulation of the source node, producing a replication event in the same brain that produced the signal in the first place. Hearing one's own voice saying something may cause the responsible neuronal node to revert to the state that produced the meme in the first place. (The novelist E.M. Forster once quipped: "How do I know what I think until I hear what I say?") Similarly, the individual who types a word onto paper or a computer screen is the first to see and interpret what the message says. These are examples of the long loop in action. The host becomes both the sender *and* receiver of messages. Indeed it has been argued that this macroenvironmental feedback loop is the reason we humans alone have consciousness: It makes us appear to ourselves as agents in the world (both "inside" and "outside" our heads simultaneously), as well as providing temporally delayed input. In effect, the returned signal presents the brain with a chance to reflect on the behavioral option it chose.

Interaction between short-term and long-term memory mechanisms is another way that nodes can be brought back into an infectious state after firing. This is because even long-term memory is constantly subjected to updating. Until recently it was thought that newly learned material is transformed into stable, solid chunks of long-term memory through a one-time process of protein generation known as consolidation. This is no longer believed to be the case. Instead long-term memories become chemically unstable every time they are retrieved. In this fragile state, long-term memories can be easily altered or disrupted. Furthermore, after they have been retrieved and are being sent back down for storage, memories must be *re*consolidated through the synthesis of new proteins. This reintegrates them with the new situation in the area where they came from, updating their consistency with any other information that may have come in during the interim. This reconsolidation occurs within the local neural system that reforms the memory.

Presumably reconsolidation occurs with sufficient precision and regularity that recall can take place without significant damage to the information contained in long-term memory. However, the process is not perfect. This is evident in the so-called false memory problem, discovered through a laboratory procedure in which information about a recent experience, made up by researchers, is fed to experimental subjects. When subjects are later questioned about the event, they tend to incorporate the false information into their account. So even relatively stable memories must be continuously reinforced—at least after each recall of the information they contain—to ensure that the information is not degraded or simply lost in the process of being used for decision-making. In effect, consistency over time is what happens when updates keep returning the memory to the same state it previously had.

Consolidation can therefore either take a node out of a memetic state or reinforce it. The Stationary and Mobile Meme Models thus differ in how the relationship between short-term and long-term information storage works out at each node. In the mobile meme case, the two kinds of memory work at odds with one another: long-term mechanisms change the infectious state introduced in some node by a short-term mechanism, rendering it un-memetic. In the Stationary Meme Model, on the other hand, short-term and long-term mechanisms are mutually reinforcing, allowing memes to remain on the spot. If the infectious state persists even through multiple episodes of reconsolidation, we can label it *meta-stable*.

The need for reconsolidation has the important consequence that memes are not restricted to short-term memory but can infiltrate long-term memory as well, since they can modify a memory *while it is in use* (after recall but prior to reconsolidation) via a fast-acting mechanism. When the reconsolidation mechanism kicks in, this memetic state then gets "frozen" in place (at least for as long as "long-term" means in the brain, which may only be a few hours). Thus the link between memes and long-term memories—if not in terms of the traditional conception as stable crystals of information safely stored away—is rescued. Infectious states can be fixed in place by protein synthesis! This obviously has important implications for the longevity of individual memes and hence for the ability of memes to create lineages.

MEMES IN MOTION

The second model of memetic replication involves the movement of the original meme from one location to another. How can this be achieved?

In only thousandths of a second, a cortical neuron's "receptive field"—the kinds of stimuli to which it responds—can shift. One moment, a neuron may be responsible for firing when a spot on the tip of your index finger feels pain; the next moment, it might be a spot near the first knuckle of that finger that is "represented" by that neuron. Scientists have for some time believed that the responsibility of an area of cortex could shift over a much longer period—say, to compensate for injury to an area of the cortex. But this level of dynamic behavior, only recently discovered and existing even at the level of individual neurons, is quite surprising because it implies that even our self-representations (at least in terms of topographic "maps" of our body that are present in our brains) are highly fluid. What this suggests more generally is that mental representations are constantly moving around in the brain.

Clearly there is a lot of change going on in the brain all the time. Just as clearly, however, there is some consistency of memory, and some sharedness to cultural knowledge. How can we reconcile the apparent chaos in the brain with the consistency and stability of social memory, with the effectiveness of socialization into the lifeways of a cultural group? After all, you remember your name from day to day, do you not? How is this possible if there is only a welter of ever-changing hubbub in your head?

In effect, the brain is faced with two largely combative goals: to learn and to remember. How can it effectively achieve both simultaneously—

and using the same equipment, since both memory and learning are accomplished by neurons, and nothing but neurons? We don't currently know the answer to this question in full. But an important way in which the brain attempts to limit potential damage to hard-won prior knowledge is by segregating the memory and learning functions spatially. The hippocampus (a small area in the middle of the cortex) apparently serves as a gatekeeper that selectively grants new impressions entry into the long-term storage areas. So tried-and-true knowledge is kept apart to some degree from more dynamic areas, while new, untested data are subjected to a variety of examinations before being admitted into the exclusive long-term memory club.

At the same time, of course, some bits of information are constantly being batted about by neurons, getting passed from hand to hand. These neurons must remain highly adaptable if they are able to hold a wide range of types of information and switch from one moment to the next. If a meme can be virtually anywhere, then the isocortex (the largely neocortical areas devoted to internal message-passing) really is just that: a vast landscape of highly similar neurons, ready to be manipulated as you please, charged with responsibility for just about anything. A neuron's connectivity, strengths of connection, relationships with support structures like glia, and involvement with neurotransmitters and other chemical partners—all must be able to change at the drop of a hat. Neurons in the isocortex must remain pluripotent. It's true that certain functions tend to regularly wind up in the same general areas from person to person, but these tend to be the primary sensory and motor areas. (The primary visual cortex is in the very back of the head, for instance.) For the rest, there is significant variation—even between hemispheres—in where a particular function winds up in a person, especially for the "higher" functions like planning and executive control that are characteristic of the isocortex.

Memes thus need not be tied to any particular location. But it isn't unexpected that memes should be mobile. Computer viruses can also move around quite freely. A comp-virus can be physically split into sections residing in multiple locations, only virtually linked into a whole by other bits of information called pointers. (This is why disk-optimizing software can de-fragment files by moving them all to one location on the physical disk to speed up their retrieval by the operating system.) Prions, our other example of a novel replicator, are also individual molecules that are free to move about their environment, the brain.

If memes are mobile, like these other replicators, then they must be adept at solving the problem of transmission. An analogy will be useful.

Think of the brain as a soup. In fact, it should be an alphabet soup, with memes serving as the letters, floating on the surface of a sea of chemicals. Each kind of meme would then be a different letter. There might be an A here and another over there, while a clump of Bs might be off somewhere else (local connectivity between neurons would probably ensure some geographical clustering of memetic lineages). The letters would bob like corks on this sea, moving back and forth, or drifting with the tide of change in the information-processing seascape. But the letters can also move about in a jerky fashion or can be thought to rapidly sink below the surface of the soup, only to reappear somewhere else a millisecond later. This is because it is, in fact, an electrical soup, and the letters are just lighted portions of a massive, densely linked network of wires just under the surface—something like a gigantic underwater neon sign. As with a neon sign, it is the flashing script that matters: The state of illumination of the sign's various sections are the memes, which can be duplicated and move about over the network of wires.

This makes it easy to see how variation and evolution can occur in a population of memes. It will naturally be most easy for the lighted message "A" to be conveyed by a section of the wiring diagram that actually bears that shape—an area of wiring that resembles two long legs attached in the middle by a belt. This would be an area of the sign with a predisposition to "A-ness." However, as with an old, disfunctional neon sign, sometimes a physical B shape will also broadcast the message "A," because the bottom length of wire in that node has short-circuited. A letter B could also display as a "C" if its right-most section of wire went out. However, it would be more of a stretch for an A diagram to display as a "C," suggesting that there are closer familial ties between memes A and B than between memes A and C. The A to C "mutation" would therefore be rather rare in the meme population. It may also be easier to go from one letter to another than the reverse. For example, it should be easier for a physical B to show as an "A" than vice versa. Thus, as various bits of the wire diagram go on the fritz, the letters displayed at a particular location of the grid can change, as can the total number of A or B letters in the electrical soup. This is the essence of evolution. The ability of the multitude of wires to transmit electrons determines what kind of illuminated state they can adopt. This analogy also suggests that each letter is composed of subunits, only some of which need to flash simultaneously to make the proper signal. Memes might therefore consist of an ensemble of components, so that a replicator node will be somewhat complex.

This is quite a different picture of the brain than we got from the Stationary Meme Model. Nevertheless both of these models may be true at the same time, although for different memes. But is one model of meme replication more likely to hold true? Are there, after all, islands of stability in the sea of moving brain waves: rocks that never move, holding fast to a moment from the past, ignoring all the stimuli beating up onto their shores, and remaining untouched by subsequent events, as expected by the Stationary Model? It now seems very unlikely that memory could cope without moving, for it would mean that some memory is—just as in computers—"addressable," with a constant physical location.

The Stationary Meme Model also requires one node to be able to regain its composure after firing, which is an added duty for a node to discharge. In this sense, the Stationary Meme Model is more complex, requiring greater sophistication in the performance of individual neurons. Although neurons are no doubt highly competent computational devices, overburdening them with tasks runs counter to the wisdom of parsimony.

Another factor to consider is the fact that the brain is a so-called "small world" network. This is a particular kind of structure that appears to generally characterize complex "real-world" networks of any kind, from chemical-reaction networks to food webs, electric power grids, networks of friends, citation links between scientists, and groups of computers. Small-world networks arise when you randomly replace some fraction of the links between nodes with random pathways. Local "neighborhoods" tend to be preserved after such a move, but at the same time it becomes possible to connect any two nodes in the network through just a few links. In effect, most power stations—or people or computers connected to the Internet—are connected together by short chains of intermediary nodes. Any person in the world today, for example, is estimated to be only about six acquaintances away from anyone else (the so-called phenomenon of "six degrees of separation"), while on the World Wide Web, one page is on average only about 16 to 20 clicks away from any other.

Connectivity in small-world networks is not perfectly regular: The physically neighboring neuron need not be the closest in terms of the length of the pathway that has to be followed to get there. (This is an important way in which the brain is unlike the artificial neural networks simulated on computers, which are fully connected—every possible link is made between nodes.) Rather, much like Internet friendships, your closest e-mail pals may be across the world, while the person in the house or apartment next to you remains an unknown quantity. Thus, in the brain, every once in a while, a single axon stretches from one side of the brain to the other, linking two otherwise quite disparate neighborhoods. This has the effect of keeping the average number of links between any two neurons in the brain reasonably low.

Models of dynamical systems with "small world" structures display an enhanced ability to propagate signals, increased computational power, and an ability to synchronize their outputs (which may explain why the synchronizing of spikes plays such an important role in the brain). In particular, infectious diseases spread more easily in small-world networks than in completely regular or completely random networks. This implies that memes should more readily survive and propagate inside brains than in alternative kinds of networks. Memes have found a kind of home, the brain, which makes movement more feasible. (On the downside, it also means that it will be necessary to observe the whole brain to know whether a replication event has occurred, because a meme that was in one corner of the brain at one moment could leap to the other side the next through one of these long-distance connections.)

For all of these reasons, I expect the Mobile Meme Model will prove much more relevant empirically than the stationary one. It is more in keeping with the truly dynamic mental organ we have come to know through recent neuroscientific advances.

THINKING IN "MEME-TIME"

Every time you turn around these days, a new discovery in neuroscience indicates a fresh kind of dynamic change that happens inside our heads.

In fact, the pictures of brains coming from the new imaging techniques are still "blurry" because the brain changes faster than pictures can effectively be taken of them.

How can replication possibly occur in the face of this tempest in a cranium? Memes are supposed to have the power to hold themselves together in the face of adversity—after all "fidelity, fecundity, and *longevity*" is the replicator motto, right? This intense mutability of memory appears to be a knockdown argument against the possibility of brain-based memes. It's a question whether any association between a certain retrievable bit of information and a particular piece of the brain lasts for longer than a few milliseconds. What could possibly be duplicated or serve as a safe location for the storage of information over time? How can we call any little piece in this stormy sea an island of stability?

Everyone intimately experiences the fact that ideas can mutate: Your beliefs change with time, just as common sense would suggest. The malleable self, something we all must deal with, is underpinned by the development of life stories we tell ourselves to give us the sense of permanence that seeing the same face in the mirror day after day seems to require. Our life narrative is simply an attempt to keep some kind of continuity in what is in fact a maelstrom inside our heads beneath our consciousness.

But the emphasis on the stability of memes within individuals—on the need to have locked-in memories—is based on the assumption that meme evolution occurs only *between* individuals. If meme evolution can occur within a host, then this perceived need for mental stability is wholly misplaced. Neuromemes *evolve* within that encompassed sphere—and *very rapidly*. The picture presented by neuroscientists of a brain constantly in flux is *not* antithetical to neuromemetics (although it is to conventional memetics, because of its exclusive tie to social transmission). Just as the HIV virus can rapidly evolve through mutation and diversification within a single host, developing an entire ecology of mutant forms side by side, so too can there be variant forms of a neuromeme, simultaneously present in one brain, that have branched off into divergent lineages. Any one of these variants can come to dominate the meme population in that brain, thanks to a variety of selection mechanisms. And any one of these forms can be the meme that makes the jump to a new host, starting the whole process of evolutionary elaboration all over again.

Meme "flexibility" within the lifetime of its host is thus not a conundrum, but an expectation of neuromemetics. Once we stop thinking about evolution from the point of view of the host and switch to the perspective of the memetic parasite, dynamicism becomes an advantage: Memes can change while their hosts stand still in evolutionary terms. This has always been one of the primary reasons for the persistence of parasites. They can hit moving targets because they are able to change direction even faster than the target can.

So we had just better get used to the somewhat counterintuitive timeframe that characterizes this new kind of replicator, with its generations that pass in an eye blink. If we are going to talk about the evolution of cultural replicators in the brain, we have to learn to think in "meme-time." Bacteria can reproduce every 20 minutes or so, but memes put the fecundity of these biological champions to shame: The rapidity of their profligacy is on the order of a thousand times faster.

REASONS FOR REPLICATING

Even if all of the foregoing is fine and good, the ideas expressed thus far remain hypothetical. What reasons do we have to even suspect that replication might be happening inside our brains? On the face of it, it seems inefficient for the brain to duplicate information. Surely evolution would select against such wasteful practices with precious resources?

In fact, there is every reason to think that multiple copies of the same information is just what evolution would produce in brains. It might be both convenient and necessary to introduce replication machinery into the normal operations, for several reasons.

First, why should bits of the brain's knowledge be spread across many cells when, as we now know, each neuron, by itself, is a pretty fancy computer and thus capable of representing a fairly complex idea or image? The answer is simple: because neurons die. There must be redundancy in a system in which any component may tragically fail without warning, and without the possibility of backup. Redundancy is fault tolerance at work.

Redundancy is also important in genetic evolution. Many genes are present in multiple copies in the same genome—sometimes hundreds of them. Some of these copies incorporate small changes in the DNA sequence. These appear to be kept in storage by organisms, ready in case their potentially different products become needed to deal with shifts in environmental conditions. Redundancy even exists in the genetic code itself, in the form of "silent" codon changes. There is a probably even greater redundancy in the neural code, in the way the brain stores information.

Someone current with the latest findings in neuroscience might want to rebut: "But we have recently learned that even *adult* brain cells can be replaced. Stem cells are constantly being recruited even by cortical areas, and wired up to take the place of dead neurons." This would indeed seem to give the old notion of the so-called grandmother cell—a single, "dedicated" neuron that fires only when the face of your grandmother comes into view—a new lease on life: If the neuron storing your grandmother's image fails for any reason, just replace it.

But, in fact, rebirth isn't good enough to fix the problem of information storage, because the information accumulated in the old cell has already been lost. If what that cell knew hadn't been duplicated elsewhere, and made retrievable, that information cannot be replaced. Sure, a new neuron could be slipped into the old neuron's spot in the network, but what would it have to say to its neighbors? It wouldn't know what the old neuron knew without being exposed to the same sequence of experiences. It could, so to speak, fill the "shoes" but not the "office" of the old neuron. The expertise the old neuron had developed over its lifetime would have died with it. And since every neuron is potentially a member of many different teams, each of these teams would have buried some of its skills with the funeral of its player.

In fact, just the opposite of the grandmother cell hypothesis is suggested by the "new" model neuron: The more complex a single neuron can be, the more important it becomes to make sure that its store of knowledge is copied elsewhere in the system. This knowledge can't be hoarded in pristine condition somewhere off-line, because the brain is a whole: Any part can be reached by any other and so is subject to constant manipulation. For true backups, neurons had to await the development of artifacts—storage mechanisms outside the body that are the brain's real "hard disks."

So it's possible that memes arose in the first place from the need for brains to keep copies around in case trauma or other trouble damaged the imprint of some crucial bit of information. From this basic backup process came abilities to copy information, parasitized by memes for their own purposes. The duplication of information within a brain may have arisen as a by-product of the growing tendency to introduce redundancy into the coding process as brains grew bigger. This pressure did not arise from memes themselves, as Susan Blackmore suggests, but from internal requirements for efficient memory representation. Our hypothesis for the origin of intra-brain memory replication, then, is that distributed representation and the multistage incorporation of memory led to replication in some cases.

Second, it's also the case that the same bit of information moves around within the brain over time. Recent evidence—thanks to the ability to simultaneously monitor multiple neurons in behaving animals over relatively long periods—suggests that as new behaviors become more habitual with repetition, the area of the brain most actively involved during such activity moves from the isocortex to more "primitive" sections of the brain. This "practice effect" is mirrored by movement of the information underlying a skill from one area of the brain to another as its performance goes from being the focus of attention to a preconscious, unthinking habit. Similarly, a meme may migrate through the brain as it goes from being a sensory stimulus to a short-term and then a long-term memory. The information may be communicated without duplication within a brain but may also in some cases involve replication, if a copy of the memory inadvertently is left behind.

There are thus a number of reasons to expect that the duplication of information, if not replication per se, is a common occurrence, especially in big brains. The step from duplication to replication may be significant but is not insurmountable, as the mechanisms I outlined for neuronal replication above build on duplication.

How Memes Qualify as Replicators

Now that I have identified a mechanism for meme replication, we can take a stab at determining just what *kind* of a replicator a meme must be. As we learned from Replicator Theory, the nature of the replication reaction determines what kind of evolutionary dynamics will follow. Will only the fittest memes survive over time, or will the first memes to arise come to dominate the population? On the other hand, perhaps every kind of meme will have a chance in evolutionary tournaments?

In fact, what we want to find is that memes can evolve toward greater complexity. Only in this way can they account for a significant portion of culture, which itself is becoming more complex, seemingly at an increasing pace. This means memes must exhibit linear, or Malthusian, growth, because only then will the fittest memes survive—those that can combine to produce more and more complex adaptations. We also know from comparisons with other replicators that the kind of replication reaction leading to Malthusian growth and the optimization of memetic features over time must not bind memes up in the process (as in the survival of everybody scenario) nor involve multiple replicators to produce each new one (as with survival of the first). Instead we must find one replicator producing one copy with each iteration of the process. So is this what the memetic replication reaction looks like?

Meme replication requires a sequence of steps. Incoming stimuli serve as the super-molecular equivalent of catalysts for reactions during which non-infectious neuronal nodes are transformed into infectious ones through a change in their "conformations." If we think of this process in Replicator Theory terms, we can say, first, that a signal enters the "containment" area (the cellular matrix of a neuronal node), setting an electrical reaction in motion. The signal works as a chaperone or catalyst to transfer memetic information from the source node to the replication assistants (such as neurotransmitters) or simply to catalyze meme functionality from local materials. When the "reaction" is finished, the meme-node is left in an electrochemical state that is infectious. In effect, then, the original stimulus sets off a "chain reaction" through the downstream node that culminates in that node adopting a new connectivity pattern and state of latent activation. Sometime later, a new stimulus is then produced that is capable of transmitting the infection to yet other nodes. It is this characteristic of the new output that makes the node suddenly memetic.

This mechanism, as described, requires one preexisting meme. Only one replicator must be present to produce another one, and then it becomes free to pursue other avenues of replication. The conclusion is thus that meme replication is one source to one copy. I therefore think we have every reason to believe that memes will function just like any of the other replicators we have studied, at least in terms of their growth pattern. They must run the gauntlet of selection, with only the fittest surviving—those memes that absorb new adaptive qualities over time.

Memes also display a number of other parallels with prions and computer viruses, the other replicators we have investigated. First, they can replicate via *conversion,* rather than construction, of the neuronal substrate. Just as prion replication depends on a preexisting, gene-produced, normally folded protein molecule, and comp-virus replication depends on preexisting artifactual memory, meme replication depends on preexisting, gene-produced brain structures. Second, memes can be reversed, or become non-infectious again, because the same neuronal node can be a replicator one minute and not the next, or with respect to one kind of input but not another, depending on its state. Third, memes can replicate through the mediation of super-molecular instigators: a signal-stimulus. So memes share a number of peculiarities with the replicators to which they are most profitably compared. And, as suggested by Replicator Theory's stringent mechanism requirements, memes exist on a single substrate (neurons), just like every other known replicator.

An important question to ask is whether memes are unlimited replicators—ones able to remain infectious even when mutations strike factors influencing their state. Such a trait is important because it means memes could take on a range of states without losing the ability to maintain their evolutionary lineage. This quality would seem to be necessary if memes are to be capable of explaining the complexity of human culture. Can neuronal nodes take on many states and still remain infectious?

The answer to this question is much less clear than previous ones, because we still don't know enough about the brain to say for sure. Perhaps, then, we should return to our comparative stance for clues, since this strategy has proven so productive in previous chapters.

We know that other unlimited replicators tend to be modular, at least based on the existing examples: Genes and comp-viruses are both modular and unlimited, while prions are neither. The correlation between these two characteristics is therefore good, based on our known examples of replicators, and for this reason probably extends to memes. Are memes modular? We did have the hint earlier on, when discussing the image of a memetic population as an electric alphabet soup, that memes might need to be made up of components, although this certainly remains a vague suggestion at this stage.

Why then should memes be unlimited replicators while prions, for example, are not? Prions are limited to one size and only a few physical conformations. Memes are not, thanks to their super-molecular scale. They are like computer viruses in this respect (since locations in computer memory consist of charged sections of a tape or disk). The conformations that molecular replicators like genes and prions can adopt are severely constrained by atomic forces. But replicators composed of multiple molecules don't appear to be subject to the same physical constraints. Their function may still be strictly tied to their form, but a new dimension of variation in form is available because the individual molecules can now orient themselves to each other in multiple ways as well. We can thus expect that memes will be able to adopt a wide range of states. Presumably a number of these will remain infectious, so that memes can be considered relatively unlimited replicators.

In the end, perhaps the most important question of all is who's in control (our *cui impello* question). Can we expect memes to be relatively *powerful* replicators, capable of taking on genes for the determination of host behavior? If not, then even if memes exist in something like the form I have suggested, they have no impact on the world and will forever remain ineffectual, invisible entities, hidden behind a genetic screen.

Essentially this is a question about how robust memes are, because their competitors, genes, are quite robust. All you need to account for some complex phenomenon is a robust replicator. Genes aren't exactly big, but they are responsible for the production of very complex things: organisms. If memes have many ways of interacting among themselves, the possibilities for combination expand considerably, and the range of things that memes can account for increases. In this way, even simple things like memes can account for something as complex as culture.

Robust replicators must be able to take on many more states than they currently do, leaving room for future evolution. To determine whether memes are robust therefore requires finding the ratio between the number of actual states a population of memes exhibits and the number of possible states they could adopt. This ratio is actually likely to *decrease* as brains get bigger because with more copies of memes around, a larger variety of states will be realized by at least one meme in that larger population. So the only way for memes to get more robust over time is to aggregate, to make the state of a meme dependent on the compound state of multiple neurons. The ability to do this, of course, depends significantly on the size of the neuronal network. Certainly the representational capacity of cortex rises exponentially with the number of neurons. Thus the number of possible multineuron nodes is several orders of magnitude higher than the number of states required to define a single node. This is how robust replicators are defined. So the bigger the brain, the more robust the memes in it—*if* they learn to organize at a higher level, making a node involve more neuronal territory.

Such a reorganization could help explain the difference between human and non-human cultures. Early memes perhaps were not particularly robust replicators because they lacked significant plasticity, being restricted to a smaller scale. With relatively few neurons involved and relatively few steps before their message was passed to motor neurons, there was little room for development. But as time went by, they found themselves in increasingly complex brains and were able to stretch out a bit themselves as well.

So, as with any evolving entity, it is probably the case that the power of memes has increased with time, in tandem with the expansion of brains. Truly effective power has perhaps only come relatively recently in the course of evolutionary history, as suggested by the considerable differences between human culture and the cultures of other species. Nevertheless it is quite clear that the complexity of human culture is itself increasing as time goes on. This kind of constantly increasing power is just what one would expect of a maturing replicator, and one whose power seems to be increasing much faster than that of genes.

Thus the mechanism I have identified, and the kind of replication reaction we have inferred for it, are consistent with what we expect of an unlimited, robust replicator. The memes suggested by this analysis will be powerful adaptive agents, sufficient to explain cultural evolution.

SELECTION *ON* SIGNALS *FOR* MEMES

It is tempting to draw a parallel between the evolution of ideas and that of the biosphere.... Ideas have retained some of the properties of organisms. Like them, they tend to perpetuate their structure and to breed; they too can fuse,

recombine, segregate their content; indeed they too can evolve, and in this
evolution selection must surely play an important role. . . . This selection must
necessarily operate at two levels: that of the mind itself and that of performance.

The performance value of an idea depends upon the change it brings to the
behavior of the person or the group that adopts it. . . . The "spreading power"—
the infectivity, as it were—of ideas, is much more difficult to analyze. Let us say
that it depends upon preexisting structures in the mind, among them ideas al-
ready implanted by culture, but also undoubtedly upon certain innate structures
which we are hard put to identify.

—*Jacques Monod*

An important component of any evolutionary explanation is the identification of what things can be selected and what kinds of pressures can be
put on those things. If signals evolve, as I have just suggested, then they
must suffer selection. If memes are replicators, then they too must
undergo a similar fate. So no theory of memetics would be complete
without a discussion of selection *on* the products of memes and, through
that culling process, *for* memes themselves (a distinction that should
become more clear as we go along).

When a student asked Linus Pauling how he got a good idea, the
double Nobel Prize winner answered: "You have a lot of ideas and you
throw away the bad ones." Francis Crick, co-discoverer of the molecular structure of DNA, said that "theorists in biology should realize that it
is . . . unlikely that they will produce a good theory at their first attempt.
It is amateurs who have one big bright beautiful idea that they can never
abandon. Professionals know that they have to produce theory after theory before they are likely to hit the jackpot." These eminent scientists are
describing the process of conscious decision-making as a choice among
alternative ideas—a form of selection. Similar accounts by scientists and
artists about the creative process are legion.

No doubt a matching selective process occurs in the purely unconscious sphere: Lots of nascent ideas get quashed before ever reaching
consciousness. Behavioral plans, after all, are rather complex creations
that must have undergone some preliminary assessment before further
mental investment was made in them. Are there memetic puppeteers
operating behind the curtain of consciousness, like the "wizard" Dorothy
found hiding behind a screen in the Land of Oz?

We can elaborate somewhat on how an unconscious process of selection among alternative ideas might operate. The brain signals responding to a single event can be mirrored at many different places (probably in the same region of the brain). It's as if the brain enlists neurons from many precincts to "vote" on each of its actions. Thus spike trains representing all of the different interpretations of a stimulus compete for the job of directing the organism's behavioral response to that stimulus. The mutual connectivity that appears so endemic in the brain—node 1 being wired up to node 2 and vice versa—may be a way in which signals from multiple information-processing streams are synchronized, to coordinate behavioral responses based on a convergence of the "best thinking" the brain has done. Synchronization could line up the alternative interpretations of a situation for better assessment of their relative merits, producing a "fair" competition in the sense that each of the spiking "racers" sets off when the same starting gun fires. This way of reining in spike trains also suppresses cheaters, who could otherwise sneak through the tournament without actually recruiting new members for their team based on merit: Head-to-head competition prohibits the raiding of competing teams. This is important because the goal of the race is to achieve the largest pool of support and thus a better chance of attracting attention. The possibility that gets voted the winner in such competitions actually gets acted upon and may also bubble up into consciousness as well. Attention is thus a kind of top-down selection system among racing spike trains.

Presumably these competitions are biased in favor of the response that will produce the behavior most conducive to the continued survival and reproduction of the organism. The resolution of the race also results in the suppression of the neuronal representations voted to be less likely interpretations of events or less relevant behavioral plans. This suppression is achieved through synaptic plasticity, which reduces the incorrect neuronal states' degree of connectedness to the rest of the network and thus their ability to conscript support the next time around. At the population level, this suppression is visible as a decay of synchronous firing among neuronal ensembles, which appears to be a reflection of the brain changing its train of thought.

In sum, a spike can be selected for its ability to maintain a message from one synapse to the next, for its ability to find the right kind of target cell in which to deposit its message, and for its ability to convert that cell to the proper infectious state. On the other hand, a state (a meme) can be selected for its ability to maintain its state (memory retention), for producing a spike train that wins in the kinds of decision-making competitions I have just described (because the spikes then gain a chance of finding a new host through social transmission), and for being responsible (indirectly) for behavioral responses beneficial to the host organism (because the meme should then not only be reconsolidated, but its ties to other nodes should be strengthened, increasing its chances of winning the next go-round). It may also need to return to its original state after stimulation (to complete the replication process required by the Stationary Meme Model). So there are indeed many kinds of selection mechanisms at work in the brain, operating on our thoughts at many points in the decision-making process.

Advances in neuroscience will likely soon provide us with a glimpse into the mechanisms that produce mental changes. What these insights may prove is that our non-random, "designed" solutions to problems are the result of a multistage mental tournament in which there is selection among blindly created variants for the strongest option, given the environment of thought at the time. In effect, decision-making is a process of Darwinian selection in the brain. It just all happens very fast. We put up alternative scenarios for action, let them battle it out for their fit to circumstances, and the candidate surviving at the end wins our vote in terms of determining what we actually do. Ideas, from this perspective, are tentative tendrils of thought, creeping forward through the neuronal network, only to be chopped off at their greenest tips by the sharp clippers of some selecting agent. Those memes that participated in the production of the winning entries into these decision-making tournaments can therefore expect to be favored by selection. They are the alternatives that go on to reproduce again—that is, they will tend to persist to the next generation of decision-making. Meanwhile others will gradually disappear from the scene. In this way, the selection *on* signals leads to selection *for* further propagation of the memetic replicators responsible for these favored phenotypes. This "mental Darwinism," if true, would bring psychology fully in line with the "hard" sciences and provide a solid foundation for a science describing social groups in terms of such Darwinian agents as well.

THE MEANING OF MEMES

What can memes be said to represent in the brain, then? What does a meme *mean*?

From a neuromemetic point of view, memes don't represent anything! Memes don't mean, they just are. Memes are states of being. In this, they are just like genes, whose expression depends on the developmental context you place them in. Genes make proteins to "express" themselves, and the consequences of gene expression can vary depending on where the gene is located. (A gene's neighbors in the genome can control the circumstances under which it gets transcribed and hence affect its impact on events.) Similarly, memes produce spikes that express their "intention." It is the spikes that represent things. When neuroscientists are interested in brain "code," they start measuring firing patterns, not the states of the underlying neurons. Neuroscientists are almost exclusively concerned with what spikes mean. Neuromemetics, on the other hand, is interested in biophysical models of how spikes are generated in the first place. What happens *before* a spike is produced is often more interesting to a memeticist than how fast or often a neuron releases a spike, or where the spike goes afterward.

Mental representation is a function of a firing pattern, then, not the structure of the neuronal network. So you can't equate meaning with memes. Meaning is always context-dependent, and so unrepeatable. There can be no replication of representations. What *can* be replicated are physical structures. As we are searching for a naturalistic theory of physical replicators, we must use a structural definition of memes. Just such a definition was provided earlier.

As we have been emphasizing, then, memes are just physical things; meaning comes in the contingencies of their *expression*. However, a state can be expressed as a likelihood to respond or fire, so states and spikes (or memes and their products) are related. In this sense, one can say that a meme *implicitly* represents something, because it has the capability of producing a spike with that quality.

What a spike represents depends on where it is found in the brain. Remember that memes probably move around a lot, so where they figure in an information-processing stream will vary, depending on who is upstream and downstream of them—that is, depending on the neurons that surround them in the chain of command. In particular, the cognitive *complexity* of what spikes represent depends on where they fall in the sequence of communications between neurons. Between a sensory stimulus and motor output, there can be numerous links from neuron to neuron. Large-brained creatures like people especially engage in an unconscious, multistep process when responding to situations. Neurons pass messages from one to another until some resolution about what to do emerges. It's like a game of 20 questions. For a given neuron, neuroscientists call the range of stimuli to which a neuron responds by firing its "receptive field." At first, very primitive distinctions are made, to preclude oceans of further possibilities by narrowing things down as quickly as possible. For visual inputs, it might be a question of whether the stimulus is light or dark. Then finer and finer discriminations are laid on the initial, generic hypothesis by subsequent decisions. The hypothesis is channeled to downstream neurons, which are tuned to fire only in response to particular combinations of features (as established by prior firings). Is the stimulus animal, plant, or mineral? Each higher level of processing encodes a more composite aspect of the visual stimulus and thus requires inputs to satisfy several conditions before firing will take place. So some neurons come on-line only when the original stimulus has already satisfied all of the criteria defining a very complex category. Neurons at the end of the chain respond only when a particularly complex representational object—like a human face—is present in the perceptual environment. What a node's output means is thus a function of where it falls in this chain of activation.

In each case, neurons are simply responding to synaptic inputs from their neighbors. A first-order sensory neuron responds to a particular rod or cone in the eye, while the proximal stimulus of the second-order neuron is the first-order neuron rather than the original environmental input. What is *explicitly* represented at each stage is only that information encoded locally, considered independently of what has gone before. But each node also contributes some "added value" to the sequence of discriminations, with the cumulative consequence that an exquisitely fine-tuned response to a complex environmental context is made possible. The resulting behavior of the organism is thus extremely flexible, thanks to the nearly infinite pathways through the neuronal network produced by the possibilities for combination inherent in this sequential process.

Nevertheless, by compressing this sequence to a single step, or decontextualizing the response by not considering "prior" information, we can argue that the later, higher-order nodes *implicitly* represent the entire set of discriminations that together define a complex object. Thanks to the sequential nature of cortical processing, the complexity of what a node implicitly represents can change dramatically.

This hierarchical picture of how the brain operates has implications for memetics. Dawkins (and everyone since) has thought of memes as concepts—that is, as complex, semantic entities like the notion of a meme itself. But from a neuromemetic perspective, memes (or, more precisely, their products) are not defined only at the conceptual level. Presumably, replicating nodes in neuronal networks can exist at any stage of the sequential process just described. If so, thanks to this hierarchical processing, the spike trains produced by memes can exhibit a *range* of implicit representational levels. So what kind of thing *can* a spike train represent? I would suggest something as small as a sound or as large as a religious tradition (such as Roman Catholicism). A spike can encode simple or complex things despite no change in the complexity of the node involved, depending on *when* the meme comes into an information-processing stream.

If memes can represent only primitive distinctions about the world, then it becomes necessary to add a number of them together to make up a proper concept. Mental representations would then be composed of many memes. Current psychological theories about the nature of concepts suggest that concepts are heterogeneous complexes made up of more primitive cognitive bits and pieces. Concepts, in modern psychological parlance, are themselves theories (so this theory of concepts is called "theory theory" of course). Neuromemetics is therefore consistent with the latest perspectives on how concepts are represented in minds.

However, no two high-level representations are likely to be exactly alike, and they will probably vary considerably from individual to individual, if they are put together from many pieces. In fact, conceptual clusters, as composites of memes and other information, are *bound* to be individualized, since they are constructed during the course of a developmental process—the complex production of a behavioral response to a stimulus. Each response is produced to suit a specific occasion. But such complex mental representations, pulled together in working memory as part of a decision-making process, are not the units of culture. The claim that representations can exist at the conceptual level and above does not touch on the qualities of neuromemes as replicators, since they can be much simpler in nature.

However, the lack of "connection" between a meme and its representation implies there may be no simple relationship between the state of a neuromeme, the type of signal it produces, and the sort of behavior that results from that. So no easy set of rules can tell you how to convert from the state of a neuromeme, which you know, to some observable behavior, which you want to predict—or vice versa. Mendel uncovered a simple way to translate from the color of peas to hypothetical genes, but the equivalent may not exist in the case of behavior. So, unfortunately, identifying how neuromemes might replicate does not necessarily provide us with the equivalent of a genotype/phenotype mapping for culture. We currently don't have many clues about what parameters in spike trains are relevant, nor how those variables might relate to traits in the underlying memes. Do we count the average number of spikes per second within a minute after stimulation? Do we wait for a longer period of salience? Or count the intervals between spikes within a train? Or the temporal correlations with the spikes produced simultaneously by other nodes? There are suggestions that *all* of these may matter and may be part of the coding system used by the brain for making decisions about behavior. We just don't know at present.

A further conceptual difficulty looms. One implication of the "moving meme" hypothesis is that a particular spike train may mean "double-handled pot'" on this occasion, but "aspect of hair-line on forehead" the next time around, and "sense of grief over a dead pet" the next. How can all these memes be members of a single lineage? If what their products represent can vary so much, how can there be a consistent relationship between a meme and its spike train? Since it's only the spike train that makes it to the site where the next meme will be created, how does the same meme get created each time around?

Steady and true input-output relations are crucial to sustaining a lineage. Such different representations would seem to preclude any such relationship developing between memes and their spikes. And we know that even individual neurons are constantly responding to different sets of stimuli from one minute to the next.

But it's actually the *meaning* that varies in each case, not necessarily anything else. Just because the meaning of the signal producing a meme varies doesn't imply that the same physical state isn't created in the node each time around. The signal can be physically the same as well each time; it just has a different meaning because it is taking place in a different context. And the circumstances eliciting the memetic signal during each iteration may share something in common that isn't obvious to a conscious mind working at the conceptual level. All of this is somewhat counterintuitive, but we constantly need to think about these processes from the point of view of the neurons involved and how they work.

Try this analogy: Just as biological twins can begin life with the same genetic complement but grow up to become quite different, so too can two memes begin their lives as identical by descent, but then, through experience, diverge in some qualities, such as what they represent. This does *not,* however, make them less "identical" in evolutionary terms. They still share the same parents and state; it's just that what happens downline when they fire off a spike may differ in some respects. It's as if twin brothers go into separate lines of work. They still look alike and call the same place "home" (the house where their parents live). So too will two sibling memes retain a physical similarity in terms of their neuronal states and keep some link (perhaps only historical) to their memetic parents, regardless of what they are up to nowadays.

THE FIRST AND LAST MEME

What does neuromemetics imply about when memes might have *first* begun replicating inside brains? At minimum, surplus neurons capable of storing information from the environment—that is, programmable neurons—were required. These only become possible with the rise of bigger brains. Nevertheless this significantly increases the length of the period during which memes could have been around, because standard memetic approaches emphasize that memes arise only when organisms have acquired an ability to imitate. This effectively restricts memes to a small number of species and predicts a very recent origin for the first meme. With neuromemetics, memes could have dawned relatively early in the growth of the cortex—certainly by the time vertebrates appeared. So memes probably began life within the brain of a relatively simple-minded organism—an animal with just enough brains for a meme to roll around in (although the 32 hard-wired neurons of the sea slug, for example, may not have been enough).

Memes, in this view, are simply physical tokens in the brain that have acquired some evolutionary agency or control over their probability of reproduction. Something about their physical nature led them to get involved in a situation resulting in their information being duplicated. Prions are similar in this respect: Before the first prion showed its face, there were lots of proteins in circulation. It's just that, at some point, proteins with certain configurations arose that, in the proper circumstances, wound up being duplicated without referring back to DNA. Memes too are Johnny-come-latelies, walking into an already complex state of affairs, able to take advantage of mechanisms that arose in the first place for other reasons having much more to do with providing advantages to genes. Memes are an example of a replicator co-opting a mechanism for its own purposes—in this case, learning mechanisms.

What was likely accomplished during this first phase of memetic evolution—that restricted to individual brains? Well, evolution is always a contest to get as many progeny as possible into the next round of reproductive competition. One way to accomplish this is by increasing one's rate of replication relative to opponents. This requires that the resources devoted to nonessential activities be minimized and that the reproductive cycle itself not take very long. The first of these requirements implies very little investment in building an interactor; the second needs a small-sized replicator that can be copied fast. Indeed, what is usually found in a race for replication is that parasites evolve which make use of the functionality of the full-fledged replicators to achieve their own single-minded goal of making baby parasites more and more efficiently. In these competitive environments, we find that increasingly simple replicators evolve: parasites on parasites on parasites (called "parasitoids"). This is, for example, the lesson from the ALife program TIERRA, recounted earlier: The ecological conditions in TIERRA favored the evolution of small replicator size and low resource use; it was a strictly competitive environment. Many later experiments in Artificial Life have confirmed this general principle. However, other simulations have been set up to show just the opposite: that more and more complex replicators can also evolve in computer memory. If transmission depends on doing some kind of useful work beside just reproducing yourself, we generally find that replicator *complexity* is favored.

Another important lesson from ALife, then, is that selection in the brain—if it favored increased complexity, as suggested by the growth of the brain and its products—must have been for functions other than the brute rate of meme replication. For complexity to evolve, there has to be a specific force promoting an increased investment in non-reproductive functions or, alternatively, a competitive environment that demanded complexity to achieve reproductive competence (people have long generation times because they aren't sexually mature until they reach the teenage years). What other functions besides reproduction might a meme have evolved to perform?

The brain is designed by natural selection to produce behavior—adaptive activity in the organism. That is what the brain does: intervene between genetic instructions and environmental stimuli to produce an adaptive response. So it seems natural to suppose that competition among memes must be to produce behavior, to be selected for the good effects they produce in the host. That is how the symbiosis between genes and memes arises: Both seek to promote "good" behavior in the organism they find themselves in. What selective force would have induced memes to invest in a more complex interactor and a lengthier developmental process to become competent in those functions—in effect, to take a hit on reproductive rate by lengthening generation time compared to competitor memes?

I think the answer is that the need to cooperate leads to complexity in this case. It's the fact that memes are necessarily social that sets them apart from essentially independent replicators like genes or prions. Memes are "born with connections": Their substrates are physically tied to one another as parts of a single, massive neuronal network. This has important implications for their evolution and, in particular, their relationship to their "hosts." Only by attracting attention, and hence directing the behavior of their host organisms, can memes induce hosts to translate spike trains into social signals, which provide them with an escape route into the macroenvironment. There the signals can find their way into another brain and help replicate the relevant memes, which thus survive into future generations. In effect, memes, to be communicated, must be able to recruit confederates. Because memes generally must collaborate with other neurons to get attention, their essential stance toward their community must be one of cooperation. In Network World, you want to have your connections strengthened, to become indispensable, a party to all computational events. The ability to slot into a wide variety of neuronal complexes, to be a good neural citizen, is a favored quality.

NICE PARASITES

This conclusion has a major implication for the nature of neuromemes. As a result of this need to cooperate, memes tend to be not virulent but rather symbiotic parasites that do better if they do no harm. In fact, gregarious, tolerant memes are likely to be most successful. Thus those who talk about "selfish" memes implicitly assume that memes are isolated, independent replicators—probably because no physical model has previously been imposed on the concept. In fact, the symbiosis between memes and organisms is real: The memetic surround in which people live, the cultural world, is to a meme as water is to a fish: invisible, but nurturing, a necessity for life itself. This implies that the environment of meme evolution is different from the standard one, where energetic efficiency is the primary criterion for selection.

The popular image of memes as virulent "mind viruses," with which memeticists are so smitten, is therefore not only repellant but wrong. To think we are antagonistic to our own brain—or that there are foreign bodies taking up residence inside us that are malevolent—is preposterous. The notion of host-destroying memes *must* be mistaken. Just look at the tremendous general adaptiveness of culture, the consequent success of the human species in conquering the planet, and the increasing complexity of cultural evolution. Indeed, the extraordinary ecological success of our species has been due largely to coevolution with memes. They're what have permitted us to have all kinds of fun other species don't have—such as recreational sex (thanks to the Pill) and virtual reality.

Memes themselves have similarly profited through their relationship with us, having become more capable replicators with time—for example, by learning to jump between hosts. Further, memes constitute (at least aspects of) the self, to which we obviously have an intimate, not antagonistic, relationship.

Thus while memes are parasites, this doesn't necessarily imply they are virulent (although some may be). "Parasitic" doesn't necessarily mean something is there, feeding off you. Considered as an ecological relationship, parasitism merely means dependent, not hostile. Actually parasites can have a symbiotic relationship with their hosts that is mutually beneficial to both parties; it's only that the parasite can't live without the protection and sustenance that the host provides. Technically both cleaner-fish and leeches are called parasites; it's just that one provides the service of picking nits off your flesh while the other sucks your blood.

Certain memes may be harmful to their hosts. But the virtue of memetics is that these conditions can be specified. This is because memetics provides a theory of meme-host conflict, leading to predictions about when potential maladaptation in the host should arise. By comparison, competing theories of cultural evolution such as evolutionary psychology are weak on this point. Evolutionary psychology argues only that it is more likely for traits in evolutionarily unusual contexts, such as modern societies, to be maladaptive. But such a blanket statement is unlikely to be true for everything from rockets to contemporary forms of the novel. Meme theory is alone in predicting *just when* cultural traits should be harmful. Memetics argues that, as parasites, beliefs will be virulent to the extent that they can replicate independently of their effects on hosts. For example, when relatively simple ideas travel well—like the notion of a ghost, or a person who continues to haunt the world after death (assuming that such ideas are harmful)—these non-gregarious beliefs can nevertheless attract attention and thereby get into the environment for further replication.

Also, memes that can't get out of their hosts are more likely to be virulent: They haven't passed the test of being adopted by numerous brains, of getting through a range of input filters. They may have arisen as a "bad mutation" through spontaneous development or bad wiring from ontogeny. Idiosyncratic beliefs, in this view, are more likely to be maladaptive to their hosts. Maladaptation is not due simply to ancient instincts gone awry in modern circumstances, as evolutionary psychologists suggest.

So memes are not go-it-alone replicators that compete with each other to minimize resource use and generation length as a means to increasing their relative rate of reproduction. In memetic evolution, *it is often more complex memes that win evolutionary competitions, not simpler, hyper-parasitic memes.* Memes would thus have certainly enjoyed more room to roam around in as brains got bigger and may have put some pressure on brains to increase in size as their symbiosis with brains became more ingrained over time.

WHY DO WE HAVE BIG BRAINS ANYWAY?

Still, there seem to be many good reasons for having big brains besides giving extra freedom to memes. So I can't finish this discussion of replication in brains without returning to the question we began with, which remains unanswered: Why do we have big brains in the first place? Here's a hypothetical scenario.

It may have all begun when the dependence of certain higher organisms on specialized foods produced selection pressure for even more neurons to remember where and how to obtain those foods. Additional selection pressure could also be put on organisms that found themselves doing better if they stuck together with others of their kind, perhaps because being on the inside of a big herd reduced the risk of being picked off by a predator, or because cooperative hunting proved more efficient. This necessitated, however, even more memory—for example, to remember whether other members of the group had proven trustworthy in past cooperative ventures.

These external pressures supply reasons for growing bigger brains. But how did the brain respond? The evolution of the human brain can be understood from the "inside" as a process in which new kinds of control over information flow were added over time. In effect, as the centralized decision-making structure of the brain became larger, it also became increasingly hierarchical. It appears that newer additions to the primate brain (such as the expanded forebrain) took over control from earlier elaborations (such as the cerebellum), in effect becoming their new masters. But all the apes share roughly the same general hierarchical organization. What has changed in humans is that the prefrontal cortex has expanded. This places the area of greatest plasticity at the top of the decision-making hierarchy, which thus allows the rest of our brain's organization to be controlled in a much more flexible fashion.

The change from ape to *Homo* was therefore about the tweaking of structure rather than the addition of completely novel information-processing modules. But this tweak led to important new cognitive abilities. Humans appear to have repertoires of responses that are, in effect, free from the immediacy of the present environment. This ability allows us to remember and to plan, and to do so in inventive and unique ways. Accordingly, the ability to look ahead, engage in conscious thought, and use language all depend on our augmented neocortical structures, centered in the prefrontal areas. The main reason for our extravagant degree of mental plasticity thus seems to be to permit these exotic psychological abilities. These abilities, in turn, allow human culture to have a maximal effect both on the shaping of our social life and, through feedback, on the instigator of social behavior, our brains. Human brains are thus forged from a mix of genetic and cultural evolution. The implications of these dual pathways for information transfer are dealt with in the next chapter.

Chapter Eight

ESCAPE FROM PLANET BRAIN

*The only way of directly communicating an idea is by means of an icon;
and every indirect method of communicating an idea must depend for its
establishment upon the use of an icon.*

— *Charles Sanders Peirce*

*Selection favors those able to use signals to manage the behavior of others in
their own interests.*

— *Donald Owings and Eugene Morton*

James Boswell famously asked Samuel Johnson, an exuberant fixture of
the literary scene in eighteenth-century London, how he could refute the
sophistry that the external world is but an illusion. The inimitable Dr.
Johnson said, "I refute it thus" and kicked a rock with his foot. However,
not even rocks present themselves candidly to the brain for inspection.
Even the most direct experience, such as pain, is only *perceived*. What is
"experienced" is some signal coming into the brain from a particular
sense organ—in Dr. Johnson's case, from his toes. Many philosophers
therefore claim we are necessarily solipsistic, living psychologically in an
ineluctable world of one, isolated from reality by an insurmountable
uncertainty about the nature of what is "outside." What if "I" am only
dreaming all this? Or what if "I" am only a virtual program running on
some massive computer?

Leaving aside such philosophical conundrums, we still face a major question from an evolutionary point of view. The main goal of memetics heretofore has been to explain how mental "secrets" can get from one human mind to another intact. The logical Mr. Spock had the ability to do a "mind meld" whenever the plot of an episode of television's *Star Trek* required some bit of information to be extracted from an alien species. He would simply place his hands on the creature's forehead (or its functional equivalent), concentrate hard, and out the required enigma would pop. Being from the planet Vulcan, he could engage in this form of direct mind-to-mind contact. But we humans lack this convenient facility. The poet John Donne famously claimed that "no man is an island," but in fact, psychologically speaking, each person is like a planet drifting through space, never being directly impacted by contact with other psychological bodies. So how do memes bridge the distance—the gap of air—between brains, each locked in a prison of its own making? If one brain is a mystery wrapped in an enigma, how could any two such things ever hope to communicate with one another? How do memes ever manage to escape their home and infect new hosts?

MIND THE GAP

Memeticists have typically answered this question by saying that memes leap from brain to brain by inducing their hosts to engage in behaviors that transmit signals to other hosts. Given the proclivity of memeticists to "think epidemiologically," this is a natural answer to make: Memes, like other parasites, must have a life cycle that includes a phase during which memes are transmitted from host to host. In effect, an infected host must engage in the production of what the nineteenth-century American semiotician Charles Sanders Peirce called an "icon." The icon is the thing that goes between. The meme simply coats itself in a vector—some kind of information-carrying icon, such as a spoken message, an observed motor behavior, a bit of text, an image, or a slab of stone. When a potential host comes into contact with that vector (that is, reads the text or hears the message), the meme leaps out of this vehicle (gets decoded), becoming active again, and infects the person, who becomes a new host. Then the infection phase inside the new host brain starts up. Even later, the meme may get encoded again into a suitable vector (not necessarily of the same medium it was originally decoded from), and the whole cycle begins over again. So what I will call the "jumping meme" hypothesis is that memes *themselves* traverse the gap between brains, in some form.

This supposition is, of course, based on the analogy between a meme and a virus. But there is a major difference between memes, as I have discussed them, and biological parasites. When DNA or RNA viruses are socially transmitted, the very same molecules move themselves, bodily, from one host to the next. But for a meme to "jump" from one host to the next, brain code must be converted to signal code and back again to make the same trip that DNA can make on its own. This presents some problems, as we will see.

SIGNALS AS INTERACTORS

The basic supposition behind the image of jumping memes is that these cultural replicators act like their biological counterparts by riding around inside an interactor (the "vector," in epidemiological terms). Is this a good image to have of memes? Can memes really jump the gap between brains by leaping inside an interactor for the trip across the gap between hosts?

This "interactor question" is something that has perplexed memeticists since day one—at least those memeticists who take the gene analogy seriously. Memeticists generally feel it's very important to identify *something* as the memetic parallel to the genotype/phenotype distinction in biology. A variety of candidates have been put forward. One might say, for example, that a recipe is the memetic replicator and the cake the memetic phenotype, or that the performance of a song is the phenotype of the remembered idea of the song. In effect, either an object or a behavior is being taken as the cultural equivalent of an organism—as the cultural interactor—in such analogies. But none of these suggestions has achieved consensus in the somewhat fractious memetic community.

Prions, on the other hand, don't produce separate interactors, but that is because they are limited replicators. Anything accounting for culture must be an unlimited replicator, and unlimited replicators, it seems reasonable to believe (based on the other examples, genes and computer viruses), must be what biologists call "heterocatalytic." The sequence information of a gene often gets read off without the code itself being replicated; instead a protein will be produced by the machinery set in motion by the genes. Similarly, a computer virus—an artifactual replicator that is basically unlimited in the number of forms it can take—can produce two kinds of signals: those that result in replication of the virus, and those that result in the activation of other programs, causing damage to computer memory or producing silly on-screen displays. These are examples of the basic distinction between phenotypic behavior and replicative behavior. The question, then, is: Do memes engage in this same division of labor? Can the same neuronal state produce a spike train with different results, depending on its stimulus and downstream interactions—either making a new replicator or making something else?

Memes are essentially memories. But memories need to be recalled, not just replicated. Recall (with reconsolidation) should be one kind of outcome when memetic neurons are stimulated. From the perspective of memetic replication, this is "off-line" activity, the kind of dead-end information that is emblematic of phenotypic production. In this case, the meme contributes to the construction of some complex of information in working memory for the organism to act upon, without the meme itself being duplicated. So it seems memes engage in phenotypic activity too. They have multiple products: not just new memes, but also messages helpful to normal mental operations. This makes memes heterocatalytic.

If spike trains are the "behavioral" products of memes, which are replicators, then they should be considered memetic interactors. And in many ways, signals seem to fill the bill, to act like proper interactors. Just as interactors can't pass on their structure directly, spikes don't make spikes; rather, spikes are generated by upstream neurons. Further, signals are degraded by noise (mutate) in the transmission channel (their relevant environment) and fade away without boosting (get selected)—all of which prove that they interact with the environment, as good interactors should. Also, by "finding" the right downstream neuron, spikes cause the next generation of memes to be born, thus determining the fate of the replicators responsible for them. Further, spike trains have evolved greater reliability (for example, by encasing axons in protective sheaths to improve signal induction) and developed new skills (such as inter-brain transmission), and so seem to be evolving new adaptations. (Interactors are the proper locus of adaptation.) In addition, there is no reason to expect successive spike trains to be similar, thanks to varying reactive fields and rapid changes in sensitivity of neurons, so spikes don't form lineages. And just as some proteins are enzymes (molecular catalysts), so too are these memetic interactors catalysts, of an electrochemical sort. All of these seem to be persuasive analogues in the brain to the ways in which other interactors work.

But the analogy is deceptive. Several problems—potentially major ones—arise from thinking of social transmission in this way, of seeing signals as meme vehicles.

First, signals are only loosely coordinated and are easily disturbed or divided into bits. As a result, they are very susceptible to corruption: They flit about the brain or macroenvironment, bumping off everything in their path, and potentially leave a bit of themselves behind at every point. This is no way for a good interactor to behave, allowing everything a chance to distort its message.

A further problem: If memes must be in both brains and signals, then memetic information must exist on multiple substrates, such as neurons and sound waves. But I've argued at length in earlier chapters that replication is substrate-specific. "Jumping" memes thus immediately violate our dictum—derived from Replicator Theory and our comparative look at other replicators—that memes exist in only a single form. I have also argued that it is the substrate neutrality of standard memetic theory that has hampered the development of memetics as a science. This is simply not an avenue we want to go down, in my view. Memes can't be translated from brain stuff to signal stuff and back again.

Even if memes *could* be reliably translated from brain code to signal code, there is typically a loss of information associated with translation between coding systems. What is well captured by one system doesn't necessarily "translate" to the next. Just think of the problems professionals have translating Shakespeare into French, much less Swahili. The problem is not restricted to concepts expressed in language either: Many non-Western cultures don't recognize the post-Renaissance concept of romantic love, for instance. At any rate, each transformation between coding systems leads to some of the information in the message being lost. How can memetics, in good conscience, contradict a fundamental result of information theory—that degradation follows transformation? I don't think it can.

This corruption problem doesn't arise with prions because they are both replicator and interactor bundled into one. If there is no transformation—no heterocatalytic loop in the life cycle—then this difficulty is avoided altogether: The replicator just makes its way around the world on its own. But memes must *necessarily* be transformed to escape from brains and make their way through the general environment and thence back into brains because neither neurons nor their electrochemical products leave their host's body.

Another way of thinking about the transmission problem is to say that signals are really just viruses—memes minimally enclosed in some defensive coating, trying to jump the gap to the next host. The meme's cycle from brain to behavior and back to brain is then something like the complex life cycle of some insects, where they begin life as a larva, then progress through a pupal stage, thence to an immature form, and finally maturity, with the creature taking on a different appearance in each stage.

But signals aren't complex enough to be meme vehicles. They can't carry physical memes around—that is, they can't hold brain cells inside them. Signals are just air-pressure fluctuations or series of photons, not bubbles floating through the air with neurons inside. This means that signals may carry sufficient information to cause a new host to produce a copy of the meme, but cannot contain the whole blueprint nor the materials with which to do the work by themselves. Not only that, but memes are defined by the ability to replicate, and signals can't replicate—or at least that's what I argued in Chapter 6.

Further, this perspective means that memes must violate the presumption—true enough of genes—that the replicator must be carried along *inside* the interactor. So if we are to go this way with the notion of interactor, it has to be generalized from its current conception—the Dawkinsian image of a lumbering robot housing the replicator inside—to admit this possibility.

I conclude from these difficulties that signals simply can't be vehicles, much as Dawkins would like them to be. Thus (with apologies to Marshall McLuhan), the meme is not the message. Memes just can't jump.

Still, it is crucial that memes make the leap between hosts somehow—otherwise, we have no shared knowledge, no culture. This process must also involve the social transmission of information through signals, or we are left with the possibility that culture is, in fact, genetic—transmitted not from individual to individual during their lifetimes, but at birth, hidden away somewhere in the sperm and egg. So with our change of focus in this chapter to *social* transmission, we have immediately fallen into a conundrum: Memes can't jump the gap from person to person, but the social transmission of information is the necessary foundation of any reasonable theory of memetics.

Signals as Phenotypes

The only option left for us, it would appear, is to consider signals as memetic *phenotypes,* or expressions of memetic activity in the world. Genes produce phenotypes, of course, like the color of your eyes or a propensity to obesity. So it's natural to think that memes must have an equivalent way of manifesting themselves, of showing their faces to the world in such a way that selective forces can then prefer or condemn them.

But we have established that a signal is the only thing that progresses from one brain to another. So according to this view, what reaches the new host is only phenotypic information, potentially modified during its journey from brain to brain. Each time a meme leaves one host brain, the next host brain would have to reconstruct the replicator from an observed behavior or a spoken phrase. If you call the bit in the middle—the information in the social channel—a phenotype, then the fact that it *does* interact with the world, and is changed by it, and then gets taken in by the next host, makes cultural evolution Lamarckian: What is inherited down the line is *acquired* variation. In effect, variation introduced into the transmission process through experience and interaction with the environment becomes the source for subsequent replication reactions.

Lamarckianism is, of course, *not* the way genetic evolution works. In the late nineteenth century, August Weismann showed that acquired changes to the body of an organism during its lifetime did not affect the germ line: The gametes, or sex cells (egg and sperm), remain isolated from any such effects. Weismann's discovery once and for all undercut the theory of "soft inheritance"—the idea that an organism could pass on its *experience* to its progeny. Information acquired during the life of the organism cannot be "reprogrammed" into the hereditary material.

We also don't want cultural inheritance to be a Lamarckian process because Lamarckian evolution has well-known problems, such as implying that the variation on which selection works eventually bleeds out of the system, leaving evolution bereft of an ability to change any further. How then could memetics account for a rapidly and continually changing thing like culture? An interpretation of signals as phenotypes seems to be precluded by the Lamarckian problem.

But what about Richard Dawkins's notion of an "extended" phenotype? Is this a way to deal with the problem of signals? Dawkins's point is that there is no good reason to suppose that phenotypes must stop at the skin and that there are many aspects of life which we can account for straightforwardly only if we suppose that they do not. For example, an organism can have effects on its inanimate environment, like making artifacts—shells, burrows, and whatnot—that are subject to genetic variation. There seems to be no principled reason not to regard the relevant genes as being "for" those artifacts, just as other genes are "for" bodily traits. If variation in the artifacts affects the reproductive success of the organisms that produce them, then those genes will be exposed to natural selection in the ordinary way. Similarly, signals don't appear to be full-fledged interactors; they don't have genes inside of them but nevertheless influence the biological success of the behaving organism. They are just a way the host is manipulated by a replicator that results in that replicator experiencing an improved ability to replicate.

In this view, memes engage in a common trick among parasites: get the host to help with your transmission by changing the environment in some way that fosters your reproduction. The proximate, or first-order, memetic phenotype would be the neural spike. The "extended" phenotype of a meme, then, would be the *result* of the spike, such as host behavior (as the spike stimulates motor neurons and gets converted into muscle movement) that goes about its business *outside* the host body.

But, again, this isn't really a solution. The problem with calling a signal the memetic equivalent of an extended phenotype is that phenotypes, even "extended" ones, are reproductive dead-ends. Dawkins's favorite example of an extended phenotype is a beaver dam. But the dam only blocks up water and provides shelter, it doesn't make new genes. Beavers do that in the ordinary way: by mating with other beavers. In effect, beaver dams don't make new beaver genes; beavers do. There is essentially no way for information acquired by phenotypes through interaction with the environment to re-enter the information stream that is used to constitute the next generation. But this is just what we are asking memetic signals to do. This makes for a fundamental disanalogy, then, between the biological and cultural uses of the phenotype concept. The informational and causal arrows in the phenotypic perspective point only one way, and it isn't back to replicators. Weismann, not Lamarck, should be our guide in understanding how memes work.

Here's our problem in a nutshell: Signals can't be vehicles (or interactors) because they are too simple. The interactor notion also can't be extended to encompass signals—it would be like suggesting there are vehicles driving around without passengers inside to direct them. But neither can signals serve as the phenotypic expressions of memes, even "extended" ones, and still produce evolutionary lineages. Phenotypes are by definition reproductive dead-ends, but signals bear the burden of carrying reproduction forward into new hosts because memes themselves can't leave brains.

Have we reached an impasse, then? Have we defined away any possibility of a social memetics? This would truly be a dire conclusion, since our whole purpose in identifying memes is to explain culture.

Signals as "Instigators"

However, we don't have to abandon all hope of explaining how memes can jump the gap between brains. What we need is an entirely new and different conception of how social transmission works. In my view, signals are not interactors; nor are they phenotypes. Instead they are what I will call "instigators." The arrival of a signal in a brain brings an influx of energy and information, sparking the crucial change in local conditions that causes a replication reaction to begin. This is an analogy neither to the meme-as-virus nor the meme-as-gene, but a return to the replication reactions of Replicator Theory, outlined in Chapter 3. This new perspective sees signals as "molecules" that can cause specific reactions, given the right preconditions. This role is consistent with the basic nature of signals as bundles of patterned energy. As packets of information, they ensure an inheritance relation by making sure the replication reaction they instigate produces the same kind of meme as made them in the first place. The physical substrate carrying the wave pattern can also serve additional roles: The signal may do the job of a catalyst, speeding up the sequence of events, or act like a matchmaker by bringing the relevant parties to the replication reaction together (and thus facilitating proceedings). But these are secondary to the primary role of getting a *particular kind* of reaction going in the first place.

What we are looking for from this switch in perspective is a mechanism that actually conserves information over time—that's what replication is all about. If we lose information, we lose everything. This conservation is accomplished by noting that memes never leave their happy situation in the brain. Rather they send forth a signal that searches for a place to create a brother meme elsewhere, while the originating meme sits there doing nothing further at all to help out.

What is new in this perspective is that signals are seen as rabble-rousers. They are projected like arrows into the environment, with which they must interact (hence the confusion that they are themselves interactors). Signals then migrate through the macroenvironment to a novel host (gaining contact through some sensory organ) and are translated back into neural impulses. Once within the brain, they are passed through neural connections to a location where they give birth to a new meme by stimulating a node in the new network, leaving it in a memetic state. This may happen a number of times as the signal moves through its microenvironment, the new host's cortex.

This implies that there is no direct meme-to-meme contact during memetic replication; memes don't go flying through the air to meet up with their brethren in other brains but stay inside their original host. Memes don't move: Signal-instigators do. This means the *idea* of a meme may be a meme, but the spoken word "meme" is not itself a meme; it is a signal. A meme can only be a state of matter coded in "brain language."

This isn't just a different way of seeing the same phenomenon of social communication. It may appear that there isn't any change in what actually happens "on the ground"—it is, after all, still the case that information goes from one brain to the next through an intermediate stage in which it adopts a different form. But the change in point of view does have real implications. Calling signals instigators rather than interactors or phenotypes is crucial because it saves us from the ghost of Jean-Baptiste Lamarck.

If signals aren't interactors, then cultural evolution doesn't involve the inheritance of an interactor's derived features—Lamarck's folly.

If signals are instigators, then replication also doesn't need to involve construction, just *conversion*. This is the crucial insight we have derived from our investigation of the other parasitic replicators, prions and comp-viruses. All the signal need do once it enters the new host is to make the conditions there favorable for a particular kind of conversion to take place. Conversion of an existing substrate is not the same thing as making a new stretch of DNA out of amino acids. Instead, for a meme, replication simply involves flipping a neuron from one state (which pre-exists the replication reaction) to another. This is not too much for a signal to accomplish, even if it *is* impoverished as a representation of the entire meme. It is still the signal that is selected through its travails in the gap between hosts, and it may even be considered to exhibit adaptations (such as linguistic features). But the meme isn't inside the signal (which preserves the interactor notion's integrity as well).

Conversion also avoids the problems of the replicator necessarily being degraded because it is reconstructed from a phenotype. Signals are not phenotypes; they are not a replicator's way of interfacing with the world. Their minimal and specialized role instead is to contain the information most likely to lead to the replication of a particular strain of memes. They must move through a communication channel and then instigate a reactive process at their destination. However, producing an infectious state in the receiving brain is not enough; it has to be the same state as produced the signal in the first place to constitute the next link in a memetic lineage.

Only a tiny bit of the signal's information may be used in the duplication process, and this may not be the memetic codes themselves. Instead replication reactions may rely on signals providing an apparently extraneous bit of information that *triggers* the replication process. The mere presence of the signal may be sufficient for the necessary memetic information to be re-derived on the spot. The result is still meme replication in the new host, but the signal need not transfer memetic information from the source to receiving brains. To ensure there is an inheritance relationship between the source and receiver hosts, it is only necessary that the signal contain information that guarantees that the *correct* nodal state is the outcome of the reaction process.

This means that the relationship between a meme and the signals it produces could be arbitrary—one of the criteria often used to define a symbolic relationship. This is handy if, as many would assert, communication systems like human language are symbolic in nature. Social memetics is therefore consistent with communication theory on this point.

However, on the downside, an observer who is trying to reconstruct a meme inside a brain from its more readily observable social signal is barking up the wrong tree. Since conversion is not the same as production, the signal is not necessarily "like" the meme in content. Such an observer would be assuming that the relationship between signal and meme is more direct than it really is. The signal need only inspire a conversion in some existing neuronal node, persuading it to adopt a different configuration, rather than making it from scratch. It is not transmitting the meme's essence from one place to another. So signals are not necessarily linked in a thematic sense to the memes that produce them.

The information brought into the equation by an instigator thus may not actually be incorporated into the final product, the new meme. This implies that whatever variation was introduced into the instigator's message by its travels does not become part of the offspring replicator. The inheritance process is thereby ensured to be Darwinian rather than Lamarckian: No acquired variation is fed back into the memetic lineage. In informational terms, memes and their signals can be completely independent.

Despite the lack of contact between a parental and offspring meme, seeing signals as instigators makes memetic replication direct again, at least in the sense that memes are not coded to make up the signal that flies between organisms. Memes themselves don't have to be translated from a mentalist code (replicator sequence) to a behavioral code (phenotypic performance) and back again during social transmission. Instead the product of a meme, a spike, is translated into a social signal. This means it's the instigators, not the replicators, that undergo translation. This implies, in turn, that it's the system that converts signals from an internal code to an external one for transporting the message through the environment that must resist the tendency to degrade, not memes. We therefore don't have to claim that information theory doesn't apply to memes; it does. It's just that memes are not undergoing the translation processes that lead to low-fidelity replication. Memetic lineages can persist over the long term because any degradation of signals is incidental to their replication.

ECOLOGICAL SELECTION ON SIGNALS

It should be fairly obvious that this is a natural extension of the ideas presented in the previous chapter about how memes get around *inside* brains. Social signaling is just an elaboration of the strategy memes learned in an earlier phase of their evolution. Neuronal signals are transduced even in inter-neuronal communication: They go from an electrical pulse within the cell to coding in chemical neurotransmitters between cells (on a very fast timescale). This cycle is simply elaborated slightly when signals are transduced for transmission outside the body, into a form (such as photon streams) that can progress through the air. Transmission, unlike transduction, does not involve a change in code or a barrier to movement.

This is a very straightforward modification of the existing system. The only difference is that now it's motor neurons that are being stimulated at the end of the cortical information-processing stream, rather than another kind of neuron. "Talking to muscles" is in fact the most natural thing in the world for an animal brain to do. It is, after all, the primordial function of the brain to produce behavior. As a result, motor programs are centered in older parts of the cortex. And communication with one's fellow organisms through behavior is also ancient. Indeed, many animals, from lowly ants on up, can produce behavioral signals of some kind, such as laying trails of pheromones along the ground for others to follow. It's just that genes control the production of these primitive signals. It's only with the evolution of big-brained creatures that genes relinquish control over signaling. So memes are merely piggybacking on a primitive function of non-memetic neurons, parasitizing an existing facility of the brain.

This suggests that it isn't just social signals that are best thought of as instigators but spike trains as well. They too are merely taking a minimal message from one place to another, where locally available resources are put to work to cause replication. No memetic information need be encapsulated in cortical signals either. It's just that the "gap problem" has brought their nature into particularly stark relief. In fact, the same problem—of leaping the gap between neurons—must be solved by neuromemes as well. The scale is considerably smaller, and the gap not so hostile to being leaped perhaps, but the problem remains nevertheless.

As soon as there was room for memes in brains in the first place, there must have been a tendency for meme-derived spike trains to wind up stimulating motor neurons. And any meme that led to good behavioral outcomes would have been useful to the brain and so become a favored guest. The only novelty here is that the meme itself actually gets spread to other hosts as a consequence. This would give the memetic lineage a whole new lease on life: The memes could survive their original host and engage in a whole new round of evolution under a new set of constraints. Any such tendency of memes to stimulate motor neurons would therefore have been powerfully reinforced.

Selection on signals in the social sphere is an extension of that occurring on signals inside brains. "Ecological" selection on the "free-living" forms of signals—those traveling through the macroenvironment—must pass the same tests as signals moving between neurons, such as the ability to find a previously uninfected host. A factor important to its success in the social sphere that was missing in the purely neuronal situation, however, is the signaling system's need to maintain the "meaning" of the message through translation from a neuronal to social signal.

There are sometimes reasons why certain classes of memes aren't well adapted to making the jump through the social environment, over and above their need to run the psychological gauntlet spikes undergo. Selection in the macroenvironment will often guarantee that actual infection rates of new hosts are considerably below their theoretical potential. Factors such as the distance to the next uninfected host (a reverse function of the density of hosts, as well as the degree to which the same meme has already infiltrated a population) and various kinds of agents that can disrupt signals (such as ambient noise, in the case of an auditory signal) will filter out those signals not up to the task of negotiating the wide open world outside the organism. So both physical and mental obstacles to transmission face social signals.

A selection process also goes on in social groups. Ideas compete among themselves for the right to occupy the mental niche that is devoted to the description or explanation of some phenomenon. This battle can be fought not just within brains but also between them, because every person will have a need to describe or explain that same phenomenon. Sometimes prospective candidates for this task are rapidly shunted aside by "right-thinking" people and may even get labeled as crackpot, which is usually sufficient to kill them off altogether. On the other hand, a "buzz" factor can lead an idea that has already achieved some popularity to become the completely dominant choice in a group through a kind of positive feedback effect: As people imitate each other, it becomes more and more likely that the person each adopter imitates will have the favored trait. This has been documented in scientific circles: If an academic paper has been cited a certain number of times, it becomes even more likely to be cited again.

Thus alternative ideas, each seeking to control one form of behavior, are culled by analogous processes happening at a wide range of scales—from the individual neuron (operating within milliseconds) to social groups of hosts (over decades). A number of these selection processes can operate simultaneously on the same ideas. For example, educational training in universities constitutes a (largely unconscious) mental selection process in the minds of students (who must consider the intellectual value of what they are taught for themselves), as well as an attempt by faculty to recruit new members for their clique (which promotes their version of thinking within some domain). The adoption of an idea by a student can be a victory for a meme, a signaling (or teaching) mechanism, and a social group, all at the same time.

THE RICHNESS OF THE RESPONSE

Researchers who specialize in the study of human language still have a major gripe against the memetic perspective on social communication. It all began with Noam Chomsky, the living author who has been more cited by his fellow academics than any other (while at the same time being a vocal critic of American foreign policy). He spurred a revolution back in the 1950s that is still continuing. His central point was the so-called "poverty of the stimulus" dictum—that what gets transmitted in signals is insufficient in content to account for what people can make of them. We recognize this ourselves when we commonly say that a single gesture or word "speaks volumes."

Think of the word "hello," spoken by the two men meeting in the street in our earlier story. "Hello" is a simple utterance that can mean many things, from "I'm happy to see you again, lover" to "I'm just making a show of being civil, even though I detest you as a person, because you're not worth violating social norms for." Not all of this information is actually present in the single word of greeting, of course. The whole field of sociolinguistics has arisen to explain how people can draw such wide-ranging implications from the social context of everyday conversation. What is actually present in the message is but a mere fraction of what can be inferred from that message by its recipient. To correctly interpret the sender's intention, the receiver must be able to fill in the gaps in the message and thus reconstruct what the sender intended to say. Essentially the "message" from communication theory is that messages don't equal meaning; they are intrinsically something less.

In fact, Chomsky's principle is true of communication in any context you care to mention: in communication between genes, viruses, cells, computer viruses, or memes. There isn't a single example of a communication process in which all the necessary information is transmitted from sender to receiver; messages are *always* depauperate and insufficient to explain how the receiver reacts. The receiver (whether a computer, a cell, or a mind) must always engage in some kind of interpretive or reconstructive process, based on the message as well as other contextual factors, to determine what the source of the message "meant" by the act of sending the message in the first place.

But how can we be sure that we reach the right conclusion based on such slim evidence? How can an information lineage persist when each receiving individual fills in large chunks of the material from his own idiosyncratic store of memories and knowledge? Memetic content might be unlikely to survive intact over numerous iterations of idiosyncratic reconstruction by different minds. In effect, this inferential model of communication could imply a relatively high rate of mutation—perhaps too high to sustain evolutionary lineages of memes. If scant messages, brief and informationally insubstantial, are heavily reconstructed by receivers to infer the sender's intent—and based on that, the meaning of the message—then the result of transmission typically might not be duplication of the original version but some new interpretation. This may even happen every time someone in the chain passes on a message.

Replication might be only at the tail end of a statistical distribution of possibilities; in many cases the meaning reconstructed by the receiver could be different from that intended by the sender.

How does real *communication* take place then? Evolutionary psychology provides one answer to the conundrum of how we can ever have replication as a consequence of information transmission: I'll call it the possibility of a "rich" response. The idea is that we should expect evolution to supply us with just what Chomsky says we need: the ability to reliably infer what's missing in messages. Evolutionary psychology suggests there are heavy-duty regularizing structures in human brains. These structures ensure that inferences take a certain form, given a particular kind of stimulus. In effect, there can be heavy channeling of mental reconstruction by receiving parties to a communication. It isn't just randomized cutting and pasting of infobits going on in there.

Since they are evolved, presumably all people share these mental structures—what evolutionary psychologists prefer to call "modules." This would seem to provide everyone with the same mental code-decode mechanism that should guarantee that, despite the poverty of the stimulus, the "correct" inference is nevertheless routinely drawn from a signal. The representation that winds up being constructed in the mind of the receiver will be very much the same as that which produced the signal in the first place, because the same manipulations are being undone at the receiving end as were done in the sender's mind to originally construct the signal. If the decoding is very close to the reverse of the coding, then we wind up with a replicate of the message in the receiving brain. The poverty of the stimulus can be compensated for by a richness in the response.

The memes that actually persist are thus those that "fit" with our evolved psychology; those that try to cross boundaries or "slip between" these modules will not survive to replicate again. But this does not discredit the general notion of replication; in fact, it provides a mechanism for predicting just which kinds of memes will survive: those in harmony with preexisting modules for processing their kind of content. One would even expect that, if brains and memes are becoming symbionts, there will be natural selection for information-processing mechanisms in the brain that ensure the survival of memetic content—even if it has to be inferred rather than simply decoded—and that specific adaptations exist (in the form of modules) as responses to this problem. Perhaps mechanisms for the repair of mistakes during inferencing could arise in this way, as they have for DNA replication.

One still might say that channeling inference—no matter how narrowly—only delays the inevitable: The memetic message must still become corrupt eventually. If some information is inevitably lost with each iteration of communication, then the fidelity of replication rapidly deteriorates, and the message soon gets lost in a welter of noise.

Determining just how long memetic lineages can persist will be an empirical question. No lineage can regenerate itself forever. Eventually some mutation will creep in that leads to an end to succession. The accumulation of adaptations by a lineage maintained over a long period is always a fine balance between mutability and fidelity, between the possibility of a better version and safeguarding the one that has proven to be the best so far. This narrow path is treaded very well by DNA replication and perhaps by neuromemes as well.

In the end, it is the combination of two factors that can produce consistent replication in memes. First, signals are instigators, which means they need play only a small (but crucial) part in the meme-making business. Second, memes work through the *conversion* of an existing substrate rather than the full-fledged *production* of new structures. Memes don't construct something complex from scratch; they merely *reconfigure it*. Coupling instigation with conversion means that the replication reaction has relatively little work to do and that signals may be quite tangential to this process. So whatever may happen to signals in their travels, replication reactions should remain capable of regularly and repeatedly producing similar memes, leading to a long-term lineage being formed.

There's also a sense in which the whole question of mutability is misdirected. Even if memes mutate each and every time transmission occurs, it's still possible for the ultimate goal of any evolutionary process to emerge: proper evolutionary lineages that exhibit cumulative adaptation. How?

One way is if the errors introduced during transmission are not biased in any way. People then are not effective at making replicas of the memes they try to copy, but they do have replication as their goal—they just happen to make mistakes. The central tendency of the resulting memes (as statisticians would say) will still float around the value of the original meme. Since this normative meme remains the dominant form in the population of learners, more people wind up using the original meme as the template for their own learning than any other value. So there is no net movement in the population of memetic forms.

What is perhaps even more surprising is that the force of evolution can still be sufficient to ensure the emergence of adaptive design when mutation *is* directed by some psychological bias. Such a bias can cause memetic norms to tend toward some new value because the copying mistakes people make are on average in the direction of this psychological attractor. Now the mean of copying errors is no longer zero. But this is still not a problem if the bias heads everyone toward the same goal—say, some functional optimum for that kind of meme. Then all the errors lead the population average toward that optimum. It's just a case of directed evolution moving the population consistently toward a new value, even if memetic replication is invariably imperfect. The only difference from the previous case is that the frequencies of alternative values in the population of memes shift along a path leading to the new norm.

In both of these scenarios, precise replication almost never occurs, but the long-term consequence is the same as in the case of high-fidelity replication among genes: cumulative adaptation over time. It's only a fixation on the way genes work that obscures the possibility of other ways to maintain the inheritance of information in the face of significant mutation.

Finally, if there is such heavy-duty reconstruction at the receiving end of a transmission, what happens to information transfer? One of the conditions for replication is that the copy must be similar to the source because information deriving from that source has been incorporated into the copy. But if reconstruction is due to modular inferencing mechanisms, and these are genetically inherited, then cultural transmission isn't the cause of similarity in beliefs and values between people. Rather each person individually relearns that knowledge thanks to naturally evolved modules for inferring things. The cultural link in social communication seems to be broken by this "on-board" mental construction.

But no one has ever argued that replication is achieved without assistance—by any replicator. As long as the process of inferring content depends on information inherited from other brains—from source copies of the meme—the transmission process is an evolutionary one and qualifies as an instance of replication. This is just what signals do as instigators of replication reactions: pass information along from the source. So the fact that cultural knowledge is inferred, based on the content of signals, does not exclude it from the category of replication events.

NEVER MIND THE GAP

The ability of some host organisms to communicate with each other through social signaling presented memes with a way to solve the "gap problem." However, you only have to remember that language is a signaling system to realize that communication systems vary enormously in their sophistication: The grunts of macaques are not nearly as elaborate as human speech. So considerable evolution has occurred in communication systems based on social signaling. This progression is bound to have implications for memes.

So how did social communication evolve? For species that lived in groups, it's possible that nervous systems had to develop the ability to learn from observing the behavior of their fellows. However, social observation can be a route through which information is transferred independently of genes. If this information is then passed along to others, a chain of inheritance in non-genetic information is established. Transmission would likely introduce noise into the chain, so we could also have descent with modification in the message. In effect, this means that genes may have permitted the development of another mode of evolution in cognitive species, bringing us quite far from genetic determinism in big-brained creatures.

The crucial step for memetics would occur when a copy of the message stays behind somewhere in the chain when another copy moves on to the next link. So instead of merely passing the message along, and then going dumb again, a node makes a copy as an interim step in its message-passing routine, perhaps to achieve some delay in the subsequent transmission, as an instance of momentary storage. This opens up the possibility of multiple copies (or near copies) simultaneously being disseminated through a neuronal population, in various, potentially crisscrossing, lineages.

If so, then genes have loosed a monster! When degrees of freedom arise in more complex networks of neurons, memes can begin to evolve. But as the number of neuronal interconnections increases, so too does the scope for action and movement by memes. The degree to which genetic control was lost—or the possibility of non-genetic inheritance increased—is greatest in species with larger nervous systems, a greater dependency on social life, and better abilities to learn from others. (Of course, humans excel in all of these areas.) Memes, according to this scenario, arose as parasites on chains of communication in such species.

Once memes evolved, what happened next to signals? Over time, evolution often produces higher degrees of complexity and greater cohesion in the objects of selection, which signals would be, as the products of this new replicator. The question is how do highly ephemeral, dynamic entities like signals become more complex? We know that signals can serve as good instigators: A computer virus can be perfectly copied through many generations, thanks to a protected transmission channel and the strict error-correction mechanisms used in contemporary computer communication protocols. Presumably the same thing happened as signals became involved in the replication of memes.

But there appears to be nowhere to go in terms of increasing the complexity of individual signals. Even language—perhaps the ultimate in communicative complexity—is not so much an advance in signaling as an advance in coding and decoding. If you look at sonograms of spoken phrases—the pictures of a phrase's wave forms in terms of amplitude and frequency—they look like nothing so much as a bunch of never-ending squiggles (mimicking the printouts from EEGs, which monitor brain waves, not coincidentally). Grunts and grammatical productions both use the same sound resources, and human language may even involve fewer different sounds than some animal sound systems. Speech is still just air molecules being pushed about, after all.

Even if individual signals are not the focus of cumulative selection, one way for signals to become more complex is to aggregate into temporally unified sequences. Individual signals may not become more elaborate, but sets of them could. Selection pressures would then be put on messages composed of many cooperating signals, leading to increased cohesion in the messages so produced. Each signal then becomes like a cell, bunched into coordinated multicellular "organisms." Certainly it appears that this is one characteristic of human language: the ability to tie signal units together into strings called sentences. Grammar, of course, is the ultimate result: Each of the units in a sentence can begin to play a specialized role (as noun, verb, adjective, and so on), greatly amplifying the range of sentiments that can be expressed in a message. Language is thus an obvious case of selection for communicative complexity, achieved in the face of little room for increased sophistication in signals themselves.

While some evolutionary advances are seen in signal streams, other advances appear in signal generators and interpreters. Both ends of the communication link have been compelled by natural selection to come up with better solutions. For example, signal receivers have had to cope with increasingly complex messages, coming in at the same speed as when the messages were ungrammatical. This has led to the need for quick comprehension, and particularly the ability to parse the continuous stream of sound inputs into meaningful units, like words. On the other hand, the generators of signals would also have needed to produce more interesting messages to parse. This required better cognitive abilities for buffering output as sentences were being put together at the originating end of the communication link, and the storing of partially interpreted messages at the other, awaiting the full sense to be expressed and received. Much of the fancy machinery supporting language is therefore in the neural mechanisms for generating and receiving spikes.

This is fully consistent with the neuromemetic perspective because it suggests that where most of the "action" has occurred during the evolution of communicative systems is in the head, where the replication of memes happens. The mechanisms that evolved for generating and interpreting complex messages should also support the reliable and accurate conversion of neuronal states, even if those states turn out to be infectious. And the merest inducement—such as the single word of greeting, "hello"—can set off the replication process, even when the source of that inducement is far away, in another brain.

Still, if each brain is an independent universe (memes being at first unable to jump the social gap), then how did an infestation of memes manage to spring up *each time* a big-enough brain was born? Would each new species of meme be able to evolve adaptations and complexities if the products of this evolution must die as soon as their host did?

It's entirely possible. Why? First, literally *billions* of generations of memetic evolution—of "meme time"—can take place within the life span of a big animal—in 20 years, say. This is a period not to be sneezed at if you're a replicator. It is, after all, the equivalent of a hundred thousand years for a bacterial colony in terms of generation time. Second, evolution just happens whenever the circumstances permit; it doesn't look into the future, starting up only if there are long-term prospects for survival. If a pool of water dries up or a computer hard disk becomes disconnected from a network, so be it; the biological or artifactual replicators just die off. Further, the basic conditions for memetic evolution were always the same, thanks to the common, evolved architecture of the brains available. Each of these newborn "universes" was, in effect, ready-made for memes. If memes were able to arise once under such conditions, they probably managed the trick quite regularly. A meme's chances for immortality were more limited in the era before social transmission, but evolutionary processes nevertheless would have occurred, over and over and over.

Of course, any communication between selective environments could have led to migrants finding their way to new contexts and provided for the longer-term survival of their evolutionary lineages. It was only late in memetic evolution that memes succeeded in reliably hopping from one island of evolutionary history to another through social communication. So the brain was not selected for size simply to support memes. We shouldn't overemphasize the importance of these brain parasites. The brain has too many important jobs to do, as indicated by the huge proportion of human genes dedicated to coding for brain function and morphology. Rather memes piggybacked on a big-enough brain to get themselves going. And once they got a handle on living inside this protected environment, they become bolder about risking life outside the brain.

In the early days, though, memes didn't even know that other islands of evolution existed outside the cortex they lived in. For them, replication meant survival to another generation. Quite simply, the meme, if it replicated, could live on in its offspring—the same motivation any of us have. The replicator's basic desire is to endure, to have another chance at the evolutionary lottery, and this desire would have been satisfied through replication inside one cortex.

If memes are basically mental entities, then the primary host of memes is not an organism but an organ, the cortex. The fate of memes is thus effectively linked to the genes that code for this organ, whichever organism the cortex happens to be in. Thus there has been a long history of interaction between these genes (which are highly conserved, at least in the mammalian line, with a consistent developmental sequence) and memes, beginning with the rise of vertebrates. Memes can thus be considered to constitute a single "species" of replicator, with a specific definitive host, the cortex.

One consequence of this longer history is that, during their long co-dependence with brains, memes have had the time to engage in significant feathering of their nests, by doing some reconstruction of the mind. If genes and culture have coevolved for a long while, then many of the seemingly "innate" traits of contemporary humans are the result of cultural pressures on genes: natural selection for genes that favor the more efficient expression of those cultural abilities. For example, considerable resources in the human brain are devoted to the processing of linguistic inputs, which can be used by memes to help guarantee their good replication. Thus big brains have evolved greater dependency on memes over time. However, meanwhile memes have sought greater independence from brains, developing more elaborate life cycles, interacting with a wider variety of things in the general environment, like artifacts (see the next chapter). Our current human "nature" becomes, in effect, a cultural artifact. Our most cherished personality traits may be ones that support our symbiosis with memes. This includes brain adaptations for language production and comprehension, for example. So we have truly entered the cultural niche as a species and can no longer live outside it. The benefits are obvious but also entail some perils since the door to a life without culture has been closed behind us.

RETHINKING COMMUNICATION

The idea that social communication involves the replication of information—a proposition central to social memetics as I have presented it here—forces us to reconceptualize what communication is all about. Communication is commonly defined as the transmission of information. However, as general as this sounds, it still isn't general enough. It leaves out quantum communication, for instance, which need not involve a channel or transmission—at least not as normally conceived—

because information doesn't move through space but rather just reappears at the destination (see Chapter 5). I therefore suggest that communication, generally speaking, is a process in which messages (typically translated into signals) are produced and then consumed. This process can occur over a wide range of scales—from atoms to organisms. The sender of a signal need not be the same as the producer, and the receiver can be different from the consumer. On the other hand, the same agent, as in the example of communication with oneself, can undertake all of these roles.

Several features of this definition of communication are worthy of attention. First, it doesn't refer to replicators. That's because the message being transmitted need not be duplicated—indeed, a message *cannot* be duplicated in quantum communication, as discussed earlier.

Another possibly surprising aspect of the definition is that it doesn't mention intentionality. In human communication, due to the complicated social cognition of our species, message production and consumption include a lot of work to infer the intention of the sender to recover the proper message (for example, getting past any potential irony or sarcasm implicit in the context, which reverses the message's meaning). But this is not a general feature of communication, which can take place without any cognition at all, as when entangled atomic particles exchange information.

While communication requires time to complete (consumption necessarily follows production), the process need not be completed rapidly; it may be mediated by artifacts, which can "hold" delivery of the message in abeyance for an indeterminate period. This makes some communication events "quasi-interactions": The sender and the receiver are displaced from one another in both space and time, as when a journalist's report is read only some days after she wrote it. In such cases, the consumed signal need not be the same energy packet as the one that helped produce the artifact in the first place, so the signal producer must trust that at some point the communication process will complete itself, making the effort of producing the signal or artifact worthwhile.

Memetics will never be a truly general explanation of the communication process because it's only concerned with explaining communication among creatures with brains. For example, only two of the three kinds of events with which we began this book can possibly be touched upon by a memetic explanation—the social and artifactual cases (Darwin's reading of Malthus and the spread of a computer-virus, respectively), but not the biological one (of *kuru*). Just what the effect of memetics will be on our thinking about the nature of communication can be made clear only by detailing how it's different from current accounts.

Three general approaches to describing communication are in vogue, each specific to one type of communicative agent. The first approach is devoted to communication among agents without intelligence, such as machines, and so is called *Mechanical* here. It's based on the mathematical model of communication, as epitomized in the work of Claude Shannon and William Weaver during the 1940s. Although this model is not directly relevant to memetics because of its restriction to communicators without brains, Shannon and Weaver's work has proven to be the foundation on which all subsequent theories of communication have been built, so we must deal with it here.

The Mechanical model was originally devised to design telephone-switching networks with certain optimal properties, such as the most efficient use of bandwidth (or the communication channel's informational capacity) with the highest possible fidelity of transmission. In this classic model of communication, a *source* selects a *message* that is then coded by a *sender* for transmission over a *channel* to a *receiver*, which then hands the message to some *destination*.

We obviously need to unpack this model a bit. The sender (which need not be the same as the source) translates the message into a signal (which can travel), so it's a signal, not the message, that is sent through a communication channel. The receiver, in turn, decodes the signal back into a message for some destination, as when a telephone converts a spoken phrase back into sound for the person listening in. The receiver must share the sender's code book and functions as a kind of inverse transmitter. Differences between the sent and received messages are due to corruption by noise or the interference of other signals in the channel during transmission. These differences are unwanted and distorting, since they add or subtract from the signal and thereby create uncertainty about what the message says.

The primary result of the mathematical approach to communication was to provide a definition of information that didn't depend on its physical substrate. Mechanical theory was concerned only with the problem of efficiently transmitting messages through channels, not with deciphering what the messages might mean. Shannon and Weaver were specific about not wanting to get involved in considerations such as the sender's or receiver's mind-set, and excluded the problem of interpreting the meaning of a message from the communication process as they narrowly defined it. The idea was simply that the message reduced uncertainty about a situation relevant to the receiver.

What could be done about noise in the channel? Shannon and Weaver's solution was to situate an observer outside the communication channel. Between the information source and the transmitter, the original message branches off toward an observation device, which surveys what is sent and received, notes the discrepancies, and transmits corrected data to the receiver. This is a cumbersome solution that few real-world communication systems exploit. Rather than adding a whole new bunch of machinery to the communicative apparatus, more popular ways of battling noise include introducing redundancy into the coding of the message or simply sending the same message many times through the channel, which the receiver can then compare.

The Mechanical approach thus sees communication largely as a process of finding the optimal coding system to compensate for the noise problem. Some ways of encoding a message into a signal will simply survive the journey through the channel better than others, depending on the qualities of that channel. The job of the sender is to find that way. Communication is social in the limited sense that it is a form of interaction between the sender and receiver, but a Mechanical communicator is not oriented toward an understanding of the machine with which it is paired in the communicative event. All an encoder has to do is produce a signal; all a decoder has to do is to attend to that signal. Thus there is no reason to credit either the sender or the receiver with any form of subjectivity, let alone inter-subjectivity. Neither communicator need be endowed with mental states and capacities. Perhaps it isn't surprising that an emphasis on the utility of different coding schemes should arise when the sender and receiver's psychology is defined out of the paradigm!

By contrast, human communication presupposes and exploits an awareness of self and others. Consequently theories of human communication reflect the very different cognitive abilities of humans as compared to machines. An approach designed to explain the peculiarities of *human* communication therefore suggests that communication involves not just the production and decoding of signals, but the inferring of intentions by both senders and receivers, based on aspects of the situation in which both find themselves. I will therefore call this approach *Inferential*. This theory is limited to intentional agents, with the quintessential example of a communicative event being human conversation. The scope of this theory is therefore more restricted than the Mechanical one, and less naturalistic.

The Inferential approach has its academic roots in the work of the influential philosopher of linguistics H.P. Grice. He argued that communicative events had to be recognized as such before they could be successful. In effect, the potential sender and receiver of a message have to negotiate a proper context for communication before exchanging messages, so that each party is paying attention to the other and expecting the exchange proper to begin. Even after the cooperative intent to communicate has been made manifest by the participants, Grice believed that human (verbal) communication can proceed effectively only if certain maxims are obeyed. These maxims include saying only what is informative, relevant, and true, as efficiently and clearly as possible. Gricean theory thus assumes that an implicit, gentlemanly agreement rules the exchange of information, and that participants share the goal to successfully complete the communicative event—that is, both desire to share a meaning.

Instead of the Mechanical concern with messages being linearly transferred from sender to receiver, Inferentialists see communication as the mutual negotiation of meaning. Inferentialism thus challenges the tendency of reductive Mechanists to equate meaning with messages (the content of signals). On the contrary, the Inferentialists assert, signals do not "convey" meanings from sender to receiver but rather constitute a stimulus from which the parties to the interaction must *actively construct* meanings. In essence, the central insight of the Inferential school—that meaning is "negotiated"—adds a step to the Mechanical model of communication. From the perspective of a Mechanist, we can say that the receiver must not only decode a signal but also infer the sender's meaning by analyzing a message *after* it has been decoded. At the other end of the communication event, the source must have some idea about what sort of manipulations the receiver is likely to make of potential messages to ensure that the correct message is sent and the meaning they intend to communicate "gets through."

Both the Mechanical and Inferential approaches have several failings. First, they don't explain *why* communication occurs or who should be able to engage in this type of activity. Studies of communication from these standard approaches are generally concerned with the narrower question of how information can be transmitted and particularly how messages are received. An explanation for the distribution and elaboration of communicative abilities among species remains missing.

There is also little theory of message content in evidence. Shannon explicitly argued that the goal of his mathematical approach was to divorce itself from the specifics of content, form, or channel; rather he sought to derive a generic concept of information through his investigations (a goal he achieved). An Inferentialist, on the other hand, seeks to typologize communicative events, particularly the kinds of events involving the exchange of spoken phrases. Unfortunately, an Inferentialist believes that most of what is inferred from messages is not transmitted— it consists of implications derived by the receiver from the communicative situation. The social complexity and specificity of communicative events therefore makes any general account of communication impossible in the Inferentialist view.

Further, the computational mechanisms by which a source selects from among all possible messages the one to be sent out into the world remain largely unspecified. Presumably not all formulations of a signal package are equally good at getting the message across. Mechanists suggest that agents maximize the efficiency by which a quantity of information is transferred. Inferentialists, on the other hand, suggest that conformity with the maxim of relevance—what will help each party to best understand the other's motivation in a given situation—is what communicators are after. While they may specify how best to send the chosen message, neither of these maxims has much to say about the need to choose among competing messages in the first place. A major part of the communicative process—how to find the best message to send, which will ensure that one's intention is properly inferred by the receiver—is left out of the explanation.

Finally, it's strange for Inferentialist theory to assume cooperation between message senders and receivers. It hardly seems likely that the interests and intentions of complex agents like organisms will always be in complete harmony. Indeed overlapping interests should be considered an unusual case since genetic endowments and socioenvironmental situations are rarely the same for different organisms. On the other hand, for the Mechanical model to become a complete description of human communication, it must show how the sender and receiver can come to have not only a common language but also common sets of premises about how to communicate messages. The communicating parties must then apply these identical inference rules to the message in parallel ways, such that communication results in the successful transmission (and hence reproduction) of the message. The Mechanical approach seems incomplete as a model of human communication; the Inferential approach seems to fill the gaps. The consequence of adopting an Inferential approach, however, is to separate the study of communication from the rest of science and make it something of an art form: the subjective study of human intentions.

A third approach to understanding communication avoids these difficulties. Because it's devoted to describing the evolution of signaling behaviors, I call it *Evolutionary*. Animals are the agents whose communication this theory is supposed to explain. The theory holds that communication is a specialized behavior involving the broadcast of information. Thus selection acts over generations to change the repertoire of displays of which a species is capable, thanks to its genetic endowments. Say an animal leaks some chemical into the environment that other animals can detect and that provides them with information about some aspect of the emitter's condition. If this information is beneficial to the perceiver—for example, in social competition—then greater sensitivity to the reception of that cue will evolve. If the receiver's response to this information is also in the sender's interest, the ability to emit that cue will also improve. So communication evolves when a coincidental association between the production of some cue and relevant information about the emitter's condition or some other environmental situation is correctly perceived by another animal. This depends on the fortunate happenstance that the animal can already sense this aspect of its environment—has evolved receptors to "taste" the chemical signal, for instance. Both the behavioral context in which the cue is produced by the "sender" and the context in which the cue is interpreted by the "receiver" are then ritualized to minimize any ambiguity in the situation, which should increase the efficiency with which each partner can fulfill their role in the communicative exchange.

The Evolutionary approach represents a very different view of communication than we have discussed before. It's a question of *dialogue* versus *dissemination*. Single events of information exchange, exemplified by dialogue between a sender-receiver pair, are the standard focus of both the Mechanical and the Inferential theories. Dissemination, on the other hand, suggests a profligate sender that tosses signals out into the environment, hoping to find one or more receivers, much as a sycamore tree throws millions of its seeds into the air or salmon broadcast their milt into the water.

But if we are to link communication to culture, it must come through a recognition that communication events are repeatable—that the same information can be disseminated not just once but multiple times, and so spread through a population. The Mechanical and Inferential approaches have ignored this fact by concentrating exclusively on the communicating dyad and on the individual communication event. What unites these first two approaches is the assumption that the sender-receiver pair, linked by an abstract channel, is all that need be considered in a theory of communication. The result is that the history of theory in both of these traditions has tended to consist largely of increasingly sophisticated models of sender and receiver psychology as earlier models have proved too simple to account for some aspect of real communicative events.

If we expand the temporal and spatial horizon to encompass a population and acknowledge that a communication event can be repeated, with the message being passed from person to person, it becomes clear that we must attend to communication *chains*. Further, the repeated communication of a message is likely to result in descent with modification in the constituent information. The dyadic communication event—with a sender-receiver pair—is still the unit of analysis for population-level phenomena; it's just that *repeated* events determine the course of information transmission through social groups. From this new perspective on communication, both biological and cultural evolution can be seen as a story of message-passing, whether through DNA or cultural messages. It's a story about stories—in particular, how individual lives and social traditions interact over time. This evolution of messages will be missed if we concentrate only on the brain (psychology) or the dyad (communicative acts), instead of society.

Recognizing the possibility of disseminating messages over a longer temporal horizon means that evolution in the mental abilities of senders and receivers can also occur through normal biological processes. New adaptations for coding or decoding messages can evolve in later generations of communicators, as can mechanisms for inferring the intent of one's partner to the communicative event. If one looks at communication chains, distortions in message content can then arise not just as a consequence of randomized noise in the channel, but because psychological features held in common among receivers introduce systematic biases in those signals and in how incoming messages are interpreted. Communication can even become an arms race between the ability among senders to deceive receivers and the ability among receivers to see through this deception. Then the transformations, distortions, and losses of information typical of social transmission are to be explained not just as side effects of "jumping the gap" between brains but as the normal consequence of "dueling" communication itself. What calls out for explanation from an evolutionary perspective is the fact that extended chains of communication can succeed in distributing throughout a population those ideas that are readily identified as a culture.

The Evolutionary approach avoids several of the problems associated with the Mechanical and Inferential approaches. First, it doesn't assume the sender and receiver intend to cooperate when they begin to exchange information. When the biological interests of sender and receiver significantly overlap, then selection on signaling abilities will generally favor increasing efficiency, as assumed by the Inferentialist's favorite philosopher, Grice. However, when interests diverge—that is, when the fitness benefits to be derived from the exchange of information are not equal between sender and receiver or do not depend on the same outcome—then an arms race can be expected between senders' ability to produce false, irrelevant, or ambiguous messages, and receivers' skills in detecting the sender's motivation for sending the signal in the first place.

Signaling theory also has a hypothesis about the conditions under which animals should form communicative intentions: Animals should seek to communicate when manipulating others through signaling promotes their own biological fitness (and that of their relatives) more than any alternative behavior they might engage in. The ability to identify a cause for communicative events is an advance for the evolutionary approach over the alternative models discussed earlier.

However, as for message content, the Evolutionary approach suggests only that signaling systems may arise as a means of exploiting a preexisting sensory bias that organisms have already developed by chance. Constraints on signal design can then be related to the environmental context in which cues or signals are emitted and received (for example, a dense jungle habitat dampens sound), as well as the physiological capabilities of the animals involved (such as body size constraints on the volume of sound that can be produced). But this is more a theory of constraints on transmission than a theory of message content. So the Evolutionary approach remains somewhat deficient on this front.

The Evolutionary approach also lacks a mechanism for selecting among possible messages to send, although presumably selection is optimized to benefit the biological fitness of the communicator. Note also that attention is focused in the Evolutionary approach on the individual organism, with models tending to take the perspective of either signaler or receiver in isolation. Thus the function of communication for the signaler is to influence the probability that the receiver will behave in a fashion beneficial to the sender. The receiver's function, on the other hand, is to acquire information that improves the likelihood of choosing an optimal response to the sender. The result, however, is that sender-oriented theory comes to somewhat different conclusions about the nature of communication than receiver-oriented models. As a consequence, an overarching perspective on communication is lacking. These remaining faults suggest there is still room for improvement.

Memetics suggests yet another theory as an alternative for communication studies, which I call *Coevolutionary*. It increases the emphasis on the hitherto ignored aspects of the communication event by noting that communication involves not merely a sender and a receiver but also a channel and a message. Adaptation in senders and receivers still holds, as in the Evolutionary approach, but can also evolve in parasites on the communication process—the messages themselves, which interact in important ways with the channels through which they travel. The Coevolutionary approach emphasizes that a consequence of successful communication can be replication of the information conveyed. Further, the evolution of human communication from simpler forms in other animals involves not just increasing cognitive sophistication, but also the creation of novel communication channels (such as mass media) and the consequent elaboration of the messages passing through those channels. A major consequence of the Coevolutionary approach is thus that *communication is explicitly linked to large-scale social phenomena such as cultural change through a physical consequence of communication by dyads: the replication of information.*

The Coevolutionary approach therefore must be seen as an addition to the primarily biological vision of the communicative act as a social interaction between gene-based organisms. This is because a parasitic system has evolved on top of that designed by the natural selection of genes in the host organisms. These organisms still use signals for their own purposes, such as social competition with conspecifics. So genetic interests can differ between individual organisms and there can be conflicts in which signals serve as weapons, as in the Evolutionary approach. But the Coevolutionary approach suggests there is an additional dynamic to the one between the receiver and sender: Both senders and receivers must cope with the rise of a new, increasingly robust replicator, the meme, that parasitizes one of their organs, the brain. In effect, there is a hidden homunculus or "ghost in the machine" behind communicative events. Communicated information can *itself* be an evolutionary agent.

The major claim of the Coevolutionary approach is therefore that *communication simultaneously involves the sender and receiver in two different relationships: first, as conspecifics with potentially divergent genetic and social interests, but also as potential hosts to a more or less robust, parasitic replicator with its own evolutionary interests.* The Coevolutionary theory thus suggests an additional relationship between sender and receiver than that of cooperators or competitors: They also share an infection. Further, these agents may be brought into contact and exchange information due to the interests of this parasite, not necessarily their own (genetic) interests.

Nevertheless there are considerable areas of overlap between the Evolutionary approach to communication and the Coevolutionary one (as the closeness of their names suggests). This is because both acknowledge the potential of conflicting interests between senders and receivers.

As a result, the psychology expected by both theories is the same in that respect. However, a Coevolutionist would also indicate that there is a potential conflict of interest between the receiver and the message. The ability to deal with messages has proven beneficial for brains in general; the relationship is thus often symbiotic. But it's in this additional level of potential conflict that distinctions between the expectations of the Evolutionary and Coevolutionary approaches can be found.

In modeling the communication event itself, the primary difference between the Evolutionary and Coevolutionary approaches is in the conception of signals. In the Coevolutionary approach, signals are defined as "free-living," extended phenotypic products of memes, rather than as manipulations of the environment by organisms. Communication can therefore no longer be considered as a process in which people play mental badminton with shuttlecocks of information. Rather communication becomes one stage in a meme's life cycle, during which it produces signals that travel through a harsh environment in search of another mind in which the meme can take up residence and reproduce. It isn't signals that are considered to be the second evolutionary agent in a communicative act. The parasite is, in fact, closer to "home," inside the brain itself.

The central question for the Coevolutionary approach, then, is whether signals are instigators of memetic replication or modifications of the environment by organisms that have no evolutionary import of their own. If it's true that memes replicate through the use of signals, then it can also be assumed that some aspects of sender and receiver psychology will reflect adaptations to this fact. In this respect, the psychological expectations of Evolutionary and Coevolutionary theories will differ, despite their common assumptions about the possibility of conflict between communicators.

The Coevolutionary theory can also fill the lacunae left by the evolutionary approach to communication. First, it provides the rudiments of a theory of message content, thanks to its greater attention to and reconceptualization of the role of signals. Signals are instigators of meme replication, which implies that they should be selected, both within the brain and in the macroenvironment, for the traits that allow them to fulfill this role. Signals will therefore be pressured to convey information vital to meme replication between hosts. This implies that signals can be expected to exhibit design features that allow them not just to survive travel through the macroenvironment intact, but also to carry those bits of information that cannot be supplied from local resources in the receiving brain. Just what these bits may be is not clear at present, due to our ignorance about brain mechanisms for information processing. But as neuroscience improves, we should begin to make empirical claims about what kinds of messages will be needed to stimulate replication reactions in the brain.

The Coevolutionary approach to communication also provides hints of a mechanism for the selection of messages. While the replication reaction occurs in the receiving brain, message selection occurs in the sender brain. So adaptations should also appear in the message-*sending* brain for this purpose. The message that conveys meaning most reliably, with the lowest probability of being misinterpreted, should be the one that the decision-making process prefers, perhaps due to its "fit" with a processing module devoted to the processing of such content in the receiving brain. Again, the inner workings of a message-selection mechanism are in need of further development. Nevertheless these ideas represent an advance over the alternative approaches. Because they largely ignore the *evolutionary* role of channels and signals, the other three approaches we have examined lack such implications for these aspects of communication.

In conclusion, models of coevolution between hosts and parasites have already provided us with insight into a variety of outstanding problems in evolutionary theory. Examples include the theory that sex evolved so that hosts could take advantage of the recombination of genetic elements that sexual reproduction provides, and that variation is maintained within species to deal with more rapidly evolving parasites. Coevolution has been shown to produce more optimal outcomes or to more rapidly achieve a given level of fitness not only in biology but also in computer science. Indeed computer scientists are increasingly using evolutionary principles to solve the problem of software design because the increasingly complex functional requirements of modern software surpass the human ability to design solutions using logical principles. Considering parasites as engines of evolutionary change has thus been illuminating for a number of disciplines. Considering cultural traits as parasites—as replicating memes—may provide insight into the communication process as well.

Imitation, Schmimitation

A further benefit of the approach to communication adopted in this book is that there is no need to mention imitation in the account of memetic replication, even though imitation is in the definition of the word "meme" in the authoritative Oxford English Dictionary. Why is this a benefit? Because the concept of imitation is problematic. Imitation is generally held to be a special type of social learning responsible for the difference between humans and other creatures. But nobody really knows how imitation happens.

Cultural selectionists and prominent memeticists (Dawkins, Blackmore, and Dennett included) admit that the psychology of imitation may remain a bit murky but argue that imitation is a peculiar *psychological* ability that has peculiar *social* consequences: When a population of imitators start imitating one another, the emergent result is culture. The appropriate question is therefore why *not* emphasize imitation as the defining characteristic of our species when there appears to be a correlation between this ability and the exhibition of complex culture? It's a very natural coupling: Culture is present in very few species; those few species have big brains; and imitation seems to require big brains. Since both imitation and culture achieve their pinnacle in humans, why not assume that each must somehow be linked to the other, especially since the development of the ability to imitate seems to be an important and early developmental goal in our species? Babies are expert imitators, after all. Presumably we devote considerable psychological resources to this ability to imitate one another. So the link between this unique learning mechanism and culture seems inexorable.

This would indeed be fine, if it wasn't possible that imitation doesn't really exist, except as a dubious distinction in the minds of social psychologists. The general rubric of "social learning" is an amorphous grab-bag category of theoretically distinct cognitive capacities, of which imitation is meant to be one. But psychologists themselves can't agree on a standard way of telling these skills apart, mostly because they haven't identified mechanisms for how they're supposed to work. Generally speaking, social learning is facilitated by an animal paying attention to some aspect of the situation surrounding another member of its species— for example, where it is sitting or the thing it is interacting with, say a plant. What's meant to make imitation different from other types of social learning is that rather than paying attention to the place or thing a novice sees some other animal involved with, it now focuses on the movements of the demonstrator—how it is engaged in eating that plant, perhaps. And with imitation, the novice acquires the rules for generating a new class of behavior—how to eat that kind of plant—rather than learning how to mimic the demonstrator's precise motions in manipulating the plant. A true imitator, then, is supposed to be able to learn how to perform some new behavior simply by seeing it done—at minimum, by seeing it only once. After this, the organism can reproduce that behavior even when circumstances are not exactly the same as those in which it originally observed the demonstrator engaging in the behavior. There is too much slop, or room for error, in the other kinds of social learning, which involve significant individualized mental processing. It is for this reason that both memetics and cultural selectionism have been based, since their beginnings, on the psychological mechanism of imitation: It's supposed to be the necessary underpinning of truly *cultural* learning—that is, learning that reflects the ability to quickly learn a repertoire of new skills from demonstrations of those skills by others.

This appears to cultural selectionists to be good enough as a foundation on which to build an epidemiological theory. After all, you don't have to have a complex appreciation of the ins and outs of each individual's tastes, desires, and knowledge to model the spread of ideas through a population.

The problem is that this definition of imitation still doesn't really tell you who should be able to do it. It doesn't mention how such mimicry of behavior might be achieved. The definition only makes reference to *behavior*, not the mental abilities needed to produce that behavior.

This vagueness has allowed two different schools of thought to spring up on the question of how much dedicated psychological machinery is required to implement this ability. We can call these schools the Mentalists and the Behaviorists. Mentalists say only those species with what is nowadays called a "theory of mind" can imitate. Theory of mind is a hypothetical mental ability that enables an animal to "read" the intention behind another animal's behavior. It allows the first animal to mentally simulate the perspective of another animal and so figure out what it is up to. Having a picture in its head of what the other animal is thinking should enable the first individual to infer how to engage in that behavior itself—imitation, in effect.

Behaviorists, on the other hand, don't care whether their candidate for imitative ability has fancy mental representational skills or not. All they require is that it display the rapid, unreinforced learning of novel behavior. It's perhaps preferable to them that the animal can do this for several types of behavior, because this suggests a general proclivity to imitate. Otherwise it might simply be a trick of the genes, an inbred program for some particular complex behavior. After all, animals with rather simple neural systems, or indeed no brains at all, can do wonderful things like build huge cities (just look at termites and ants). This is the primary reason that Mentalists restrict imitation to species with brains big enough to be credited with complex internal representations—typically apes, humans, and perhaps dolphins—while Behaviorists argue that less well-endowed species, such as rats, birds, and fish (or even some invertebrates such as the octopus!) can also be admitted to the club, as long as they can "do what they see."

So the question is: Does imitation require a lot of brainpower—which would mean that only a few select species are capable of it—or can humble animals with no cortex also imitate?

The cultural selectionist is likely to be unconcerned by this question of who exactly is "allowed" to imitate. This may be an empirical question, but it doesn't affect the selectionist's models, after all. These models can still churn out predictions about population dynamics from different regimes of cultural transmission, whatever creature turns out to be acquiring information through imitation. So cultural selectionists can pursue their program of research unperturbed by these empirical niceties.

But, in fact, this business of cognitive prerequisites is vitally important to the general question cultural selectionists are interested in: cultural evolution. This is because one particularly glaring consequence of imposing a big brain limit on imitators is that birds then get excluded. Learning birdsong becomes mere mimicry—the matching of a vocal behavior to particular sensory inputs: Hear a song; reproduce it. This means singing behavior is not sensitive to the situation and involves no understanding of the message-sender's motivation in transmitting the call in the first place—in short, it isn't full-fledged imitation. However, this introduces a definitional problem. Birds may be exempted from the "imitation club," but their songs have been well documented to exhibit all the characteristics of cultural evolution, like mutation, drift, and directional selection. The birdsong case therefore implies that cultural evolution can occur without the ability to imitate, contrary to what both memetics and cultural selectionism have so long assumed. Complex population-level effects, such as fully cultural traditions, can arise despite limited cognitive abilities among members of that population.

Cultural selectionists can always retort that they don't mind if birds have culture, even if this places them with the Behaviorists in saying that "lower" animals can imitate too. Imitation would still remain *the* mechanism of cultural learning. This move would allow cultural selectionists and prominent memeticists to continue to found their theories of cultural evolution on imitation. And all of them do so for the same reason: because it's the only form of social learning likely to produce the reliable inheritance of information. Richard Dawkins has certainly continued to identify imitation as the sole means by which memes can replicate.

However, it isn't clear how imitation is going to obey the requirement of producing an evolutionary lineage, of achieving fidelity in *chains* of transmission. Culturally acquired ideas are often lost or transformed in memory *after* an individual acquires them. Just think of the children's game of Chinese whispers, in which each child tries to tell the kid on his left what the kid on his right just whispered into his ear. What went into the front of the chain as "Sally slept at her friend's house last night" comes out the end as "Sammy wept when his aunt groused at him." Chains of imitated information can produce consistent cultural traditions only if imitation is buttressed by additional psychological processes that insulate or reinforce the acquired ideas. If imitation is to be the mechanism by which ideas are preserved across space and time, then it must provide psychological services that insulate socially transmitted information from being modified by individualized learning along the way. Otherwise the information won't be replicated but rather continually modified. Imitation simply may not be able to ensure sufficient fidelity in the transmission of information to effect cultural replication. We just don't know because the basic question of how imitation occurs remains only vaguely understood. Strangely for such a central question in the social and psychological sciences, relatively little empirical work exists on this topic to guide us through the variety of possible interpretations.

Some might argue that the recent finding of "mirror" neurons solves the problem of delineating a mechanism for imitation. Neurons in a monkey's prefrontal lobe have been found to respond not only when the animal is poised to grasp a piece of food, but also when the human experimenter in the monkey's field of vision is about to grasp the same piece of food. Mirror neurons seem to provide a straightforward mapping between one's own actions and the observed actions of others. The same regions of the brain that send commands to our muscles when *we* act also seem able to recognize the same action when it's performed by others. This implies that perceptual and motor abilities are intimately connected and that imitation may be as simple as "monkey-see, monkey-do." This would represent a drastic reduction in the psychological underpinnings necessary to produce imitative behavior. So it could solve our mechanism problem: It provides an explicit neuroscientific underpinning for the ability to imitate. It also favors the Behaviorists, which is necessary if we are to admit songbirds into the imitation club.

However, mirror neurons don't really solve the mechanism problem. The result in monkeys has been seriously overinterpreted. Mirror neurons are likely to be found in any system that contains feedback loops within the cortex. So they may be all over the brain. For example, the same neurons fire when someone is viewing an image of the Mona Lisa and later recalling that experience. Now it's perception and *imagination* that involve the same neurons, firing at nearly the same rates; no behavior is involved at all. The neurons involved in this match are also spread all over the cortex. So this doesn't prove anything except that mirror neurons are part of the vast collection of neurons responsible for either action or observation. If imitation were really so simple as the mirror neuron story suggests, many more animals would pass the behavioral tests for imitation, which they don't. So learning novel things isn't just a matter of linking perception to behavior. This relationship necessarily holds for any animal. The mere fact that the same areas of the brain light up when two kinds of tasks are performed does not limit information processing to a simple loop between them; many variables probably still intervene between perception and action in the case of imitation. Whether such a neurological system can be the seed for the human capacity to imagine someone else's state of mind thus remains an open question. Those who are presently looking to mirror neurons for a viable theory of imitation will have to look elsewhere.

There's another problem with the evolution of imitative ability that also suggests we may not be thinking about this whole problem of social learning in the right way. To efficiently imitate others, an organism probably must invest in extra neural tissue, which is costly in terms of the energy required to keep it going. But the first individual who happens to get stuck with a mutation coding for the expensive network supporting this psychological ability will find herself in a population with no complex cultural traditions on which to draw. Carrying around a big brain in such a situation is simply a hindrance: The individual is handicapped by having to find extra food to feed the extra brain-stuff. Such a mutation provides neither the individual nor the social group with any benefit. So there's this Catch-22 in the evolution of the complex cognitive machinery underlying the imitative ability: It isn't any good unless everyone has it, but it won't evolve unless it's reasonable for the first person to develop it. Imitation has few advantages when rare, even though it can be very beneficial to a group when common. This constitutes a considerable barrier to the evolution of imitation, since any gene-based ability is likely to start as a single mutation and hence will be very rare at the beginning.

How then did humans manage to hurtle this start-up cost and actually begin to evolve this trait? Lots of "just so" stories could be told, but the bottom line is that no one really knows at present. So we have purely theoretical reasons to think that there might be something wrong with seeing imitation as the basis for something as fundamental as cultural evolution.

One still might retort that even if there are major problems with imitation, no viable alternatives exist. If we want to explain what might be special about human culture, it would appear that we're stuck with imitation, which remains the best candidate for a psychological mechanism of quick, reliable social learning.

But we need not be led inexorably to a reliance on imitation. By emphasizing imitation as the mechanism for transmission, you are almost forced to think that cultural evolution begins—and ends—with social learning; that the only important thing in the life history of a meme is transmission *between* brains. However, experiments in Artificial Life such as TIERRA tell us that a lot of evolution can go on inside a single intelligence. So too must early memetic evolution have been a story of populations of memes *within* a brain. A considerable proportion of memetic history thus went by before the first meme learned to escape its birth-host. This means that memes have been evolving for a long period of time, and not merely for social communication. This underscores the irrelevance of social learning to memetics. Any kind of self-promoting behavior leading to the replication of memes within a brain would have done to get the ball rolling. It is true that *social* replication places other kinds of constraints on memes, and these will influence the kinds of memes that evolve once social communication has become their primary mode of replication. But in their early period, the flavor of memetic evolution can be expected to have been quite different.

By acknowledging this history, the kinds of psychological explanations you can draw from to explain memetic evolution opens up. For example, many genes survive in the genome because they are good at manipulating the system, not thanks to the vital proteins they produce. Take so-called "junk" DNA—meaning DNA that is parasitic on the genetic system, interested only in making more copies of itself, even within the genome of one individual. Just as not all genes produce proteins, not all memes need to induce behavior. They could be "junk" memes, lying in wait for a useful job to do but meanwhile being parasitic on the memetic system. At least as long as there is selection for faster reproduction, some rogue memes will feel the pressure to parasitize the memetic reproductive system by going it alone. (There is no sense of guilt in the ranks of replicators.) There might even be a kind of division of labor among memes. Some memes, like some genes, probably engage primarily in the regulation of other memes' activity—"busybody memes," in effect, interfering in the affairs of others.

All of this gets missed with the emphasis on imitation. A review of the early history of neuromemes may even be able to explain how memes solved the Catch-22 conundrum noted above: how social transmission became common while being costly at the same time. It seems doubtful that the highly inefficient and haphazard process of early social communication would lead to the replication of a meme in another brain—at least at first. Thus early selection pressures for signal production would have been weak, perhaps insufficient to result in a major transformation of the memetic system or to have bootstrapped modes of social communication into existence in the first place. Rather it is more natural to argue that an efficient replicator system was already in place within the brain, having evolved over a long period. In effect, memes at first did not live or die by their ability to send social signals. Nevertheless any success at jumping the gap between hosts would have been rewarded and so fed back into the mechanisms for memetic replication. In this way, memes could have persisted without signal-based life cycles for a long period, quietly evolving a variety of social skills, during which time random successes at social communication became more regularized.

With neuromemetics, the whole imbroglio surrounding imitation can be avoided. It would be hard to build a structure as sophisticated as a science of cultural evolution on the shaky foundations of contemporary social learning theory. Luckily it's not the foundation on which this book is based. The "electric meme" is grounded in a more fundamental science than social psychology: It draws its inspiration from neuroscience.

Chapter Nine

THE TECHNO-TANGO

Technical civilization, and the human minds that support it, are the first feeble stirrings of a radically new form of existence, one as different from life as life is from simple chemistry.

— *Hans Moravec*

As computers become increasingly intelligent, and as genetic engineering and bioengineering begin to produce true cyborgs (people whose bodies have been augmented with artificial components), our general ignorance of what drives improvements in technology is rapidly becoming more dire. Strident forecasts by reputable figures in science and public life suggest that human beings could become second-class citizens in a twenty-first-century society dominated by intelligent machines. To address such possibilities, it's vital that a readily understandable framework be established for dealing with the evolution of technology.

One reason for the lack of a ready-made stand on this issue is that traditional evolutionary theory doesn't concern itself with artifacts. But evolutionary theory *should* have to account for the existence of things like computers, which are merely fancy artifacts. Why? Because computers and other artifacts show evidence of complex design and inherited features; successive generations of computers—or tennis shoes or even clothespins—appear to form lineages. This strongly suggests that artifacts are subjected to an evolutionary process. After all, there is only one currently accepted explanation for the rise of complexity: natural selection, as limited by intrinsic physical constraints on organization. So do artifacts evolve? I suggest they do. But it's important to recognize that they evolve through interaction with mental "artifacts"—in particular, ideas, which can be memes. *Co*evolution between memes and artifacts is at the root of cultural change. My claim, then, is that memes alone are not enough. The incredible dynamicism of cultural modification in modern Western societies, driven largely by technology, can only be the result of "parallel power": two lines of inheritance working together, feeding off each other in a positive fashion. Providing a framework for understanding how memes relate to artifacts is the goal of this chapter.

How can we think of artifacts? Not all artifacts fit into a single category recognized by evolutionary theory. There isn't a truly general way to speak about these "made things." They can be extended phenotypes, interactors, templates for signals, hosts for replicators, parasitic replicators, *or* true replicators, depending on which artifact you refer to. Some artifacts can potentially be many of these distinct things at the same time. Each of these perspectives is only a partial view of what an artifact can be.

Artifacts are thus a heterogeneous group sharing only the quality of being produced from environmental materials through the activity of organisms. And even this relationship between the organic and material worlds is being severed as factories become automated and machines begin to make machines. With computer viruses, the divorce between human activity and artifact production becomes even more complete, as the virus can make copies of itself with only the most incidental involvement of any biological agent (usually a seemingly innocent mouse click).

Further, new kinds of artifact are invented as time goes by, with more complex ones coming later in the sequence. The types of artifact that arrive later have internalized the ability to change themselves in more fundamental and powerful ways. In effect, new, more dynamic categories of artifact are continually being added to the existing roster; each adds a new evolutionary dynamic to the overall picture of modern life. I therefore conclude that artifacts are evolving greater *evolvability*, which accounts for their ability to occupy multiple evolutionary categories. Accounting for the development of this evolutionary process—and the role of memes in that process—is our task here.

ARTIFACTS AS PHENOTYPES

The first artifacts were phenotypes. You have an idea for a better hammer, and the "expression" of that idea is the made thing, the new hammer. In effect, we can say that artifacts originate in the human mind as *mental objects* and afterward are turned by people into *physical objects*. The idea that was born in its inventor's mind is thus like the artifactual equivalent of a genotype, while its phenotype (or physical manifestation) is the actual object. More precisely, we might say the idea (a new meme?) is the genotype and the activity of the individual in making the artifact is the phenotype, while the secondary consequence—the *extended* phenotype, if you will—is the artifact itself. For example, the idea of a stone tool manifests itself directly as the knapping or flaking of a flint, with the end result being the Acheulean hand ax characteristic of an early human species, *Homo erectus*.

Artifacts like hand axes are inanimate objects because they are "pure" phenotypes. They exist as independent physical objects in the environment, as the *embodiment* of an idea. They thus become divorced from the vital force behind their creation: the idea that made them possible in the first place. They are also dead-ends in an evolutionary sense: They cannot make copies of themselves. This makes such artifacts lifeless and inert in character.

These abiotic substrates are created (in some cases) by cultural parasites, memes. Hand axes produced by subsequent generations of *Homo erectus* may be quite similar, just with changes in form or style, which gives them the appearance of being created through a process of descent with modification—and makes the sequence appear to constitute an evolutionary lineage. But, in fact, artifacts are extended phenotypes of memes, created through an intermediary agent, the host organism.

However, artifacts *are* the focus of selection pressures. They are the things that can acquire and exhibit adaptations, thus becoming more complex with time. But what *evolves* is the idea behind the artifact. The artifact can, in some instances, be "upgraded" to the new model or archetype for that class of artifact. But in these early stages, when artifacts are simple, you simply make a new one from scratch to reflect the better idea you just came up with. This year's hunting bow is not, for example, a recycled and retooled bow from last year; it has instead been made "fresh" from raw materials. So there is a characteristic production cycle for new artifacts in this early phase of human evolution: idea, product, use, selection; then new idea, new product, new round of selection. All the novelty arises first in the idea. The physical objects themselves—the artifacts—do not evolve because there is no direct line of inheritance from one artifact to the next. Hammers don't make hammers; people with hammer-making knowledge do.

Since making artifacts involves transforming "natural" bits of stuff in the environment into something new, people are, in effect, reforming aspects of the world in which they then have to live. Any such factor with which a species must deal to survive and reproduce constitutes part of what ecologists call that species' "niche." The making of artifacts can therefore be called "niche construction" or "ecological engineering." Such reworking of one's surroundings has been going on for a long time. Even lowly ants, termites, and bees reconstruct their niche by making massive homes for themselves out of mud. Worms also chew up their surroundings, making it a more congenial place not only for them but also for a wide variety of microbes and plants. We therefore shouldn't think of niche construction as something that requires brains, although if you have brains, the number of ways in which you can modify your environment increases considerably. And humans are obviously the niche constructors nonpareil, since we are the species that has transformed the face of nearly the entire planet through generations of cumulative, ever more powerful activity.

Once you've engaged in this kind of activity, the environment has new kinds of objects in it with which you can interact, to your benefit or harm. Niche construction creates a new evolutionary loop in which the transformed environment begins to put new selective pressures on the organisms that made those modifications. Every new technology introduces new capabilities as well as new hazards. For example, wagons or cars not only can make people more mobile but also can run over innocent bystanders and kill those riding inside them as well. In any case, the existence of mechanical vehicles may promote any genes that help in the production of such artifacts, such as the genes that increase the skill with which new (and safer) vehicle designs can be made. Littering the landscape with such knickknacks, adding artifacts to what nature has already provided, can thus significantly influence the way in which natural selection goes to work on a population. This is especially the case if these modifications persist, as artifacts often do, while the organisms that make them come and go. Much of what we call the "built environment"— roads, cathedrals, sewer systems—has been sitting there, as part of the landscape, for generations. In effect, such artifacts are inherited by subsequent generations as an aspect of the ecological world they face.

However, this seems to be only part of the picture, because artifacts, which include things like paintings and fashionable clothes, are generally felt to be part of our *cultural* environment. Shouldn't there be some consideration of the fact that creating such artifacts feeds back not only onto our genetic makeup but onto our cultural lives as well? Niche construction can have an impact on the speed and direction of biological evolution because artifacts are physical objects "out there" in the environment, just like mountains or trees. Making hammers can be a "natural" consequence of genetic programs for that kind of behavior. But skills for modifying the environment can also be transmitted through populations in an epidemiological fashion—that is, through social learning.

What can now be called *culturally learned* niche construction, as you might expect, introduces new dynamics into the relationship between clever organisms and their environment. Learning how to make artifacts from others can introduce quite different dynamics into a population than would be observed had the same routine for making artifacts been inherited from parental genes. This is because it takes a long time for biological evolution to select among naturally occurring mutations those few that would enhance our ability to modify the environment in some particular way. But once we developed a rather general biological mechanism for learning new things socially, the universe of possibilities opened up, because an individual could then acquire many new abilities for making things within a single life span. Our abilities to clothe, house, and medicate ourselves have enabled us to respond to environmental challenges not so much through genetic mutation and change in our body's shape, but rather by reducing or eliminating those aspects of our environment that we find unpleasant, thanks to these newfound abilities—abilities coming not from a shared genetic proclivity, but from knowledge held in common by cultural groups. In effect, we have been liberated from the strictures of our bodies by our ability to remake the world in our own image. Humans are thus not helpless victims of circumstance or mere objects of selection, but rather the potent constructors of highly modified, culturally inherited worlds.

ARTIFACTS AS INTERACTORS

If artifacts are part of culture, and specialized knowledge or know-how is necessary to make artifacts, then perhaps the ideas going into the making of artifacts are memes. When a woman knits a scarf with a certain pattern, does the pattern reflect a meme (or meme complex) in the mother's mind? Can we use an evolutionary perspective to illuminate the relationship between memes and artifacts?

A hammer doesn't augment human musculature and achieve a widening of the human niche just by lying there on the ground. It must actually be put to work building a chair or a skyscraper. Artifacts are thus linked to memes by the fact that cultural tools, to have real force, must be wielded appropriately. This requires specialized knowledge, acquired by individuals through practice using those artifacts. It's people with expertise that make artifacts an important component of culture. This is made obvious by the science fiction scenario in which some social group crashlands on a new planet, blessed with a wealth of technological gismos beyond their wildest imaginings, only to perish because they can't figure out how to make anything work, not being the originators of that technology. Know-how has to match with artifact.

So people must interact with technology. Does this make artifacts what evolutionists call interactors? Answering this question will require a bit of analysis.

Take the example of a vehicle, like a wagon. A wagon is a machine made up of some wheels, axles, and a platform, combined to carry loads efficiently. If function follows form, perhaps as a person constructs a wagon, his idea of a wheeled wagon is bodily transferred to that artifact, in its shape. The wagon thus becomes a good Dawkinsian vehicle, carrying the wagon idea "inside," as the artifact moves about or otherwise interacts with the environment. The wagon, with its every movement, sheds wagon-memes like manna in all directions, just waiting to be picked up by passers-by, because the idea of a wagon is embodied in its outline, in its basic design. The analogy between artifacts and biological organisms, both of which carry replicators around inside them, should be clear as the motivation for this argument.

How, then, does a wagon-meme create the next generation of wagon-memes? A wagon may incidentally communicate the *idea* of a wagon, but can it directly transmit to the perceiver the recipe for duplicating the wagon? A person who sees a wagon going by won't necessarily learn how to build one as a result of that experience. Neither can the wagon physically replicate itself. Since wagons don't fabricate wagons, the meme must first get out of this "vehicle" and back into a mind, so that this active, organic agent can create the next wagon. But the meme's problem is easily solved. A wagon can also communicate the idea of motion and the possibility of transportation through its form to anyone who sees it going down the road. That is, the very act of perceiving the wagon alone causes the meme to leap off the wagon and into the perceiving mind. The recipient acquires a memetic copy of the inventor's original idea of efficient transportation of heavy objects via rolling wheels by virtue of absorbing this visual signal. In effect, perception can separate idea from instantiation, or extract the essence of "wagonness," by engaging in a kind of deconstruction of the wagon. The act of looking at something is thus not a case of a stimulus coming from the world into the eye. Instead it is as if the eyes send magical darts at the wagon, which cause the wagon idea to be chipped away from its material context and then sucked back into the perceiver's eyes. The meme then proceeds into the brain, to lodge itself there as a fully functioning duplicate of the idea in the mind of the wagon's inventor. One replication of the wagon-meme is complete.

In this example, memes can be found in signals, minds, and artifacts—a good example of their much-vaunted ability to jump from substrate to substrate, and the value of defining memes abstractly. The wagon idea, as a meme, physically moves from someone's mind out into the world by inspiring the constructive activity of the host individual. Once there, it latches onto an artifact (the wagon) that has been made; then it moves into someone else's mind after a ride on a signal when the wagon is perceived. From its base in the perceiver's mind, it can cause more meme-embodying artifacts to be made. High-fidelity replication is assured by such a procedure because the meme itself—its whole information packet—makes the entire convoluted journey from one mind to the next through this roundabout journey via artifacts and signals.

But this explanation is wrong. There are several problems with it. First, perception is, in fact, the passive reception of a signal by an organism; eyes do not zap X rays or anything else at objects. Any view that requires "active" eyes, ears, or noses is unscientific. Second, this viewpoint introduces a major asymmetry between the way in which memes get into artifacts and the way they get back out of them again. Memes are created at the same moment as the artifact itself, as a kind of side effect of that production process, in the form or function of the artifact. But memes are then extracted from their physical home without harm to the artifact—or indeed any kind of involvement of the artifact at all—through a kind of magical process. A signal making its way through the air (or some other channel) has to bounce off the artifact, acquiring meme-hood in the instant of contact, and then continue on its journey, thus transformed until it is perceived. However, no physical change in the nature of the signal has occurred, so how are we to account for its change in status from mere signal to meme? I don't think there is any coherent way to see how signals work as long as you think of artifacts as standard interactors with replicators inside.

Also note that in biology, genotypes are inert stores of information (DNA molecules, wrapped up tight inside cell nuclei), while organisms, the interactors, are dynamic agents active in the world. In the case of artifacts, we get the reverse: creative, dynamic genotypes (ideas) and static, uninvolved interactors (artifacts) that move only if some external force is applied to them. Isn't there something wrong with this picture?

It's clear from this analysis of wagon-memes that artifactual "interactors" are not like their biological counterparts in several key respects. The replicators (if any) responsible for their construction are not inside the artifacts themselves. Instead the ideas that make new kinds of artifacts possible are located at some remove, in someone's brain. So the memes are not actually housed by the artifact itself, as DNA is held in the nucleus of each cell. Artifacts also have no ability to repair themselves, and so are more liable to degrade without outside assistance than organisms. Further, the mechanism for converting their energy supply into useful form may not be inside them either, like the step-up or step-down transformers at power stations that adjust the voltage for electrical equipment inside households. In sum, artifacts are not independent reproductive systems. Artifacts are only part of a larger system required for making copies of themselves: the activity of a knowledgeable individual or, in a more modern context, a factory working from blueprints and raw materials, must also be added to the mix. Artifacts are not self-contained reproductive systems because the replicators and catalysts that produce them are "off-board" or "outsourced." Artifacts are zombies; if they move, it's because they're under remote control. For all these reasons, we would have to broaden the notion of an interactor significantly to include artifacts. We can't really use Dawkins's original idea, that interactors are "vehicles" for replicators, to describe this class of made things.

Can we legitimately expand the interactor notion to cover artifacts, then? I think the answer is a definite "sometimes." Simple artifacts like hammers are tools, which are useful ornaments for expanding the niche of their users. For example, stick tools make otherwise inaccessible food items (such as underground termites) available to chimps. But tools have influence on the environment only through use of an organism's muscle-power. Tools are just the "extended" traits of those organisms to which they become attached: A stick wielded for defense or for poking at the ground is (literally) an extension of one's arm. For this reason, I suggest that relatively simple artifacts like tools cannot profitably be seen as interactors; they are extended phenotypes. For the same reason, I argued in the previous chapter that signals can't be interactors: They're just too simple.

But some artifacts are truly complex. More complex artifacts became possible after big-brained creatures figured out how to combine tools as means to a more powerful end: machines. Machines have internal structure: multiple, heterogeneous parts amalgamated to serve functions that none of the parts, taken individually, might have succeeded in addressing. Nothing in the animal world is equivalent to a pulley, for example, much less a computer. Some animal-built artifacts can process information—termitaria (the towers of clay that termite colonies build) can regulate the air temperature inside themselves, for instance. But computers are capable of much more, thanks to their modifiable innards. Their insides can take on new states under the command of instructions from elsewhere inside the machine. This is possible only because they have complex internal constitutions, unlike any animal-produced tool.

An important revolution in artifact production came when machines were linked with power sources independent of human musculature, such as steam engines and electric generators. This invention provided such artifacts with the mechanical equivalent of a metabolism: the ability to change internally through the incorporation of energy. Steam made possible factories with greater powers to shape metal and other raw materials. It also allowed artifacts to move on their own: Steam enabled people to step aboard trains for trips to distant destinations. Powered machines are therefore like cells in having not only considerable internal differentiation but also the ability to use (if not exactly acquire) their own source of power. This independence enabled powered artifacts to interact with their environments in new ways; they were no longer just slaves to organisms.

Complex artifacts are therefore good interactors. Given that computers can become hosts to parasitic replicators like viruses, they must work something like organisms, which are similarly susceptible. So perhaps we can legitimately call these "metabolic artifacts"—the ones with their own power sources—interactors. Indeed it's hard to know what else to call them, even though they can't reproduce independently, like biological organisms.

The development of machines occurred only in the past few hundred years and in association with a single species—humans. It must therefore be the case that, at some point in history, some classes of artifact became sufficiently sophisticated that they moved into a new evolutionary category. Since this was a historical development—computers followed long after hammers—an evolutionary process must be responsible. Although these more sophisticated artifacts remained (extended) phenotypes, because they were still the result of human activity, they could also now be called interactors. But since we can't apply the same evolutionary title to all of them, artifacts must be a grab-bag category: Pick an artifact, and sometimes it will play one evolutionary role, sometimes another.

ARTIFACTS AS SIGNAL TEMPLATES

So far, we have concluded that some complicated artifacts, like wagons and rockets, can be interactors. Some artifacts move around under their own power; they can even be "smart," like computers. But what role do memes play in this gradual evolution in artifacts? If these are physical rather than biological interactors, then where are the memes? In particular, what is "inside" complex artifacts, making them "go," if it isn't memes? What allows them to be part of an evolutionary lineage? How can we rightfully say that information passes through artifacts on its way from one mind to another if memes themselves aren't present the whole time?

In fact, nothing need be inside an artifact. Instead interactor-artifacts can be thought of as templates for signals. Before investigating how artifacts act as templates, what they are templates *for* must be clarified.

Signals are patterned streams of particles flowing through a channel. Examples of signals include spoken words, a light beam (stream of photons), or a digital message (a stream of electrons). They can be called "natural" if an organism produces them (as in the sounds of speech) or "technological" if a machine does (as in our example of a photon stream). Artifacts, on the other hand, can be templates for generating signals. Such artifacts consist of a substrate with information inscribed as structures in it or on it (as color, shapes, or patterns on a background). Examples here include tattoos, the famous paintings on the cave walls in Lascaux, France, books, magnetic tape, and DVDs. Artifacts are things that can choke the transmission channel by sitting there inertly, waiting to serve as the template for a signal.

In contact with an artifact, a signal can start to reflect a new pattern, which changes its amplitude and frequency, for example, to "reflect" the fact that it is now carrying information about the nature of the artifact it bumped into. For example, even though Shakespeare has been dead for nearly 400 years, his speeches are still present in current-day brains, having been stored in widely available books. Thus ideas can make their way from one mind to another through the mediation of artifacts, carried along their way, both before and after contact with the artifact, by signals.

This process of communication through artifacts can, but need not, result in the replication of some idea, making it a meme. In most cases, perhaps, the signals that project from an artifact don't lead to the replication of any meme associated with that artifact. As I argued above, seeing a wagon doesn't necessarily lead the perceiver to recreate the wagon inventor's idea in his own mind. People often see things without fully examining and understanding them; they are concerned only with using the artifact to achieve some other goal, such as getting somewhere.

Nevertheless it seems undeniable that at some point in time, the social transmission of memes moved from being mediated by signals to also being mediated by artifacts. But just as signals aren't memes, neither are artifacts. Both are *instigators* for replication, not replicators themselves. Signals, remember, are mere scraps of information, which must be augmented to constitute an idea. There is still much work for the brain to do once the signal has been received, just as cognitive scientists and linguists have argued.

In some cases, then, the progress of information into artifacts and back again through signals was good enough to result in information being replicated between two different brains. This evolutionary development—messages hopping a ride on artifacts in their progression from host to host—is a "natural" outgrowth of an existing capability among memes. As organisms learned to modify their environments in specific ways, messages came to get buried, or embodied, in artifacts, waiting for later release when struck by a signal-bearing medium. Eventually memes learned to exploit this in-place mechanism for their own selfish ends.

This is what makes physical artifacts different from their biological counterparts: the role of signals in the replication of memes. Memes can't exist in both brains and artifacts, any more than they can exist in both brains and signals. Either case violates the single-substrate rule. But when artifacts are seen as signal templates we no longer face the multiple-substrate problem. Memes stay in brains, where they belong. Artifacts join signals as intermediaries to replication, achieved somewhat indirectly through their agency.

This may seem counterintuitive. Think of the Rosetta Stone, for example. How is it possible that memes aren't stored in the surface of that piece of rock? The writing on it unlocked a whole dead civilization, that of the ancient Egyptians. It therefore seems to speak to people with the voices of those long dead, millennia later. For dead Egyptian minds to be able to communicate with us, memes must have been stored in that slab of stone, you might think.

But, in fact, rocks don't talk. Their virtue—and the reason someone took the time to carve similar patterns on that particular stone three times over—is that rocks form a relatively inert and durable substrate for signal templates. There's no meaning—no memes—on them. The Rosetta Stone, despite redirecting photon streams into the environment for thousands of years, was "mute" until the Frenchman Jean-Francois Champollion took seriously the idea that the same text might be written there in different languages—as a kind of translator's table—and worked out the meaning of the Egyptian hieroglyphs from the Greek words.

If memes are not lying dormant in the stone, if they reside only in brains, then how do dead memes miraculously appear again in the brains of people generations later? The link is necessarily indirect, through the artifact, which achieves this miracle of transmission from the dead. This transmission involves a number of steps: The stone functions as a long-term storage device for the written templates that, when "read off," produce a catalytic signal, which then helps in the re-creation of the meme in a brain. But such a pathway can be as efficient as direct transmission of the catalyst from a source brain, if conditions are right. Remember that, in my book, brains are preconditioned to construct memes, given a chance. So there doesn't have to be much of a stimulus, and the stimulus can just as easily be a signal derived from a rock as one projected by a mouth.

How does the artifact production process work now? Using the host organism as an instrument, neuromemes can produce both signals and artifacts. For example, a meme can cause someone to say something or to type a word into computer memory. However, as in this example, it will generally be the case that a meme must decide whether to spur its host organism to produce an artifact *or* a signal. Different motor programs will typically be involved in the production of one or the other. To say something, the mouth must be moved; typing requires fingers to peck away at a keyboard.

So memes have an either/or choice: Either produce a brain spike that escapes into the macroenvironment as a social signal (in the form of photon or phoneme streams), or produce a physical template—an artifact—that can catalyze signals at some later time. It's basically a choice between current or later transmission (and perforce, a decision between replication today or tomorrow).

The direct production of a social signal is easier. Essentially the move is from one dynamic coding system to another—from spike to speech—

whereas the production of an artifact requires the coordination of a complex sequence of host activity, like hacking away at the surface of a stone. The direct, signal-mediated route therefore arose earlier historically. Adding another step in the transmission process—the "layover" on a stone—also introduces greater uncertainty into the memetic life cycle. In particular, replication is delayed compared to the relatively quicker, guaranteed result of a signal. This need to find a way to secure a return on the investment in a stone template also caused this mode of replication to arise only after brain-to-brain signal mediation.

However, artifacts potentially represent an improvement over signals in one respect: Many signals never reach a new host and thus constitute a reproductive dead-end for the memes that produced it. Producing the Rosetta Stone, on the other hand, could potentially harvest a much better return on investment because it sticks around, giving many later signals the opportunity to establish a whole new life for the responsible memes. The all-important evolutionary goal of continuing the memetic lineage may be achieved more readily by the artifact constantly generating signals that reach another brain than host-generated signals would.

So artifacts are much more likely to produce multiple copies of the meme than a signal. This may compensate in some circumstances for the meme's reduced degree of control over the fate of the signal eventually produced from the Rosetta Stone, hundreds of years later. Thus there may be a net selective advantage to artifact production in some circumstances. Robinson Crusoe on his desert island should definitely forgo yelling for help and instead put a message in a bottle, for example.

COMMUNICATIVE ARTIFACTS

My claim is thus that more complex artifacts can work as physical interactors through their ability to serve as templates for signals that transmit memetic information between minds. We can draw further lessons from this new view of artifacts. In particular, we can distinguish two major classes of artifactual interactors.

The primary purpose of a wagon is not to communicate the wagon-meme to those who happen to observe it going by. Its basic function is to supplement human muscle power by moving loads about efficiently.

That is what the lineage of wagon designs gets better at doing over time. Indeed, as wagons get more complex, the mechanism that makes them move may get buried inside other machinery, so that the ability to perceive how they work becomes more difficult. The spread of the meme for "rolling transport" may therefore be hindered, rather than furthered, by technological developments in wheeled vehicles.

The main purpose of books like this one, on the other hand, is to transmit the writer's message. Books exhibit a number of adaptive features designed to facilitate this end: the stark contrast of black type on a white background for easier reading, a clear typeface, a durable substrate—all of which are placed inside a protective cover. As the printed template produces visual signals that someone can interpret, some memes may be replicated in the process. And as improvements in printing have taken place, it has become easier for the messages in books to be disseminated. Developments in the form of books appear to reflect a selective bias toward the more efficient communication of any embodied messages held within their covers.

So successful wagon designs are *not* those that communicate "wagonness" most readily, but those that "communicate" loads more efficiently from here to there. Nevertheless any artifact that functions as an interactor allows signals to reflect off it, which can lead to memes being reproduced; the artifact becomes a memetic interactor in such cases. As this class of interactor-artifacts developed over time, however, some of them came to be explicitly *designed* to serve as signal templates and hence serve better as disseminators of memes. Various classes of artifacts nowadays—like this book—are therefore made to transmit information or to hold messages. Artifacts whose primary function is to serve as a signal template can be called *communicative* artifacts.

After eons of directing hosts to produce signals, memes therefore devised a new strategy for reproduction in new brains. In the first instance of such activity, a meme produced a signal that caused muscular or other activity by the host organism. This activity, in turn, resulted in the production of an artifact, probably through the relatively minor modification of existing environmental materials (as in the case of early hominids bashing at stones to create primitive tools). These tools are the extended phenotypes of those memes, as suggested before. But these artifacts also included two-dimensional templates on their surface, embodied in the form of the artifact itself, from which signals can be "struck off." A symmetrical stone with a sharp edge, unusual in nature, communicates its function by its very shape, telling all who see it: "Use me to make cuts in things!" A signal thus produced, "reflecting" information from that template, can then make its way through the environment, eventually to become a stimulus to a new brain. On making contact with some sensory organ such as an eye or ear, the social signal gets translated back into a neuronal spike train. This train then searches for a location suitable for a replication reaction inside that brain, which acquires the idea that stones can be modified to suit human purposes.

Once the ability to produce artifacts evolved, elaborations of this basic strategy for the social replication of memes naturally arose. The next type of artifact ensuing from this development encoded messages as a patterned structure on the artifact's surface, using the physical substrate as a "background." Examples of this novel type of artifact include paint on rocks, ink on paper, and magnetic charges on celluloid tape. This is an outgrowth to be expected only after the first, simpler route to meme replication had been well established.

A subsequent step in the evolution of interactions between memes and communicative artifacts results in an even less direct process of memetic reproduction. Now the neuromeme produces a signal that induces host behavior which produces an artifact (a machine) that in turn produces a memetic artifact with a signal template. Examples include the printing press, which produces books that can produce photons (visual signals), and the tape recorder, a machine that can encode tapes which play musical sounds (aural signals). This additional complexity in artifacts is worth noting because, from the perspective of the neuromemes, the nature of the memetic copies eventually produced is more uncertain as more steps intervene. So this lengthier process of replication should arise later—as it did, historically speaking. But it's just an elaboration of the original move from signals to artifacts in the first place in the sense that it required a "decision" by memes about which route to use to achieve reproduction.

The advantage of this strategic move away from host muscle power in favor of more powerful machines, and therefore the reason such production processes probably evolved, is that machines can produce many more copies of the communicative artifacts than would otherwise be possible. Just compare the number of popular novels printed for businesspeople on trains or planes to the number of medieval *incunabula* hand-written by monks slaving away in their monastic cells (which can now only be found locked away in museums). The increase in artifact population size (millions of copies in the case of pulp-fiction titles, for example) in turn significantly increases the pool of potential hosts for a given set of memes. The trade-off, from a meme's perspective, is between the greater uncertainty and roundabout replication arising from machine-based artifact production versus the greater potential number of copies that could be produced through this route. The balance must have eventually tipped in favor of using machines to make templates to make signals.

Artifacts that have some control over the production of signals, rather than being passive, inert templates for signals, come next in the evolutionary sequence. Electronic artifacts (tape players, television sets, virtual reality machines) generate signals themselves rather than relying on ambient radiation to be reflected off one of their surfaces. There is thus a "generation gap" between books and television. The advent of electronic artifacts also introduces greater control over signal content, destinations, and distribution, and simultaneously reduces the uncertainty associated with the original production of such media. The potential population of hosts also increases dramatically again. Each of these is obviously of benefit to the memes interacting with such artifacts. Just think of the number of people in the world who know the theme songs from popular television shows, compared to the number that have learned the tunes from an original, locally produced play.

The relationship between a template and a substrate becomes even more elaborate when the substrate wraps in on itself to create an entire microenvironment for signal templates. This makes the artifact a snug home for the templates, to some degree independent of the larger environment with which the artifact itself must cope. The obvious example here is a computer, which houses sophisticated templates for the dispensing of signals: memory banks in the form of two- or three-dimensional arrays of signal templates. Further, we are no longer restricted to static templates; these are active ones: These registers can be readily modified, so that messages can be updated. This is a major step forward—making computer memory more flexible, like human memory—with obvious results. Moreover, a computer's templates are not restricted to storing just one piece of information; many millions of bits of information can coexist inside a single machine. This is a throng of templates that memes (and other types of information) can use.

In this way, artifact-based signal templates have evolved, step by step, from being coded in the shape of an artifact (such as a stone tool) to a pattern on its surface (ink on paper) to a structure internal to the artifact itself (a computer program). This sequence requires an increasingly convoluted production process to make the template. In the first step, both the substrate and the template are produced in the same action (since they are the same thing in a hammer: its external form); later a substrate suitable for more sophisticated templates must be prepared before the code is added. For example, clay tablets must be molded before symbolic shapes (cuneiform) can be pressed into their surface. Even later, with more highly evolved types of artifact, a complex artifice like a computer housing must be created prior to the addition of the template inside.

Also there is an increasingly intimate connection between the host and the template over time. The template moves from being an *ecto*parasite—that is, a parasite clinging to the outside of a host (like the shape of a hammer)—to an *endo*parasite, one living inside the protective womb of the host's body (like computer memory). From the perspective of both the artifact and the template, then, the observed progression is a natural coevolutionary sequence for a parasite and host: one of increasing dependency on each other.

This trend suggests that the relationship between artifacts and memes is one of symbiosis—*both* of the involved species use their association to regulate interactions with the environment. As the degree of dependence between them increases—that is, as the relationship "tightens"—either the usefulness of each partner to the other tends to increase or one becomes progressively more exploitative of the other.

What advantages accrue to memes by progressing through this sequence? Ectoparasitic templates on artifacts are essentially naked and remain exposed to the environment, which means they can be readily degraded. The messages they can produce are also very limited as a result: Only *transcription* of the template structure into signals takes place, typically through reflection off that surface (as when photons are deflected from the shape of a hammer). The form of the artifact can become increasingly complex (progressing, for example, from a two-dimensional surface to one with some relief or from a black-and-white to a color palette). But there is still only a longer string of signals to be read off the surface of the artifact; there is no increase in the sophistication of the signals.

By moving inside the artifact, the template gains not only shelter from environmental forces that would weaken and weather it, but also an opportunity to feed off the artifact's resources, if any, for the creation of more complex structures of its own kind. For example, with the introduction of computers, the template gets both a comfortable microenvironment in which to live and the possibility of developing into a new code, thanks to the provision of a substantial and concentrated energy source and room to grow, with sophisticated, looping execution structures for the development of computer programs. Further, as computers are networked, signals can be transmitted between hosts, resulting even in the replication of programs (template states) along the way, as in the case of computer viruses.

Thus, although artifacts are physical substrates, not organic ones, they still behave in many ways *as if* they were proper biotic hosts for parasites. Since artifacts are not biological things, however, the same kind of relationship cannot be expected to hold in all respects. An organism-as-host-environment is special because it can *adapt* to the presence of the bioparasite. The continued presence of the parasite is therefore dictated not only by its own morphological and physiological characteristics, but by its ability to defend itself against the host's response to its presence. Host and parasite thus become a coevolutionary system.

But a physical host cannot respond adaptively when a parasite decides to manipulate it; the host has no ability to learn new responses; it is not behaviorally plastic (although it may be *made* of plastic). The relationship between an artifact and replicator is always one engaged "at a distance," through the mediation of a template. Although the two may still coevolve, this leaves memes with only indirect, and relatively feeble, control over their artifactual "hosts."

ARTIFACTS AS REPLICATORS

The foregoing does not end the litany of evolutionary roles that artifacts can adopt. An earlier chapter provided a long discussion about computer viruses. They are able to duplicate themselves in electronic forms of memory, and thus constitute an instance of an artifactual replicator, even if they are restricted to manipulating an existing substrate. In effect, this makes them parasitic replicators. However, an even more novel kind of artifact looms on the horizon. Nanites—nano-scaled machines that can presently self-assemble into almost-visible objects and that may perhaps soon be able to self-replicate—are envisaged by nanotechnologists as mechanical replicators that do not rely on hosts. Given the right conditions, they will be able to make use of some form of energy to reproduce themselves from raw materials, independent of any human activity. Therefore such nanites would not be parasitic, they would be true artifactual replicators. Naturally any such replicator would begin its career as a relatively primitive form but then could evolve greater complexity without human interference.

With truly replicating artifacts, we clearly would find ourselves in a new situation, because memes would remain physically independent of such artifacts (back in the brains of those who invented the new replicator), although the replicating artifact and replicating neuronal node could clearly interact with one another—as when the inventive nanotechnologist, under the influence of memes-for-nanites, sets up the conditions necessary for the first nanite to be created. From that point forward, however, we could have dueling replicators: memes competing with the nanites themselves for control over the course of nanite evolution, because it's possible that the interests of physical replicators and biological or cultural ones would not be the same. For example, most nanites today are made of carbon and might seek to use a major source of concentrated carbon—biological organisms—as fodder for their own "bodies."

If some artifacts do cross this final Rubicon, becoming fully independent evolutionary agents, they will have run the gamut of evolutionary roles, from simple phenotype to full-fledged replicators. This would reverse the message from the previous chapter on social memetics. There I argued that signals are "none of the above"—at least neither replicator nor interactor (they are phenotypes, of course). But artifacts, on the contrary, can be "all of the above"—from simple phenotypes to full-fledged replicators. We can only await future developments to tell where the evolution of artifacts will go from here.

The Machining of Culture

So how do memes fit overall into this picture of artifact evolution? I believe the relationship between memes and artifacts has itself evolved as the nature of artifacts has changed.

Artifacts like hand axes and wagons are simple because their function follows directly from their form. In this, they are like prions. Therefore any memes associated with such artifacts will be limited replicators, because their ability to vary and still perform their function must be severely constrained. Wagons with square wheels don't roll and so can't transport anything. Such wagons don't last long, and neither would a meme "wagon with square wheels." Simple artifacts also embody only one idea. In the case of the wagon wheel, we may also have just one meme—the idea that "round objects roll"—embodied in the form of this class of "made round things." This meme is replicated when someone innocent of this idea acquires it through observing the wagon go by.

More complex machines, on the other hand, are capable of performing multiple functions and therefore of being involved in the transmission of multiple memes simultaneously. We could say that the axle connecting the wagon's wheels is an important independent technological innovation and must have required a new meme to have evolved. Putting the platform on top of two sets of wheels connected by axles was a further development of the basic idea of rolling transportation. A wagon may thus be associated with several memes, each embodied in a separate portion of this relatively primitive machine.

Thus artifacts can evolve multiple functions by developing component parts, each of which performs an independent job. They can also become multifunctional by dividing into a background and template, with the template evolving the function of producing multiple signals that convey separate meanings. A major break arises with this physical division because form no longer automatically indicates function in template artifacts. Merely observing an artifact's use no longer communicates the idea behind its creation. When you see a hand ax hacking at something, it's fairly easy to infer that the ax is designed for chopping things up. But if you see someone reading a book, you won't be able to figure out why the reader is laughing or crying unless you can get a glimpse of the page yourself. This is because the template contains information that must be interpreted to have meaning. Black-and-white patches on a piece of paper must be translated from their two-dimensional state to make a three-dimensional impression, a true picture in the mind. Once we leave wagons and such behind, we get to communicative artifacts in which the material object itself becomes in a sense secondary; it remains important only to the extent that it serves as a good support for the template. For this reason, it doesn't much matter if words are written on stone, paper, or a computer screen.

The development of more sophisticated artifacts like nanites does not mean that simple phenotypic artifacts are made obsolete or fade into obscurity. Instead they continue to have their purposes, if highly specific ones. But in the course of modification, an artifact may cross the "border" from one evolutionary category to another—for example, from phenotype to replicator. This kind of major leap in evolutionary "status" is not unheard of in the biological world: Prions are replicating proteins, simple molecules that acquire the ability to replicate via a physical modification of their shape. This single act transforms them straight from genetic phenotypes to independent replicators. Prions don't become interactors first, nor hosts to parasitic replicators; prionic proteins become replicators themselves in one fell swoop. Similarly, particular kinds of carbon, like the soccer-ball-shaped Carbon-60 molecule (called a "buckyball" after the inventor of geodesic structures, Buckminster Fuller), can be induced to assemble themselves into aggregates with regularized shapes such as tubes. Aggregates of C_{60} can then form micromachines, like tiny transistors or molecular rotor wheels, or different-shaped C_{60} aggregates can be combined to produce macroscopic (visible) structures. Such self-assembly is a major step on the way to replication, so C_{60} is potentially on the road to becoming not just another "inert" molecule but a replicator.

A DARWINIAN DUET

Since both memes and artifacts obviously evolve, and also interact, it is fair to say that memes *coevolve* with artifacts. This is certainly the case with communicative artifacts, because the idea that made the artifact-as-substrate—for example, the *idea* of a book (as a two-dimensional surface for encoded symbols)—is different and unconnected in an evolutionary sense from the memes associated with the template—that is, from the *content* of a book (as a particular sequence of symbols). It may also be true that the earlier generation of simple artifacts coevolved with memes. But the idea that gives one wheel some peculiar feature (say, multicolored spokes) may not be memetic at all, even though the basic idea of a wheel-as-round-thing *is* a meme. Thus *aspects* of an artifact may be produced through the replication of a meme, while other aspects of the very same artifact may be non-memetic in nature. This creates a coevolutionary relationship between artifacts and memes of a different sort. In such a case, memes and other ideas combine in the creation of a compound artifact, with the memes being responsible for only certain aspects of its design.

One way in which coevolution between memes and artifacts can occur is through the avenue of use. With practice at hammering, a clever person might find that his particular hammer is poor at some aspect of the job it has been designed to do. Perhaps the hammer's head keeps falling off as it bangs on things. In a few cases, the user may think of a better way to keep the hammerhead on the shaft tightly. This may just involve the hammerer changing his behavior in interaction with the tool, such as using shorter strokes from the elbow rather than swinging the hammer from over the shoulder. Or it may require a design modification to the tool itself: Perhaps the user devises a new means of affixing the head to the shaft. This design innovation may take place through trial and error or through a mental simulation of that trial-and-error process ("What if I wrapped some cord around this bit here or notched the end of the shaft here?"). Either way, the process of design modification—or artifact mutation—can be thought of as being Darwinian in the sense of involving selection among random mutations.

In fact, technological innovation can be described as a Darwinian process of descent with modification *operating in parallel* between ideas and their implementation—that is, in both the technical knowledge base of artifact-making or artifact-using and in the artifacts themselves. The inherent dynamism of technology can then be seen as due to the *interaction* of these coevolving processes. Inventors feed off existing artifacts as somehow faulty models, experiment with many options for their modification, and then create improved ones, which are descendants in a lineage of similar artifacts. These new exemplars of what an artifact can be in turn stimulate novel thoughts about how the next generation of improvements can be wrought on this raw material, creating the next generation of ideas related to such artifacts. It is this back-and-forth between thoughts and their manifestations in the world that makes technology particularly dynamic.

More formally, here's how I propose technological change occurs. New artifact types are created through *invention,* or random mutations in form. This starts a new evolutionary lineage. *Innovations,* on the other hand, are modifications of these inventions through the recombination of parts. Add a tool to a tool, and you get a machine. Combine a machine with another machine, and you get a more complex machine. Such single-step recombinations between artifact lineages ("combinatorial chemistry") can rapidly produce complexity. Over time, an artifact lineage can therefore show evidence of cumulative selection (variation with descent) and manifest an adaptive design with greater and greater power to transform the environment.

Simultaneously there is a process of mental evolution in know-how that can also be described as Darwinian. The production of new artifacts is first simulated in the mind, perhaps through an unconscious process, in which various amendments to the artifact are "tried out" for their competitive advantages before the first thingamabob is ever physically attached to a thingamajig. This process of mental trial and error may recur at the level of research and development within a firm, in which potential emendations to a product are selectively tried out by a team, perhaps through "brainstorming," rather than being simulated within one individual's mind. Yet a third level of selection among artifactual forms can occur in the marketplace, as when one brand of toothpaste is chosen over another by a consumer. In any case, innovation can be described as a Darwinian process of iterative modification, not a directed one of "intentional design."

This means that to explain cultural change we must deal with the evolution of both technology *and* cultural knowledge. It's the interaction of these two Darwinian processes—of descent with modification in the body of knowledge available to a society relevant to the production of some artifact, as well as the embodied modifications in the artifact itself—that must be modeled for a complete understanding of techno-logical evolution. And only a unified Darwinism, using a single set of principles for both memes and artifacts, can place such disparate phe-nomena into a common analytical framework.

Who's in Control Now?

I have established that there is a parallel between evolution in memes and evolution in artifacts: They are two Darwinian processes working in tan-dem to achieve the phenomenal rates of cultural change we see in the world around us. How, then, do these two processes actually interact? Both memes and artifacts have an independent evolutionary dynamic associated with them. But if they are mutually involved in cultural change, then there must be *co*evolutionary dynamics between them as well, caused by their relationships with each other. This complex rela-tionship can really be pursued in detail only by modeling the process. The formalism devised by cultural selectionism (derived from popula-tion genetics) is admirably capable of dealing with this modeling prob-lem, after suitable modification. But, as noted above, the nature of the interaction between a meme and an artifact also differs depending on what kind of artifact is being considered—is it a simple phenotype, an interactor, or a replicator? So an entire class of models will have to be developed to deal with the varying kinds of relationships that memes and artifacts can have together. This simply hasn't been done yet.

Why are we left hanging here, at this somewhat unsatisfactory point? Why has there been little theoretical attention to developments in technology from memeticists? One reason is that the relationship between memes and their hosts—including (by extension) artifacts—is symbiotic: What is good for one is generally good for the other. Thus there are relatively few areas in which interests conflict: What benefits hosts in their battle to stay alive and prosper has also proven to aid the career of memes as mental parasites. And over a substantial range of phenomena to be explained, there has been considerable overlap between the expectations of standard social science and the memetic kind of host-parasite approach.

Still there are bound to be situations in which conflict arises between memes and artifacts, especially when (and if) they become full-fledged replicators themselves. Who's in control of technological evolution? For much of their history, both memes and artifacts have been relatively impoverished replicators, with little ability to affect the behavior of the other. This again implies that there are few domains in which their unique interests can express themselves or make themselves felt in the world.

At the same time, the story of culture turns out to hinge much more significantly on artifacts than on memes. Memes began life as replicators and have achieved certain advances, especially through their interactions with signals and artifacts. But over the same course of time, artifacts have rapidly progressed from simple phenotypes to replicators, thanks to their involvement with ideas. This rapid evolutionary succession makes it even more imperative that we understand precisely how fast-changing artifacts evolve.

The lack of attention paid to artifacts has meant that there has been a tendency to make too much of memes, to think that they play too great a role in human culture. It's a natural mistake. If memes are like genes, which drive biological evolution, then surely memes must play the same central role in cultural evolution. They must be the agents that make us fully human, unique in all the world. It has therefore been necessary here to cut memes down to size by showing that even if they are *a* leading player in the tale, the drama turns out to be a love story. And the actor playing opposite memes in this romance turns out to be artifacts. Memes must dance a tango with technology to make the curtain come down after the third act. This "demotion" of memes to shared billing should cure meme enthusiasts from making ridiculous claims about the ability of memes to cause everything seen during human evolution, from big brains to language to consciousness. In fact, there is more to culture—and human psychology—than the replication of memes. This extra bit is artifacts, and all that goes with them.

Certainly it will be fascinating to live in a world with replicating proteins, replicating organic and mechanical memories, and replicating miniature machines. But to understand how replicating artifacts might come into being, we need to figure out how novel ideas become novel objects—and how these inventions then become the foundation for a successful line of innovations—if we are to dodge the more frightening scenarios being put forward about the new millennium (like man-eating nanites). The concepts put forward here are only a first step in this direction.

RETHINKING CULTURE

What even these tentative steps make clear, however, is that the viewpoint developed here has major implications for our notion of culture. As I observed earlier, culture is most commonly seen these days as the information in individuals' heads that they have acquired from others. I want to take issue here with this formulation. The definition is too narrow.

Let me first note, however, that this cognitivist definition of culture has many virtues: It's clear, simple, and comprehensible. You know what to count in order to take stock of culture: ideas in people's heads. It also automatically identifies a mechanism—social learning—as crucial to the dissemination and replication of culture. One problem, at the empirical level, is that it may be difficult to tally up the frequency of ideas in some group of people. It's easier to do that with readily observable phenomena like behavior or artifacts, but cognitivists define the object of analysis as something inside the mind. This difficulty is why some memeticists suggest we make the observably duplicated elements of culture, like behavioral patterns, the replicators—in effect, reversing the perspective we have adopted here.

The connection between culture and cognition is also powerfully supported by the circumstance that human beings—the planet's sole possessors of full-blown culture—are those creatures with brain sizes that seem disproportionately large for their body size. Surely these two singularities are linked, the reasoning seems to go. The dominant theory of human brain evolution—the "Machiavellian intelligence hypothesis"— suggests that it's the complexities of social life that have caused the ballooning of our brains, which then allowed cultural accretions to creep in. Calling the need for impressive social cognitive abilities the cause of big-brain evolution seems to work: Human groups are considerably larger than those of other primates, so remembering more names and faces may have spurred the growth of cortex in our species. And other creatures with big brains—such as apes, elephants, and dolphins—also seem to have flexible, complex social lives.

Within the currently dominant cognitivist camp, culture is generally defined as either widely shared traits—a view held by most cultural anthropologists—or as socially learned traits, a conviction common among cultural selectionists. These definitions are not identical. For example, some socially learned beliefs can be minority values and so still would be called cultural by cultural selectionists, but not by those who emphasize sharedness.

There are also problems with the "widely shared" criterion. First, what is rare today can be universal tomorrow, thanks to the social diffusion of information. At what degree of commonality does information become cultural? Second, some beliefs may be shared as a result of similar but independent experiences, rather than through the social transmission of the idea. So for the cultural consensualists, the cultural and the communicated do not overlap completely. But this breaks the link between culture and society, which most social scientists would want to keep.

I therefore argue that the criterion of social learning is more crucial than the actual commonality of beliefs in defining what is cultural. The implicit objective of recognizing culture is surely to distinguish it from biological inheritance, including the information that evolutionary psychology suggests has been placed in the mind by natural selection in the form of specialized abilities for information processing. So it's best to acknowledge this explicitly by including the acquisition mechanism in the definition of culture. The true criterion of what is cultural underlying the cognitive revolution is how such information is acquired: through social learning, rather than genetic or mental inheritance. Culture is something individuals acquire during their life spans through social living.

But if you "follow the information" as it loops between heads, often through various artifactual incarnations, it becomes clear that you can't divorce cognition from technology. You can't really separate the production of communicative artifacts on the factory floor from their role in the dissemination of information. It's the state of the art in communication technology that determines what artifacts can do as manipulators and purveyors of information. What is used today as a "mute" piece of inert technology can be fit into a communicative artifact tomorrow, thanks to the ease with which inorganic materials can be recombined. So recognizing that the cultural knowledge in people's heads has origins in interactions with artifacts necessitates adding technology as a dimension of the culture concept.

But perhaps we can divide "social" learning into two categories: information learned directly from people, and that acquired from artifacts. This would keep the original mode of social learning as cultural, while relegating the other to "technological learning." However, such a division is unlikely to work because much of the information in artifacts has been put there by people with the express purpose of communicating it to others. Artifacts are just intermediaries. The routes through which information circulates simply involve too many interchanges between people and artifacts, at least in modern societies, to keep the distinction between direct and indirect transmission clear.

Recognition of this mediation has been delayed, I think, by the fact that the discipline traditionally designated as official guardian of the culture concept—anthropology—has been preoccupied until recently with traditional societies. In these groups, technology is highly limited, especially with respect to communication, which remains almost exclusively face to face. But this ancient social context is rapidly disappearing. For the culture concept to remain relevant, it must adapt or die.

Luckily there *are* alternative evolutionary approaches that do not see culture in strictly mentalist terms. These alternatives are actually more consistent with the general objective to explain what is in people's heads. In fact, change in the pool of non-genetic information in a population is better described, especially in modern societies, by paying attention to the sources of that information. And to the fact that much of this information is stored, in various kinds of artifacts, in the environment.

Somewhat ironically, recognition of the fact that culture is partly environmental reasserts the uniqueness of the human species. As it has become clear that animals can do most of what we used to think was specific to humans—like make tools or use symbolic communication—the so-called "cultural adaptation" that separated humanity from the rest of creation has tended to dwindle in significance. Chimpanzees are now well known for having a variety of cultural traditions, or patterns of behavior passed from generation to generation via social learning. They even have group-specific techniques for using artifacts. For example, two chimpanzee populations living in similar forest environments in nearby African countries exhibit distinct styles of the same behavior. Tai Forest chimpanzees from the Ivory Coast dip a stick into a termite nest and then lick the termites off with their tongue. In contrast, Gombe chimpanzees in Tanzania use a more efficient technique: They use their free hand to strip termites off the stick-probe and then shove the mass of insects into their mouths.

What is less well known is the fact that such behavioral traditions also exist in "lower animals" like the humble guppy, in whom arbitrary and maladaptive social customs for finding food can be maintained through social learning. The rudiments of culture thus exist in a range of species, but the number of traditions in any one of them is few and their speed of change slow compared to the human case. How can this variation in the "strength" of culture between species be explained? In particular, what *is* unique about human culture?

The argument now—and one that will stand the test of time—is that what distinguishes humans is *cumulative* culture. Human culture includes traditions that build on their earlier accomplishments to reach more sophisticated culminations, either in thought or in material form. Thus the key to human cultural evolution is that a new "ratchet" effect arose, in which cultural advances were built upon progressively in a way not seen in the social traditions of other animals. Human cognition would thenceforth become increasingly complex and differentiated, eventually achieving modern levels of sophistication without the need for further biological change.

Of course, calling this process "ratcheting" merely gives a name to what we already surmised was important. What we really need is a mechanism to drive the ratchet round and round. What really separates human culture from the culture of other intelligent creatures?

A prominent point of view suggests that cumulative culture is possible only if people can learn lots of new things before they die, not just the few new tricks to which monkeys and apes are restricted. This limitation consigns monkey culture to the continual reinvention of simple novelties (like the famous invention of potato-washing by a Japanese macaque) and ape culture to a relatively small number of cultural traditions. Fully robust culture is presumably restricted to species with well-developed abilities to imitate others in the social group. Otherwise individuals spend their lifetimes reacquiring the cultural basics of their parents, and no accumulation is possible. General intelligence and a rich social life are not enough.

According to this argument, the crucial human adaptation is the ability to understand others as intentional agents like oneself. This psychological change enables people, and only people, to truly imitate others: to copy not just what someone else does but what they *intended* to do (even if they don't, because of some fumbling, actually achieve their goal in practice). This facility is the only means to transmit information socially with any reliability, because it permits learners to ignore errors in performance by teachers.

An implication of this line of argument is that the difference between cultural learning (which only humans do) and social learning is imitation. The problem here is that the experts are still divided about whether we should credit apes with the ability to imitate. What chimpanzees, for example, appear to learn from observing tool use in others is that certain kinds of effects on the environment can be produced; they don't learn how to manipulate the tool itself from seeing their fellows in action. In other words, apes can't "ape" one another. They learn from others through what is called "social priming": Their attention and exploratory trial-and-error learning is directed in part by what they see others do. Apes can learn what things in the environment to pay attention to, and what constitutes a resource or a danger, by observing their compatriots. But the observing animal is left to draw his or her own inferences about how to make use of that knowledge and, in particular, must learn to make use of resources like tools through personal experience. Apes thus cannot reproduce the action patterns or learn strategies or goals by observing others. Hence insight—the discovery of a new way to extract nuts, fish for termites, or reach inaccessible food—cannot be directly transmitted to other members of the group. According to this view, each generation is condemned to recreating the cultural techniques and knowledge of its parents because of its own psychological inadequacies.

This small difference in psychology between humans and apes is then supposed to blossom into big differences at the social level, especially over time. The result of our ability to share knowledge through true imitation, iterated over time, is the ratcheting of the pool of cultural knowledge toward greater complexity.

This is a highly plausible account of what separates us from the rest of creation. But the real cause of cumulative culture may be the ability to share knowledge *across generations* through the manufacture of artifacts.

In my view, culture can acquire complexity through a *material* history. Culture thus has a body as well as a mind; it's just that the body of culture is made of materials like metal and plastic rather than flesh and blood. Our intelligence is both molded and augmented by our environment, and our environment in turn has been modified by the intelligence of our parents. We are educated in part by our own artifacts. The result is that we live in, and have become increasingly dependent on, our constructed environment. Taken together, genes plus memes and artifacts create a life-form of increased complexity, similar to symbiotic relationships in the biological realm such as lichen (the combination of a fungus with algae). Human culture is a phenomenon that emerges from the interactions between humans, memes, and their constructions.

Taking cognition out of the brain/body and putting it in the environment has a number of advantages: You don't have to carry it around with you; it becomes a public good that everyone can use without using it up; it can endure after your death and so assist your descendants; and it enables you to engage in complex behaviors that would otherwise exceed your brain's "bandwidth" (temporal information-processing capacity). Certainly the team of one computer plus one human can solve problems that neither in isolation could attempt. Further, by making knowledge public, you can begin to engage in the *social management* of information. Some "intelligence" can be distributed through the organization of the social group itself and so can compensate for some lack of processing ability at the level of individuals. That is, certain kinds of social structures are better able than others to actually *process* information. These organizations can themselves produce more optimal solutions to group problems (such as the distribution of food) and then get these solutions into the right hands.

Culture, in this view, is not just the information each generation acquires through direct teaching and sophisticated social learning. There is a built environment to account for, with its own causal role to play. For the truly exorbitant increase in cultural sophistication characteristic of the modern age, the development of extra-brain storage capacities seems to be required.

Making artifacts also depends on having precise manipulators: something like hands, for making things. Dolphins don't have technology, despite their big brains, because they don't have fingers. Presumably it takes some intelligence to produce information worthy of being off-loaded into artifacts, but it also requires the ability to do so effectively. So dolphins, despite all the cognitive prerequisites, have no technology to speak of, for lack of dexterous appendages.

What do apes lack then to achieve cumulative culture? They have hands with opposable thumbs and big brains just like us, yet they exhibit only a few relatively modest cultural traditions to show for it. Further, if we have memes, then surely so do apes (and dolphins): Their brains are big enough. So the activities of this kind of replicator can't be the crucial factor distinguishing the other big-brained creatures from humans.

What is missing, quite obviously, from the forested surroundings of apes is a cache of complex artifacts. This deficit may have a psychological cause, such as an inability to imitate one another's behavior or to engage in the multistep planning associated with the construction of complex artifacts. Alternatively apes may simply never have lived in circumstances competitive enough (with other species) or severe enough (in terms of survival) to have made invention a necessity. For whatever reason, the real correlation we can observe in the world today is not between big brains and cumulative culture (because apes and dolphins have one but not the other), but rather between sophisticated artifacts and cumulative culture. This combination is found only in humans.

So "big culture" is not necessarily a function of big brains per se but rather of the ability to produce complex artifacts. This is not to deny that cumulative cultural traditions depend on a reasonable amount of intelligence as a precondition. They do. Artifacts are, after all, physical objects made through effort and concentration. They need complex planning to produce. But cognition alone is not enough. At some point, there simply isn't any more one animal can learn within its finite life span, even from other clever animals. Some of the accumulated wisdom of the species has to be shunted into the environment for it to endure. It's the fact that the artifacts are there and can be made use of that influences subsequent cultural developments.

Humans thus engage in a distinctive way of life: We live in a "culture niche." This term acknowledges that the ability to share knowledge for coordinating human social activities is important. But plain, old-fashioned social learning may be enough to accomplish this sharing. Cumulative culture need not depend on sophisticated, poorly understood skills such as imitation. What I mean to emphasize with the term "culture niche" is that our way of life is also defined by material adaptations: our physically constructed environment as a storehouse of cultural information.

Culture must therefore be considered information circulating in a population through social communication, which may be mediated. Since mediation can involve not just signals but artifacts—information stuck in the communication channel—not all culture need be in peoples' heads. Some will reside in the environment, in these artifacts. The mediation of communication introduces a new complexity to the transmission of cultural information and hence to cultural dynamics, which now depend on innovations in the means of storing information in the environment—that is, on the evolution of artifacts. This makes the culture concept more messy than a purely cognitive appraisal. But that's the price we have to pay to "follow the information," which is the main job of an evolutionary account, because that is where the dynamics are determined. And to repeat what has become, in effect, a mantra in this book: If the dynamics are different, then you have to recognize that the underlying mechanisms are novel. Indeed, in evolutionary models, cultural dynamics *are* different if you admit artifacts into the definition of culture. Having all those signal-producing templates lying about makes a difference to how fast and how far a species can evolve.

In particular, it's becoming more and more clear that the increasing complexity of twenty-first-century Western culture cannot be occurring strictly in response to what individual people are learning from the other people around them. Where exactly, for instance, is the information collected by the much-vaunted Human Genome Project? The billions of As, Gs, Cs, and Ts are certainly not in anyone's head—and arguably not in any *collection* of people's heads. Instead it is an extremely large database stored on centralized computer hardware, connected to the Web. Much of what we "know" nowadays is like this: information warehoused only in quite un-fleshy human institutions. To account for the rate and direction of change in the cultural knowledge of "modern" societies, we must look at these technological information repositories and how people interact with them.

The standard cognitivist conception therefore ignores an important part of culture—the evolution of artifacts. Artifacts are inherited, which means they fit in with the general spirit of the cultural selectionists; they also influence human psychology and so should be important to those with a mentalist bent, like evolutionary psychologists. Further, they figured significantly in the earliest definitions of culture, and so including them would represent a return to an earlier, perhaps "golden" age of omnibus definitions, in which everything from harpoon heads to religious symbols was thrown into the category of culture. We must include in our definition of culture not only what we learn socially, but also the implicit knowledge we acquire or simply make use of that derives from artifacts, perhaps placed there by earlier generations. Such artifacts range from skyscrapers to cars to "personal organizers" to the piece of string you tie on your finger to remind yourself to buy some milk on the way home. These bits of "crystallized intelligence" influence cultural activity because they serve as tools for thought and behavior. So culture is not just what you learn from others; it's not just the pooled information in a networked group of heads. Much of what we rely on, what makes us modern humans, is "out there"—in the environment, in the seemingly menial stuff we use every day to get things done.

This perspective suggests that human culture really *is* something remarkable. After all, we produce more sophisticated artifacts than any other species, by a long shot. We will almost certainly underestimate the power of culture and mistake our own place in the world, if we do not take into account the reality of our highly reconstructed environments—the cultural niche—into which we have put much of our "intelligence." And some of the "intelligence" required to maintain and reproduce these complex artifacts may be selfish memes at work in the world.

Chapter Ten

RETHINKING REPLICATION

Repetition is the only form of permanence that Nature can achieve.

— George Santayana

We have had to expend a lot of effort to reach the point where real empirical work can begin on memes. In particular, I have been anxious to establish just how replication could occur in the context of social communication—and in a way consistent with what we know about the process of replication in general. Particularly important was the discussion of other known replicators, keeping in mind the principles of Replicator Theory. Comparative investigations have often proven fruitful and proved so here. This investigation allowed us to infer new propositions about the general nature of replication.

First, any mechanism achieving such a precise, rather magical result is likely to be restricted to one kind of physical substrate. This is because many precise conditions are involved in the replication of material things, and all replicators—memes included—are material things.

Second, defining similarity was shown to be crucial to determining which replicators can be classed together. For this purpose, we inferred that a Same Influence Rule is what is relevant to evolution. This condition is satisfied if two replicators have the same job with the same overall consequences in some larger context. A class of replicators can thus be identified that are in potential competition, thanks to sharing a similar physical form and role. They use the same strategy for combating selection. Even though they may vary in certain respects physically, these aspects are irrelevant to their likelihood of success, so these replicators can be tagged with the same label. Genes are defined, in effect, by having the same influence on their surroundings. They don't have to be exactly the same physical sequence of bases. For example, there can be changes in the third base defining an amino acid without affecting the protein produced by that molecule. This is all that is relevant in many regimes of selection, including "survival of the fittest." In prions too, sections of their structure are irrelevant to their ability to infect other molecules. And certainly the "same" comp-virus can have different physical realizations in different computers, depending on the specifics of the local memory "ecology." For example, they need not be in the same location on the hard disk of two different computers but still are called upon by the operating system under the relevant circumstances and are copied into other locations just fine. A comp-virus program need not even be stored in the same series of physically contiguous locations in memory. Just as a gene might include an intron, or non-coding region, so too can a comp-virus, like any program, be stored in fragmented form, as long as the pointers to the fragments all work properly. The intervening memory locations are not part of the functional program but are nevertheless there in physical terms, just like the intron in the case of DNA. Similarly we can expect there will be irrelevancies in the physical instantiation of a meme. Some aspects of the neuronal network will have no impact on their functioning as members of the information-processing cooperative. So for all of these replicators, there are physical aspects that must be the same, and others that do not matter for performance of the crucial evolutionary roles they must play as replicators. This is all the Same Influence Rule requires.

A third major insight into replication comes from recognizing that there are a number of obvious commonalities in the ways the non-genetic replicators we looked at work. This is, I suggest, because they are all members of what I believe is a newly identified class of replicators, which I call super-parasitic. The super-parasitic class of replicators includes (but may not be limited to) prions, computer viruses, and memes. These replicators share one feature with "normal" parasitic replicators like viruses: They cannot replicate on their own. Instead they rely on machinery evolved to benefit *other* replicators (like DNA) to duplicate them. This principle can be iterated. For example, in the biological world we have viroids, which are parasites on parasites (bioviruses). They are even more primitive bits of matter that co-opt the mechanisms developed by the virus which they depend on to help *them* replicate.

But *super*-parasitic replicators not only need such machinery; they also depend on another evolutionary agent to make their substrates for them in the first place. They are like hermit crabs: They turn the empty shells left behind by another species into their own survival machine and main abode. Super-parasitic replicators essentially adopt substrates that are the prior work of other replicators. What is more, an *interactor* for the other replicator is turned into a *replicative* substrate by these super-parasites. For example, prions turn an ordinary protein into an infectious agent for their own reproduction.

From the perspective of Replicator Theory, these replicators engage in a kind of third-order replication process, which determines how fast their populations can grow. First, the growth rate of a super-parasitic replicator (like a meme) is a function of the rate at which the replicators responsible for its substrate (in this case, genes) are produced. Second, memes must also wait for these newly made genes to produce organisms, which the super-parasitic memes then draft for their own purposes. On top of these two constraints comes the replicator's own ability to convert this interactor (in particular, the brain tissue of these organisms) into a replicator like itself. This too can occur with greater or lesser speed, depending on what kind of replicator it is. So any increase in the population of a replicator like memes comes only at the end of a long chain of preconditions; it's a highly *derivative* process.

All of the replicators in this class of super-parasites are also similar in another way: They all use the *conversion* of an existing substrate to an infectious conformation as their primary mechanism for replicating themselves. This follows naturally from their dependence on the substrates of others: Without the ability to build their own homes, they are reduced to renting or buying an existing property and then converting it to suit their needs (often with the guarantee that they can convert it back to its original state once they decide to sell or move on). This grants them greater flexibility in the ways in which they can arise, since a bit of matter can be rearranged into the proper conformation through a number of routes but can often only be made in the first place by one method. Some of these routes are quite different from the single method characteristic of DNA replication. The "routes to replication" through conversion include:

— *Mutation* (or spontaneous conversion)
— *Transmission of signals* through protected channels to new hosts, which catalyze conversion, or
— *"Natural" bias* (a structural tendency to adopt the relevant conformation)

Comp-viruses illustrate these causally distinct avenues of replication. First, *mutation* can occur through file corruption, perhaps due to a power surge in the computer. The second route, *transmission,* can happen through instruction messages being sent to other nodes in a computer network, which results in switched register values in the new host computer's memory banks. Third, a comp-virus can be acquired if a user happens to copy a host program already infected by the viral code, which can happen when an infected software program gets downloaded over the Net or gets shipped out on "hard" media from a factory. In this case, some sectors of the user's computer memory will automatically get "tuned into" states that define a comp-virus by previous computer activity (such as file copying). This constitutes an example of conversion through a *natural bias.*

Since memes are our primary focus, let me also explain how analogous processes result in memetic replication as well. First, a meme could arise through *natural bias* if the "natural" state of a node in a host's neuronal network supports a particular infectious state. For example, during a brain's development, some bit of memory may acquire a certain configuration that ensures it will return to that particular state, even after many kinds of disturbance. If this state happens to be an infectious one (capable of generating a spike that converts other nodes), then this node's natural bias is toward a memetic state. The tendency to adopt a memetic state could be due to genetic causes manifesting themselves during the growth and maturation of the brain, which is consistent with the notion from evolutionary psychology that there is significant innate structuring of the brain. Such nodes might represent fundamental cognitive distinctions such as the conceptual primitives of "over" and "under" or "inside" and "outside," which can be applied to a range of cognitive phenomena.

Language suggests such primitives exist through its wide application of these terms to things besides relationships between physical objects—like "*under* the gun" and "*inside* knowledge."

Second, a neuronal node can be stimulated to adopt a memetic state as a consequence of processing some external stimulus, received in the form of an action potential or spike. Only this process requires the involvement of the hosts' social environment and an act of information *transmission,* such as reading a book or listening to someone talk. This works via the mechanism of signal instigation outlined in the chapter on social memetics.

Like the other parasitic replicators, new memes can also arise through spontaneous *mutation* as a result of internal mental activity. The difference from the case of transmission is that the original stimulus is now internal to the brain, rather than coming from the social environment of the host. Via its connections to other neurons, a memetic substrate may suddenly be stimulated to adopt a new configuration that can replicate itself. In effect, incidental activity by related neurons connected to the node flip it into a new state, to which it's then likely to return.

Memes (along with the other super-parasitic replicators) can thus arise in various ways, whereas genes are restricted to one. This takes us some distance from the traditional view of memes as the necessary product of social learning, particularly imitation. The upshot is that memes should exhibit altogether different kinds of evolutionary dynamics than traditional models of social learning would suggest. In particular, elaborate regimes of evolutionary change in memes can occur within individual brains, with a number of implications detailed in earlier chapters.

It's also true that all of the super-parasitic replicators duplicate themselves at linear rates: They are "Malthusian replicators" that undergo selection such that only the fittest variants survive in their host populations over time. They share this characteristic with genes, and so coevolution with genes is likely to be a true power struggle. Who's in control when there is conflict between super-parasitic replicators and genes is not obvious but will have to be determined on a case-by-case basis. The outcome will often come down to a footrace, with the winner being the replicator that can replicate faster (assuming they can stay on course—major differences in mutation rates can spoil the chances of fast-moving memes, for instance).

This is a fascinating result that has immense implications for a general understanding of the evolutionary process going on about us, in this modern world of replicating genes, artifacts, and memes. Outbreaks of both prion-based diseases and new computer viruses are headline news, so these super-parasites are not mere curiosities that have no bearing on our everyday lives. These two replicators are forces to be reckoned with. The same may be true of memes: Our worldview will be substantially altered if they are proven to be (at least partially) responsible for cultural evolution.

As with any other form of parasite, these three replicators can largely be distinguished by their life cycles. They are particularly identified by the substrate on which they primarily depend, the nature of their relationship with that definitive host, and the kinds of interactive products they make. In comparing the super-parasitic replicators in this way, this last quality becomes particularly interesting, because it's clear that both of the super-molecular super-parasites use signals as intermediaries. Memes generate both inter-neuronal and inter-personal signals. Like memes, computer viruses don't physically move from one host to another through the air or through electric cables. They use signals sent between memory locations, typically with the aid of operating system instructions that the virus must be able to call on for assistance. Thus memes are most similar to comp-viruses, because they are both super-molecular and both use signals as instigators. Both memes and computer viruses are also defined in electrochemical terms, as states of potential change on electronic, magnetic, or cellular substrates. The primary difference between them is that memes use a biological substrate provided by genes, instead of an artifactual one produced by people.

There are also some remarkable parallels in the *evolution* of comp-viruses and memes, as we have analyzed them. Both have gone through three major phases, in which new complications to their life cycles have been introduced. DNA has gone through a long history of additional complexities as well, in terms of the development of more and more sophisticated interactors. But as parasitic replicators, this avenue of evolutionary elaboration is not available to comp-viruses and memes. Their development didn't come in terms of spawning more complex interactors; instead they have gone from relatively simple reproductive cycles to ones with additional stages appended to them. (Complex life cycles are common in bio-viruses as well.) Both of our replicators began replicating only within single hosts (either machine or organic) but then developed the ability to distribute themselves among hosts in a second phase of evolution. This ability started with a rather primitive mechanism for jumping the gap between hosts. (The mechanisms were different, of course: Comp-viruses piggybacked on intermediate disk-hosts, while memes used signals as instigators.) But then this life cycle became further elaborated into a more sophisticated, flexible mechanism (direct electronic transfer for comp-viruses, the use of artifact way stations in memes). This commonality of trajectory suggests it may be a "natural" one for super-molecular, super-parasitic replicators. Apparently signal-*assisted* replication becomes signal-*mediated* replication if given a chance.

This analysis also suggests there is a relationship between a replicator being super-molecular and its ability to use signals for the mediation of replication. Why should this be so? Because with an additional dimension to play with (the interaction between molecules), the tight constraints imposed on a replicator of adopting but a single physical conformation are loosened. This appears to lead to the use of one or more of these molecules as a signal. This in turn leads to some flexibility in the kinds of substrates a replicator can adopt (as in the example of comp-viruses moving between magnetic and optical memory thanks to the common use of the same binary code for signals).

I have also argued that the notion of a replication reaction, derived from Replicator Theory, can be applied broadly to all instances of replication—even to those replicators (like memes) that are super-molecular in scale. I believe it's useful to think in these terms because it reminds us that the replication process is a physical event located in space and time. The replication reaction is that nexus where information transfer and duplication take place. It's the linking point in the chain of events defining an evolutionary lineage. Replication at any scale requires an input of energy and produces something that wasn't present before: a second instance of the replicator. So these events are reactions in the traditional chemical sense. In the case of signal-mediated replication, everything is still required to be in the local situation from which the copy springs into existence, except that the source is only present—by proxy, as it were—

in the form of the signal. Contact between the source and the copy is what is lost in such cases. (This is what necessitates the need to divorce the replicator notion from that of an interactor, although the one is not always carried around inside the other, as Dawkins thought.) Information transfer is achieved, despite the inability to "read" that information directly off the physical source, through the signal carrying this information to the site where the copy is produced. That is, in fact, what the signal is evolved to do.

Biology versus Culture

Our comparative examination of these replicators also provides us with a clear indicator of just what separates biological from cultural replication, for these are two different arenas in which evolution can take place. At first glance, it might appear that the distinguishing feature of biological replication is physical overlap between the parental and offspring replicators. For example, in the defining case of DNA replication, one strand of each copy's double helix is donated by the source molecule to its offspring. In effect, each of the progeny molecules shares a half of the "mother" double helix. So the source and copy share some physical material, inherited from the previous generation of DNA. On the other hand, think of a truly cultural type of replication mediated by a copying machine: The new piece of paper coming out of the machine on the left contains no part of the original, which is regurgitated off to the right. There is no material contact between the copied piece of paper and the original: An intermediate step in the copying process holds the relevant information electrostatically on the copier's rotating drum. So this is a reproduction process in which there is a *physical* link between the original and the copy—the original must be there on the drum or nothing happens, a blank page comes out—but no *material* overlap between the original sheet of paper and the copy.

However, prions, which are arguably biological in nature, also replicate without material overlap between the source and the copy. After contact, the source prion goes on its merry way physically unchanged, while the copy is simply a molecule that has changed its shape, but that existed prior to the interaction. It isn't created through a process in which the original prion sacrificed some part of itself to its "offspring." No exchange of atoms has taken place; there is physical *contact,* but not overlap. The prion case seems to preclude material overlap as a clear divider between the biological and cultural realms.

What seems to more effectively distinguish cultural from biological replication is the use of signals as mediators of the replication process. This characteristic holds for both comp-viruses and memes, but not prions (or DNA). Such a distinction is quite interesting because it suggests that culture really is intimately related to the process of communication (as noted above), while biological evolution is not. It's true that biological *interactors* such as cells and organisms have developed the ability to communicate as a mechanism that facilitates their survival and reproduction. It's also true that messenger RNA can be thought of as carrying information from nuclear DNA to the cytoplasm, where proteins are synthesized. This could be considered a kind of mediation or message-passing during biological replication. But RNA is *not* a signal: It's a static molecule, not a wave-like pattern of molecules. So even if we admit this generous interpretation of mediation, it still isn't mediation *per se* that distinguishes cultural from biological replication, but rather *mediation by signals.* Cultural replication reactions are always instigated by signals racing in from elsewhere, whether this transmission of information derives from another computer in the network, photons reflected off the page of a book, or as a spike sent by another neuron. Signaling is not a central feature of the replication process in biology, but it is for culture.

Mediation by signals means that cultural replication always involves code-switching. In DNA, replication occurs through complementary template-matching: There is a one-step relationship between one base and another on the opposite strand of the molecule. Similarly, making a new prion is a matter of direct, contact-based information transfer between the surface of two proteins, making them the same. One code suffices in both cases: The template being created is either the same as, or the molecular complement of, the original template. In cultural replication, by contrast, the inheritance process always requires multiple steps, because it must solve the problem of the physical "gap" between a replicator and its copy. With replication mediated by signals, information is necessarily transferred from replicator to signal and back again. In social memetics, these two steps, switching from one substrate to another, suffice. In coevolutionary memetics, there is yet a third step: coding the artifact-based template as well. In the coevolutionary case, the progression of inheritance is therefore from meme to artifact to signal: three changes in substrate, and hence in information coding.

These switches in coding systems from one medium to another allow an arbitrary relationship to crop up between a replicator and its products—something not seen in the biological world. The information structures or codes in signals and artifacts are not necessarily "attached" to particular physical replicators. For example, comp-viruses can use multiple computer operating systems and communication protocols to generate the signal that makes its way from one site of replication to the next. Similarly, by cooperating with other mental machinery, a meme can generate any number of artifacts to get the same message across to the next potential host. So this arbitrary (some would say "symbolic") relationship holds whether the cultural replicator is itself artifactual in nature or not.

Dependence on signals may also be a more fundamental quality of cultural replication than the other quality shared by comp-viruses and memes: being super-molecular. This is because it is possible, in the future, that computers will be molecular in scale, as with the prospect of DNA–based computers, or even smaller, with quantum-scale computers also on the horizon. Since these computers will still need programming to do complex calculations, viruses will doubtless still be able to arise on these new substrates for computation. Replication may also spring up in artifacts other than computers in the future (like nanites), with no necessity of being super-molecular in scale. So being super-molecular is probably not so essential to the nature of cultural replicators as the use of signals to breach the distance between the source replicator's abode and the site where the copy is produced.

The Big Picture

Can we visualize what a replicator must be like, then, independent of its particular substrate? I think so. All replicators—genes, prions, memes—are things that can be represented by *ball-and-stick constructions*. Indeed this is how Watson and Crick originally worked out the peculiar shape of DNA, so crucial to its functioning—by fiddling with a big construction kit of wire and plastic balls. For genes and prions, which are molecules, the "balls" are associated with atoms, the "sticks" with electrical bonds. For memes, which are super-molecular, the balls are cells (neurons), while the sticks are synaptic connections. For comp-viruses, the balls are memory registers; the sticks are pointers from one register to the next. So all of the replicators we have identified can be thought of in these terms.

Another way of thinking about ball-and-stick constructions is as *networks*. Networks, after all, are just nodes connected by links. And networks can be represented mathematically as matrices, or rectangular collections of boxes (or cells) filled with numbers that represent the nature of the connection between the node identified by its row and the node identified by its column.

An interesting property of matrices is their reducibility—that is, the number of rows or columns in the matrix can sometimes be eliminated without losing information. For example, given knowledge of the laws of chemistry, the three-dimensional shape of DNA (the double helix) can be reduced to two dimensions and represented as a matrix with two rows, as if the twisting helix were stretched out into a straight line (as it is when DNA is transcribed). And thanks to the strict complementarity of each rung of the DNA ladder, you actually only need write down the sequence along one side of the ladder; the order of the other is completely determined by its complement. Thus geneticists commonly represent a gene sequence by a one-dimensional matrix, or vector. This is the "string" of letters that you will find representing a gene in any gene bank or library.

Prions and memes, on the other hand, are unlikely to be so readily reducible to a single dimension, because the rules for converting from a lower-level representation to a higher one will be too complex to justify leaving out the intervening information. Nevertheless prions can certainly be represented by the values in a matrix with some small number of rows and columns. I think memes can too, although this remains to be empirically ascertained.

I think we can safely conclude from this discussion that we have learned many things about the nature of that most fundamental of evolutionary processes, replication, from our attempt to locate another replicator: memes. This project has led us not only to a better search image for the ostensible cultural replicator, but also to a better understanding of what is happening in other evolutionary domains—and even with respect to the oldest and best-known replicator, DNA.

Chapter Eleven

THE REVOLUTION
OF MEMES

Culture is not inherited through genes, it is acquired by learning from
other human beings. In a sense, human genes have surrendered their
primacy in human evolution to an entirely new, nonbiological or
superorganic agent, culture.

— *Theodosius Dobzhansky*

Genes began life as strictly functional entities. As late as 1933, the eminent geneticist T.H. Morgan could claim that "there is no consensus
opinion amongst geneticists as to what the genes are—whether they are
real or purely fictitious." To many, genes were only a name for whatever
was responsible for the Mendelian patterns of inheritance people were
observing in biology labs. No one yet knew just where to find genes, nor
how they fulfilled the statistical regularities in the proportions of traits
that appeared in each succeeding generation first identified by Mendel. It
was only in the 1950s that genes became incontrovertibly real, material
entities—the biological analogue of the atoms in physical science. When
Watson and Crick constructed the famous tubular model of the double
helix in a Cambridge laboratory basement, genes "took on" a material
structure. Scientists had identified the molecular substrate that held them
and the mechanism that enabled them to make copies of themselves.

The basic machinery of genes has thus been known for 50 years. The
nature of prions is also coming rapidly into focus, thanks to the urgency
of curing the human diseases they cause. Various other simple, self-

replicating proteins have also been manufactured as special-purpose, toy replicators by chemists working in laboratories. However, the replicator underlying the more recent transitions in evolutionary dynamics—those associated with human cultural inventions such as language and writing—has not yet been properly identified.

Currently, then, memes exist only as hypothetical functional entities, with the express purpose of explaining observable similarities in cultural traits over time. I argue that developments similar to those that occurred in genetics must now take place for memetics to be established as a science. We must find out where memes are hiding. The difference from the situation of the 1930s is that genes have already been identified and may be responsible for the statistical regularities we see in the field, in the "cultural" traits human populations exhibit over time. In effect, what we are calling memetic effects could be due to that other replicator being at work. For memes to become real, we have to find out just how they operate and, if possible, see them in action. It's time for a new scientific revolution, for the beginning of a "molecular memetics" to mirror the revolution in biology that occurred with the identification of physical genes.

The Evolution of Memes

In this book, I have used information gleaned about the nature of replication to explain what memes are. I began by suggesting where we should look for memes, where our "molecular" memetics must start. Artifacts and behaviors fail the replicator test. Only brain states have the necessary qualities to replicate: They can cause similar entities to arise through information transfer. So memes should be in the brain. Does the brain have the qualities to harbor replicators like memes? Yes. It's an isolated, energy-rich environment housed inside the braincase, holding a soup of cells and chemicals. Memes could even persist in a brain in a vat, if that vat contained the requisite chemical bath as an energy source.

But what inside the brain could replicate itself? The brain is so complex that it's difficult to know where to start the search. Since the earliest times in Western intellectual history, the brain has largely been thought of in terms of an analogy to something easier to comprehend. Many of these analogies happen to reflect changes in the concept of electricity. The classical Greek conception of the brain was as a *pneuma,* a kind of nervous fluid. This became animal electricity with Galvani's late-eighteenth-century studies of jerking frogs' legs and more recently morphed into spiking action potentials. Now, with the rise of molecular studies of the brain, we can more precisely isolate electrically charged ions passing through gates and jumping across microscopic gaps like electric sparks. If electricity is a stream of electrons—small atomic particles moving quickly through a channel—then perhaps memes are small conceptual elements transmitted through a particular channel, a linked chain of neurons. The suggestion is that a meme must somehow be electric too: An *electric meme* is what we should look for.

The next step was to identify a temporal and spatial scale within which a new replicator could gain control over brain activity. This was necessary because genes are known to be responsible for aspects of long-term memory. It turns out that memes, if they exist, must work very quickly indeed—within milliseconds. If they are to replicate despite this very rapidly changing substrate, then they are also likely to be quite small—just a node in the network, perhaps one neuron, or a few of them acting in concert. The only thing that can replicate quickly enough to be a meme is the state of such a node; changes in the physical substrate, in the network of connections and nodes itself, take too long to qualify. So a meme must be the *state* of a node in the neuronal network. This state is electric because it determines the node's electrochemical propensity to fire an action potential (or spike) at other neurons in the brain. And this memetic behavior is itself electric, since spikes are discharges that change a neuron's polarity.

The important point is that a state, unlike the physical network itself, can be duplicated elsewhere in the network. And the state of a node becomes independent of genes to the degree that its current qualities have been produced by a history of stimulation by previous meme products, as well as other stimuli. Although the basic physiology of neurons and their connections are determined by genetic evolution, the state of a neuronal node at any time is determined by the sequence of events experienced by the host organism and endogenous memetic evolution.

A meme, then, is essentially the state of a node in a neuronal network capable of generating a copy of itself in either the same or a different neuronal network, without being destroyed in the process. Memes have gone through several stages of evolutionary advance. Such replicators must have first evolved *within* brains. Even before this, however, the states of neuronal nodes were able to change location within a brain when a stimulus altered its original conformation, turning it "off" in one place while simultaneously turning it "on" somewhere else. This is "normal," non-memetic information-processing in the brain. The brain-as-network simply flexes and distorts as a given state changes location.

Then the states of nodes began—perhaps by chance at first—to replicate within a brain. A signal stimulated one node to fire, causing another node to acquire the same conformation or electrical potential, while itself returning to its original state. Then two copies of the state existed, probably in similar locations. This state had proven to be infectious: It could be "caught" by another node in the network. A new replicator had emerged. One copy of the state might not be functional (a "junk meme") in terms of providing spikes beneficial to the host organism, but nevertheless was there, lying in wait for events to make it useful.

A number of things play important roles in the process through which this parasite on cortical activity replicates. First, the definitive host to neuromemes is the *physical network* of neurons itself. That is, the substrate on which parasitic memes "feed" is the neuropil—the suite of neurons and their support structures that participate in, and maintain the potential for, conducting information between neurons.

Second, the replicator itself, an electric meme, is an electrochemical potential, existing in a neuron or set of neurons. This *state* of a node in the network can be measured in terms of sensitivities to incoming stimuli, which influence the likelihood the node will fire. The node on which the meme "sits" constitutes the *effective* network, the set of linked neurons likely to fire simultaneously.

Third, the *output* or *product* of the node (a temporal set of "spikes") is not the meme's phenotype, strictly speaking, but a curious kind of thing: a signal. Signals are perturbations of the environment, often focused in a channel, that propagate information from one place to another.

After neuromemes got the ball rolling, the next step in memetic evolution was the beginning of social memetics. In this development, memes began to replicate *between* brains. An incoming signal still stimulates a memetic node to fire, as before, causing another node to acquire the infectious conformation, while itself returning to its original state. But this second node of neurons now exists in another brain. How? The memetic spike simply has to stimulate a motor neuron to engage the host organism in a behavior that produces a social signal, such as a stream of speech, that can be consumed by a second organism. Sensory receptors in the receiving organism then convert this signal back into a spike train that can instigate the local replication of the memetic state in *its* brain. As a result, there are two copies of the meme, probably in similar locations in the two different brains. Evolution within a memetic lineage, given this development, became independent of the fate of any one host organism. In effect, a meme had jumped the gap from one host to another. This led to new kinds of incentives for the sharing of information in social species coevolving with their memetic parasites.

Think of what this new perspective means for our view of social transmission. Just as computer memory can be turned into an alternative Earth in Artificial Life programs, so too is the brain a new kind of spherical "planet" on which an ecology of replicators can evolve, with their own unique characteristics that reflect the strange mix of selective forces they face. New person-planets can be seeded by signals. This seeding starts off a new colony of memes in a potentially hostile environment. These memes can then evolve into new species of ideas. Further, there is a constant barrage of message-meteorites bombarding the surface of the planet (through eyes, ears, touch), each carrying the spores of new meme-species. So the mental ecology is never able to settle down into a stable equilibrium but is always being invaded by potential new members of the memetic community. This produces the expected degree of dynamicism in the picture of what happens through the social exchange of information.

The fact that memes are parasites on this brain-planet, and produced only by a very specific mechanism, saves us from major error. Some memeticists have said that the brain, and even the sense of self, is composed entirely of memes. This evokes an absolutely visceral response from many people, and rightly so. Just as people are beginning to crawl out from under the rock of genetic determinism—a view promulgated by sociobiologists and others—prominent memeticists seem determined to slide people back under it. But this time, the rock is to be called "memetic determinism." You are slave to your memes—you *are* your memes, they say. If one replicator isn't the cause of our behavior, then it must be another, these memeticists seem to feel. But this position is both unnecessarily alarmist and almost certainly untrue. It is a piece of hubris and a major impediment to the memetic perspective being generally adopted by the lay public because it leaves no room for human agency. Luckily this position simply isn't consistent with the idea that a meme is a highly specialized kind of neuronal state, and not all forms of intercellular communication within a brain will produce it.

This doesn't mean that the ancient philosophical notion of "free will" can survive the coming onslaught of neuroscientific advances, however. It is still likely that we will have to recognize that the mind is an emergent property of the brain, and nothing more. It may even be the case that our conscious awareness of "making a decision" comes only *after* the real neuronal work has already been done subconsciously (as suggested by some important experiments carried out by Benjamin Libet over the past 20 years). What we have in our heads is a complex mixture, caused by an interaction of a couple of replicators (genes and memes) as well as regular forms of independent learning. No single agency is in charge of anything we do.

Signals began as neural communicators, then social ones. Most recently, however, they have learned to involve artifacts in their travels. Artifacts are abiotic substrates for signals located in the macroenvironment. They are a curious collection of objects because, from an evolutionary perspective, they can be anything from beaver dams to computer memory. It's also important to distinguish signals from artifacts. Artifacts can interact with signals, as when electrons circulate through a computer or photons glance off the page of a book. So artifacts and signals have independent existences. One is static, the other dynamic. They also have different roles: Artifacts can store information safely, while signals are designed for moving information around efficiently. Given this role as a storehouse, it's natural that artifacts, and not signals, should in some cases develop the ability to replicate. Signals are, after all, about communication, not replication.

So naturally, with their ability to use signals for replication between brains (developed when social memes evolved), memes came to involve themselves with artifacts, either by helping construct artifacts or by using them as templates for the signals they produce. This instigated a new coevolutionary process between artifacts and memes. The evolution of culture has come to depend more and more significantly over time on the evolution of artifacts. My earlier discussion of how memes coevolve with artifacts provides the basics for a working theory of cultural change in societies that have significantly modified their niches through the accumulation of technology. Artifact evolution is itself a Darwinian process with great speed and power, and human culture can be thought of as the product of complex interactions between organisms, memes, and artifacts.

So in neuromemetics, which covers the first phase of meme evolution, information transfer is satisfied by the production of a signal spanning the gap between one neuron and the next; replication is based on the mediation of signals *within* organisms. Social memetics (the second phase) depends on signal-mediated replication *between* organisms. In coevolutionary memetics (the third phase), information transfer becomes even more indirect. First, a meme-inspired behavior produces, or modifies, an artifact, which in turn does the job of producing or modifying a signal that eventually reaches a new host. It is a case of *artifact- and signal*-mediated replication between organisms. Each of these steps, while representing an evolutionary outgrowth of earlier steps, also manifests itself as a stage added to the earlier, simpler life cycle of a meme. (Such growth in the complexity of life cycles is a standard pattern in the evolution of parasites.)

This conception of memes maintains the focus on the specificity of replication reactions, which is necessary because cultural replicators are still physical things that must duplicate themselves through some particular mechanism. All the complexities in the evolution and life history of memes arise in the mediation of inheritance, not in the replication reaction itself, making the cause-and-effect relationship between parental and offspring replicators more and more tenuous over time, and hence more treacherous.

THE REVELATION OF MEMES

One of the primary intellectual benefits of neuromemetics is that it makes a naturalistic theory of both communication and culture plausible. The primary long-term benefit of the approach developed here, then, may be that it directly connects advances in the biological and particularly neurological sciences to the explanation of social phenomena. As a result, the ability of such a social science to tie into results from these other sciences is automatically ensured, since its principles must be continuous with knowledge about workings at the lower levels of organization covered by the biological and physical sciences. In return, the social sciences can not only make use of, but build upon, these discoveries by pointing out the recursive nature of causation, in particular, the "top-down" factors that influence goings-on in the brain, as when pheromones wafting between individuals lead to the activation of mate choice mechanisms. The result could be that social science becomes as progressive as the biological sciences on which it depends.

Neuromemetics thus provides us with a testable mechanism of replication that is consistent both with other sciences and with the expectation that culture evolves through the descent of replicators. Many memeticists will no doubt resist this conceptual move and dislike the alterations in perspective it requires, preferring the old-style "epidemiological" social memetics. However, I think the current lack of progress in empirical or theoretical memetics belies the charm of continuing to talk in abstractions. Memeticists like Susan Blackmore say that only social transmission is included within the compass of proper memetics, because Richard Dawkins and the *Oxford English Dictionary* define the meme in terms of a social learning mechanism: imitation. But there are theoretical reasons to override the etymologist's concerns: If replication within a brain is part of the *evolutionary* history of the meme, then it must be included in the project to explain the social phenomena that Blackmore and others believe memetics is about. Even if this extends memetics into the domain normally given over to psychology, such an assignment of territory is an accident of academic history, and not a proper theoretical reason for excluding such events from the domain of memetics. If we take as the central feature of a meme's definition the fact that it is a replicator, and evolves, then we must follow that fact wherever it leads, regardless of disciplinary boundaries or dictionary definitions. In fact, quite a different picture of memes arises if we do acknowledge the psychological part of the meme's evolutionary history and trajectory. Most importantly, it gives memes a more plausible beginning and, coincidentally, a longer and more interesting history.

The difference between neuroscientific and Dawkinsian memetics is crucial, and many conceptual changes are entailed by it. First, my definition of memes performs the same kind of shift as occurred in the sociobiology revolution. Where sociobiology represented a shift in the focus of selection from the organism to the gene, neuromemetics signals a similar downward shift in perspective from social to inter-cellular communication. Memes are now defined at a scale lower than that which Dawkins and others have suggested: at the inter-cellular level within an organism, rather than between organisms. It's true that the life cycle of a meme sometimes involves the use of signals that work at the inter-organism scale, but memes *per se* are not making the journey between hosts. A meme will still care, of course, about where its relatives are. Hamilton's insight, the foundation for sociobiology, forced us to acknowledge that a gene should seek out its brothers, regardless of which body they might be in. Neuromemetics forces on us the same switch in viewpoint, but in an opposite direction: a recognition that a meme should care about its relatives, but might have its whole family right there in the same brain with it!

It's even possible that replicators can single-handedly achieve a significant unification of the sciences—that elusive, but much-to-be-desired goal of philosophers and practitioners of science alike. How? Replicators are states of matter and therefore can be described using the language of physics, at least in principle. But they also catalyze reactions among bits of matter—this is the replicator dynamic mentioned earlier. At this point, we have crossed over to the domain of chemistry. However, resulting from the catalysis is a new packet of information that bears a resemblance to the packet that created it—that is, replication—with a mutation or two thrown in perhaps, due to the transmission process. In effect, the replication process, when iterated, generates a chain of objects exhibiting a pattern of descent with modification. Since this is the criterion defining an evolutionary process, we have now entered the traditional domain of biology. Next we face the prospect of explaining sociocultural phenomena as the outcome of a replicator new to the scene, the meme. This is where the social sciences normally come in. And psychology is required to understand the replication mechanism used by this new cultural replicator. So new replicators, new sciences. Since each science deals with the same kind of generating process, just at different scales of organization, both method and substance are unified. Each replicator builds on the work of previous ones; each new science is devoted to understanding the products of more fundamental ones.

If replicators play such an important role in the general scheme of things, it seems important to locate them and to understand how they perform their useful work. For example, getting the right picture of how *kuru* is caused will lead more rapidly to a cure for related diseases having the same symptoms (a patchwork of holes in gray matter), such as Alzheimer's disease. Once you get a grip on a replicator, the dynamics it produces are much more likely to become yours to control. To help us gain this control over memes, I've presented the best clues I could for identifying them. In this book, I've offered a theory of how the replication of information may occur at the neurological level and at the social level—in effect, a model of what memes might look like.

Only concerted empirical investigation can determine if these ideas are true. Luckily, within the next five to ten years, neuroscience should be in a position to provide empirical evidence that the kinds of entities described in this book exist. However, neuroscientists are not explicitly looking for replicators nowadays; it isn't a project on their agenda. This is because very few neuroscientists care about evolutionary issues. Certainly the problems of the social sciences seem very far indeed from their lab benches. The study of neurophysiology is so fast-moving on its own, and seems to depend so little on the "big questions" of why the brain is good at processing some kinds of information but not others, that it appears the evolutionary context of psychological processes can be safely ignored. Neuroscientists can only be expected to adopt the goal of finding memes if the theoretical interest of such a discovery is amply demonstrated. But if I have made the prospective kudos seem valuable enough, a race to find the first meme will be spawned among the more adventurous members of the neuroscientific community. (Remember, the last person to discover a replicator in the brain, Stanley Prusiner, won a Nobel Prize for his work!)

On the other hand, should it turn out there are no such things as memes, the consequences of that discovery for our understanding of the world around us will be equally profound, for it will mean that we observe an evolutionary process involving the transmission of information—cultural change—that is not underwritten by replicators. In this eventuality, we would have to explain how people come to have similar ideas in their heads without being "infected" by others. Remember that the entire biological world is the result of activity by genes, replicators *par excellence*. Even the infections that computers can "catch" evolve because the agents that cause them, computer viruses, are replicators. Without memes, we would have, for the first time, a pool of things—ideas and values, in this case—that exhibit similarity, duplication, and inheritance without a replicator being involved. Such a discovery would perhaps be even more magical than uncovering a new form of replicator, especially as common perception and thought appear to occur regularly from person to person and generation to generation. How can we get something like descent through the constant *re-creation* of beliefs and values in each person's head? Such a phenomenon would cry out for explanation, should memes be proven impossible.

No doubt such a discovery would also leave each of us feeling a little bit more alone than we did before—and alone in a new way. The essential *psychological* fact of a world without memes would be that we each live in a mental box of our own construction, assembled from the bits and pieces of information we get from others. How does successful communication take place, then? How is it we can understand each other? Attention will immediately focus on how the brain can reliably remake the meaning of messages from signals sent through the air. The power to communicate will fall to evolved neurological structures for assuring that human brains correctly interpret these signals. Institutions to regularize social interaction, which assure a familiar context for the transmission of signals, will also become important components of any explanation of cultural evolution. This is probably why we have ritualized routines for greeting one another, and for presenting ourselves to others (a separate public "face" or "persona"). Culture somehow becomes even more wonderful and amazing, if perhaps more inscrutable, once memes are gone.

Establishing whether memes exist is thus a scientific project of primary importance. Its outcome will give us a new kind of awareness about the nature of evolutionary processes, how social groups work, and our own place in the world. I will accept the conclusion of this project either way: memes or no memes. That's the only justifiable attitude for a scientist to take, in my view. But regardless of how the meme story turns out, the study of the most complex evolutionary process we know—cultural evolution—looms as one of the most significant and exhilarating undertakings in science today. Understanding ourselves, and the societies we make, remains one of our greatest challenges. Electric memes may yet play a crucial part in the eventual solution of these mysteries.

Notes

CHAPTER 2: A SPECIAL KIND OF INHERITANCE

24   Laufer 1918.

24   Sahlins 1960:44.

24   *Memes as explanation for cultural evolution:* I will take it for granted both that culture exists and that it evolves. Even though both of these claims remain contentious—disputed even by some evolutionary approaches to human behavior such as sociobiology and evolutionary psychology—I don't have space to address the arguments of such naysayers here. An excellent book by Boyd and Richerson (in preparation) accomplishes this job handily in my view, so I don't need to tread over this much-trodden ground again.

25   *Definition of heredity:* Maynard Smith 1987.

25   *Replication unnecessary for evolutionary process:* Godfrey-Smith (in press).

29   *Barometer not a cause of rainfall:* Sterelny and Griffiths 1999:101–2.

30   *What is culture?:* I will restrict myself to evolutionary explanations as the only kind fit for historical phenomena like culture and will not deal here with the early history of thought on cultural evolution in the social sciences, which has been admirably detailed elsewhere. See, for example, Ingold 1986; Richards 1987.

30   *Cognitive notion of culture is pervasive:* One would think archaeologists might represent a stronghold against cognitivism. After all, they necessarily deal with artifacts as their "raw" data. However, they too have largely bought into the cognitivist paradigm and often consider their "material culture" as only indirect evidence of what is truly important: inferring the mental representations of people long dead.

31   *"Modern" evolutionary theories:* By restricting my attention to modern theories, I exclude from this review evolutionary approaches that are no longer active, such as those developed by earlier generations of anthropologists like C.H. Morgan, Leslie White, and Julian Steward.

32   *Application of sociobiology to humans:* Alexander (1979) is the acknowledged fountainhead. See also Flinn 1997.

32   *"I, personally, find 'culture' unnecessary":* quoted from Betzig 1999:17.

33   *Genes plus environment can explain human behavior:* for example, Hill and Hurtado 1996:13–14.

33   *"To whatever extent the use of culture . . .":* quoted from Alexander 1979:78–79.

34   *Compendia of human studies showing fitness maximization:* Betzig et al. 1988; Winterhalder and Smith 1981.

35   *Evolutionary psychology:* Proponents are now legion. John Tooby, Leda Cosmides, and Donald Symons (called the Santa Barbara School because they are all centered at the university there) are the theoretical avante-garde; their primary theoretical statement is Barkow et al. 1992. Steven Pinker is the popularist extraordinaire (see especially his *How the Mind Works*), with Wright (1994), Gazzaniga (1994), and others as close seconds. Textbooks are also available; see, for example, Buss 1998 and Gaulin and McBurney 2000.

35 *Psychology necessary to explain behavior:* I hasten to add the proviso that the phrase "evolutionary psychology" is currently being used in two different ways by intellectuals. There is a sense in which it has been widely recognized that psychology can no longer remain out in the cold, a holdout from the Darwinian revolution. The brain is an evolved organ and deserves to be treated as such. This is the general sense of the term and would include, potentially, much of neuroscience, the study of neuroanatomy from a comparative perspective, and so on.

But a more specific use is actually more prominent these days, especially in popular scientific presentations. This is the one to which I refer in this book. It is a particular flavor of evolutionary theorizing centered around the work of a married couple presently at the University of California, Santa Barbara, Leda Cosmides and John Tooby, a psychologist and an anthropologist, respectively. To ignore this distinction would be to mistake a number of corollaries identified as important by this group as fundamental principles of evolutionary thought applied to decision-making. They are not. Modularity, the concept of the "environment of evolutionary adaptation," and psychic uniformitarianism are all part and parcel of the "Santa Barbara view," derived from specific assumptions about how evolution works on brains. This group takes its inspiration largely from the cognitive sciences, and more particularly from Chomsky. But alternative sources of theoretical insight are available. So even if this brand of psychologizing does not last, such a failure should not invalidate the more general project to understand the brain as a functional part of evolved organisms, and psychology as the scientific study of that evolved organ. The discipline of psychology should never return to a pre-Darwinian era.

35 *Evolutionary psychology as "white knight" for psychology in general:* Tooby and Cosmides 1992.

35 *Aversions in rats:* Garcia and Koelling 1966.

37 *Reconstructing ancestral environment as task of evolutionary psychology:* Symons 1987.

38 *Modularity claim of evolutionary psychology:* Tooby and Cosmides 1995:1189.

38 *Modules as mental "instincts":* Pinker 1994.

39 *Evolutionary psychology provides laws:* Cosmides and Tooby 1987.

40 *Darwinian algorithms provide evolved but contingent responses:* Buller 1999.

41 *"Jukebox" model:* See Tooby and Cosmides 1992:115–16. This analogy was used by Sir Peter Medawar in a 1959 article called "The Future of Man" <http://www.santafe.edu/~shalizi/Medawar/future-of-man.html> to make roughly the same point.

43 *Brain too small to hold necessary cultural variation:* Peter Richerson, personal communication.

43 *Human Relations Area Files (HRAF) repository:* <http://www.yale.edu/hraf/>

44 *Evolutionary psychological studies of mate preferences:* See Buss 1998 for a discussion of many of these studies.

44   *Homicide study:* Daly and Wilson 1988.

45   *Cultural ecology:* Steward 1955; Vayda and McCay 1975; Netting 1986.

45   *Cultural variation within eco-zones:* Examples are provided by Boyd and Richerson (in preparation). See also Guglielmino et al. 1995.

45   *Ape and cetacean culture:* Whiten et al. 1999; Rendell and Whitehead (in press).

47   *Diffusion of innovation studies:* Rogers 1995.

47   *Models of the psychology of innovation adoption:* Henrich (in press).

49   *Cultural selection and epidemiology:* Cultural selectionists are explicit not only about the structural similarities between cultural and genetic transmission, but also about the relevance of epidemiological approaches. As Cavalli-Sforza and Feldman (1981:54) acknowledge, "epidemiology is an important source of inspiration for the study of cultural transmission" in these models, so the two perspectives are tightly linked.

49   *Common versus idiosyncratic beliefs and their spread:* Sperber 1996.

49   *"Cultural selectionism":* This position is expounded primarily by academic biologists and anthropologists. The key figures are the Stanford biologists Luca Cavalli-Sforza and Marcus Feldman, who originally developed the formal approach, the University of California biologists Robert Boyd and Peter Richerson, who popularized it and extended it, and the Stanford anthropologist William Durham, who has provided the most compelling case studies, as well as some theoretical wrinkles. I will follow the formulation of Boyd and Richerson most closely. The most important works by these authors are Cavalli-Sforza and Feldman 1981; Boyd and Richerson 1985; and Durham 1992. A related approach using a different formalism that has not been further developed, even by its proponents, is Lumsden and Wilson 1981. For general reviews, see Durham 1992; Feldman and Laland 1996; and Flinn 1997. A variety of monikers have also been assayed by various commentators for this somewhat heterogenous approach. Durham (1992) prefers "evolutionary culture theory," Cavalli-Sforza and Feldman (1981) "gene-culture coevolutionary theory," and Boyd and Richerson (1985) "dual inheritance theory." All of these are perhaps more precise descriptors but are too unwieldy.

50   *"Unit of evolution":* Maynard Smith 1987.

50   *Definition of culture:* Boyd and Richerson (in preparation).

51   *Importance of "population thinking" in cultural evolutionary studies:* Boyd and Richerson 2000.

52   *". . . an accounting system for keeping track of information . . .":* quoted from Boyd and Richerson (in preparation).

52   *Agnosticism about the processes underlying cultural evolution:* Boyd and Richerson (in preparation).

53   *Genetic changes dependent on cultural preferences:* Feldman and Cavalli-Sforza 1979.

54   *"a means of exchanging knowledge . . .":* quoted from Pinker 1997:190.

55   *Cultural "badges" erect barriers to transmission:* Boyd and Richerson 1987.

56   *"Natural origins" of culture:* Boyd and Richerson 1985.

57   *Genes and culture coevolve, with either party determining changes in the other:* Kendal and Laland 2000.

59   "Brainworm" fluke story: Dawkins 1982:218; Dennett 1991.

62   *"Poetry can be recognized by its ability . . .":* quoted from Valéry 1989.

P<small>AGE</small>   **C<small>HAPTER</small> 3: A<small>DDING</small> R<small>OOMS TO</small> D<small>ARWIN'S</small> H<small>OUSE</small>**

65   Head quote: Dawkins 1976:192.

68   *"if it could be proved . . .":* quoted from Darwin 1909 [1859].

68   *"one special difficulty . . .":* quoted from Darwin 1909 [1859].

70   *"Limited" Darwinian theory:* Griesemer (in press b).

72   *Darwinian "extrapolations":* For Darwinian immunology, see Edelman 1993; for business models, see Koch 2001 and Schwartz 1999; for Darwinian cosmology, see Smolin 1997.

72   *Computer viruses and memes as replicators:* Dawkins 1993.

73   *Replicators as the basic units of evolutionary processes:* Hull 1980.

73   *Meme as replicator associated with culture that could evolve like gene:* Dawkins, in his Introduction to Blackmore 1999.

73   *Four minimal conditions for replication:* These criteria are closely related to those found in Sperber 2000, with supplementary input from Godfrey-Smith (in press) and Maynard Smith 1987.

74   *Duplication as intrinsic to replication:* Duplication is not a condition of replication in Sperber 2000 or Godfrey-Smith (in press), but Eigen (1992:161), for example, defines replication as a process that results in the doubling of molecules.

77   *Replication limited to "direct" causation:* Hull 1980.

77   *Photocopied paper as replicator:* Dawkins 1982:109.

77   *Replicator "power":* Dawkins 1982.

78   *The term "interactor":* Hull 1980.

78   *Interactor determines fate of replicators:* Hull 1980.

79   *Evolutionary concept of lineage:* Hull 1988:409.

80   *Tape recorder sequence example:* Sperber 2000.

80   *Replicators, not interactors, as elements of lineages:* It is a bit confusing that systematists—those who draw pictures of evolutionary history in the form of branching tree diagrams—seem to suggest that things like species are elements of lineages, or nodes along a single branch, with such drawings. But, in fact, what is being traced with a line of ancestral and descendant species is typically some gene variant, or set of variants, that distinguishes that line (not lineage) of organisms from others in the same evolutionary picture. So trees of species, if they are properly drawn, are really trees with genes underneath. See Nei 1986 for further explanation.

81   *"evolution is the external and visible manifestation . . .":* quoted from Dawkins 1982:82.

81   *Lewontin's strategy for generalizing Darwinism:* Lewontin 1970.

81   *Lewontin's strategy as "levels abstraction":* Griesemer (in press b).

83   *Formal theory of replicators:* You can't have a theory of something with only one member in the class of things being explained, as would be the case if only genes were being studied.

83   *Manfred Eigen's collaborators:* These include Peter Schuster, Peter F. Stadler, Robert Hecht, Walter Fontana, Robert Happel, and others.

83   *Replicator Theory:* For an overview, see Hofbauer and Sigmund 1988 or Eigen 1992.

83   *An artificial chemistry:* Fontana and Buss 1994.

87   *"Survival of the first":* Michod 1999.

89   *Artificial replicators:* Rebek 1994 or Lee et al. 1997.

89   *Oligonucleotide replication:* Szathmáry and Maynard Smith 1997.

89   *Cost of being common:* Szathmáry 1991. Michod 1999 calls this same regime the "survival of anybody," but that appellation doesn't imply an equilibrium in which many different replicators can coexist.

90   *Darwin's inspiration by Malthus:* Richards 1987.

91   *"Bottom line" from Replicator Theory:* Michod 1999:46.

91   *DNA as wads of genetic spaghetti:* Deacon 1999.

PAGE   CHAPTER 4: "THE REPLICATOR ZOO

93   Head quote: Darwin 1909, Chapter 15.

93   *DNA not first replicator:* See, for example, Kaufmann 1993.

94   *"once 'information' has passed into protein . . .":* quoted from Crick 1958:153; italics in the original. See also Crick 1970.

95   *IBM petaflop computer:* Service 1999.

96   *New methods for understanding protein folding:* Baker 2000; Van Gunsteren et al. 2001.

96   *The three primary TSEs: Kuru,* the disease found among the Fore people of New Guinea (discussed in Chapter 1) was probably also a disease caused by prions.

97   *Prions inserted into the brains of living laboratory mice:* Telling et al. 1996.

97   *Prion disease strains breed true:* Chien and Weissman 2001.

98   *Carriers of prions:* Hill et al. 2000.

98   *We may begin to find prions all over the place:* Tuma 2000.

98   *Sup35 causes gene transcription:* True and Lindquist 2000; Partridge and Barton 2000.

99   *Prion replication is two-step process:* Saborio et al. 1999.

99   *Body can clear some prions:* "University of California–San Francisco Researchers Report Test That Detects Prion Diseases, Illuminates Novel Findings About Infectious Prions," Science Daily report, posted 9/30/98. <http://www.sciencedaily.com/releases/1998/09/980930081855.htm>

100   *Prion replication kinetics:* Eigen 1996.

100 *Prions and species evolution:* Li and Lindquist 2000.

102 Quote: Dick 1982:214.

102 Quote: Gordon 1993.

102 *People treat computer faces like people:* Reeves and Nass 1996.

102 *Self-replicating computer programs:* Ralph Merkle ("Self replication and nano-technology," <http://www.zyvex.com/nanotech/selfRep.html>) and Mitchell (in press) are examples; the one I use here comes from Merkle; Merkle, in turn, took it from "Self-reproducing programs," an article in Byte magazine, August 1980, p. 74.

103 *Cohen's definition of a comp-virus:* Cohen 1994:1.

103 *"a program that can 'infect' other programs . . .":* quoted from Cohen 1994:3.

103 *"its ability to attach itself to other programs . . .":* quoted from Cohen 1984.

103 *Types of comp-viruses:* Spafford 1992:731–32.

104 *Basic parts of comp-virus life cycle:* This section draws from Ludwig 1990:16–17.

106 *"Attack" phase detrimental to comp-virus:* Ludwig 1990:16–17.

109 *Von Neumann architecture:* Actually the same physical and analytical division is present (100 years earlier) in the designs left behind by Charles Babbage for his Analytical Engine, which had both a "store" for memory and a "mill" for the calculating unit. But Babbage's machine was not built until recently.

109 *Inside a computer:* Excellent background on how computers work can be obtained from two special issues of Smart Computing magazine: Vol. 3 Issue 4 (November 1999) and Vol. 4 Issue 3 (August 2000), accessible from <http://www.smartcomputing.com/editorial/sTOC.asp?guid= gb48e3fo&type=5&vol=3&iss=4>

112 *Types of comp-virus life cycles:* Two kinds of logical cases are ignored here, because they haven't been relevant historically (at least yet!). The first is the possibility of replication *within the same disk* in one computer. Viruses can conceivably replicate within magnetic memory, taking up multiple places on a single hard disk. For example, this can happen when a (rather stupid) RAM–resident comp-virus, lacking a signature or signature-recognition algorithm, keeps on reinfecting the same file, or similar files, in long-term memory.

The second possibility we will not examine further is that of RAM–to–RAM replication *between* computers. In such a case, a RAM–based repli-cator produces a signal that goes through the virtual operating system (OS) to the normal OS of the source computer, then through a wire to another computer, where it is interpreted by the normal OS, passed to the virtual OS, and written to RAM in the receiver computer. This has been put forward as a proposal by some observers but has not, to my knowledge, been carried out. (An example is "Network TIERRA," which builds on the case study TIERRA discussed at length below.) There doesn't seem to be much point to either of these types of scenarios, mostly because they don't extend the evo-lutionary dynamics of the viruses in any way, fruitful or otherwise.

114   *There is no such thing as a "good" comp-virus:* Bontchev 1994.

114   *TIERRA:* Ray 1992.

115   *Legendary first experiment with TIERRA:* This account was adapted from
      Kelly 1994:285–89.

115   *"I never had to write another creature":* quoted in Levy 1992: 221.

117   *Maynard Smith convinced:* Maynard Smith 1992.

118   *Core War:* The game was invented in the mid-1980s by A.K. Dewdney and was
      popularized by its publication in his "Computer Recreations" columns in Sci-
      entific American magazine. Core War is a descendant of Darwin, a game writ-
      ten by Victor Vyssotsky, Robert Morris, Sr. (father of the famous 1988 Internet
      Worm's author), and Doug McIlroy in the early 1960s. Darwin ran on an IBM
      7090 mainframe at Bell Labs in New Jersey. By the late 1980s, Core War had
      become a cult pastime among computer programmers. See the following
      sources: Core War FAQ, <http://www.stormking.com/~koth/corewar-faq
      .html>; John Perry's "Core Wars genetics: The evolution of predation" <http:
      //www.ecst.csuchico.edu/~pizza/koth/evolving_warriors.html>; and "Core-
      wars for Dummies: An introduction and tutorial" <http://www.koth.org/info
      /corewars_for_dummies/dummies.html>. The classic reference remains, how-
      ever, Dewdney 1984.

119   *Ray told to put TIERRA in quarantine:* Levy 1992:218.

120   *Numbers of wild network comp-viruses:* Alan Solomon, "A brief history of PC
      viruses," <http://www.cknow.com/vtutor/>

121   *Brain Virus:* White et al. 1995.

122   *Viral epidemiology in the 1990s:* White et al. 1995.

124   *The chance of your computer being infected:* Kim Zetter, "How it works:
      Viruses" PCWorld magazine, October 13, 2000, <http://pcworld.com/fea-
      tures/article/0,aid,34551,00.asp>

124   *Love bug virus:* Kim Zetter, "Viruses, the next generation" PCWorld mag-
      azine, November 1, 2000,
      <http://pcworld.com/features/article/0,aid,34551,00.asp>

124   *The virus writer profile is much broader:* Kim Zetter, "What makes Johnny (and
      Jane) write viruses?" PCWorld magazine, November 15, 2000,
      <http://pcworld.com/features/article/0,aid,34551,00.asp>

127   *Comp-viruses engage in simple manipulations of their environment:* Spafford 1992.

134   *Viral programs can express—or compute—anything:* Cohen 1986.

135   *Replicators seek to "convert" you:* Ridley and Baker (1996) apply the term
      "conversion" to all forms of prion creation; here I'm generalizing it to comp-
      viruses, because they too alter the state of a preexisting substrate as their
      mechanism of replication.

PAGE   **CHAPTER 5: THE DATA ON INFORMATION**

136   Head quote: T.S. Eliot (1963 [1934]) "Choruses from 'The Rock.'" In *Col-
      lected Poems 1909–1962.* New York: Harcourt Brace.

136   Head quote: Crandall 1996.

137 *"information can be shared at negligible cost . . .":* quoted from Pinker 1997:190.

137 *"mental states are invisible and weightless . . .":* quoted from Pinker 1997:329.

137 *An idea doesn't have mass or charge or length:* paraphrased and adapted from Williams 1992:10.

138 *Only the information embodied in the nucleotide sequence gets from one generation to another:* Deacon 1999.

138 *Evolutionary lineages are limited to one time and place:* Hull 1982.

138 *"these [computers] — the most complex things produced by the human mind . . .":* quoted from George Johnson, "First cells, then species, now the Web," New York Times, December 26, 2000.

139 *"Information is physical":* Landauer 1991.

140 *Physical reality includes not just matter and energy, but information too:* Jacob 1973:95.

140 *"Information is information . . .":* quoted from Weiner 1961:132. Francis Crick, in his statement of the central dogma of biology, also suggested that three factors were involved in the translation of DNA sequences into proteins: "the flow of energy, the flow of matter, and the flow of information." So this important founder of molecular biology was attuned to the physicalist interpretation of biology. See Crick 1958.

141 *Erasing data consumes energy:* Landauer's perspective suggests that the commonsensical T.S. Eliot got it exactly wrong in the quotation at the head of this chapter. We do not lose knowledge in the conversion to information. Instead the proper question is: What do we have to lose, to forget, to get from information to knowledge? And how can we filter out only the most salient aspects of knowledge to achieve wisdom? The relationships with respect to amounts of stuff are just the reverse of those suggested by the poet.

141 *Reversible computing:* Bennett and Landauer 1985; Bennett 1995.

142 *Quantum teleportation:* Bennett et al. 1993. See also <http://www.research.ibm.com/quantuminfo/teleportation/>

142 *Quantum teleportation demonstrated:* Furusawa et al. 1998. See also Bouwmeester et al. 1997, 1998.

144 *You can't xerox a particle:* Wootters and Zurek 1982. This quantum limitation is on duplication, not on replication. The method being discussed is one that involves some outsider to the system scanning an object for information so as to engage in a reproduction process. This is not what happens in replication, where no observation need take place: It all happens within the circumscribed circumstances. "Particles" in the form of complex molecules can, of course, replicate (just think of DNA).

145 *"No information without physical representation!":* Landauer 1991.

146 *Staircase and DNA have same physical structure/information:* Hull et al. (in press).

146 *Biological and physical information inseparable:* Mahner and 146 1997:339.

148 *Proteins "recognize" DNA:* Alberts et al. 1994:408.

148 *"Structure determines function" as dominant principle:* Sarkar 1996.

149 *Lock-and-key flow of information:* Lowenstein 1999.

151 *"difference that makes a difference":* Bateson 1972.

153 *Stories remain the same, regardless of substrate:* Dennett 1995:353–54.

154 *Structural equivalence:* See, for example, Wasserman and Faust 1994. This principle implies that two structurally equivalent memes might be found in the same processing stream—one early on, another one later. The first on-line meme might then implicitly represent some gross distinction, while the later one determines whether the stimulus fits into a complex class (i.e., is tuned to fire "in response" to human faces). On structural terms, these two memes should be considered the same; on representational grounds, they are quite different. But on the basis of this difference in implicit representational quality, these two memes might have different functional roles in the sense of determining host behavior (the meme's phenotype). This suggests that structurally equivalent memes can produce different phenotypes. This is called pleiotropy when genes do it and is not considered a problem: The gene simply has multi-functionality. So too here.

156 *Information content defines a replicator:* Williams (1995).

156 *"Archival" forms of genes:* Richard Dawkins in the Guardian–Dillons Debate between Richard Dawkins and Steven Pinker on the question: Is Science Killing The Soul? at the Westminster Central Hall on February 10, 1999, <www.edge.org/documents/archive/edge53.html>

PAGE   **CHAPTER 6: STALKING THE WILD MEME**

159 Head quote: *The Winter's Tale,* Act III, Scene 1.

159 *Bowl of soup example:* Blackmore 1999:63.

159 *"memes are not magical entities or free-floating Platonic ideals . . .":* quoted from Blackmore 2000:61.

159 *Memes are both in the head and out in the world:* Dennett 1991, 1995.

160 *We don't need Mendelian laws of culture for memetics:* Blackmore 1999:56.

161 *Obesity caused by viruses:* Dhurandhar et al. 2000.

163 *Mentalist redefinition of memes:* Dawkins 1982:109.

164 *"A meme should be regarded as a unit of information . . .":* quoted from Dawkins 1982:109.

164 *"propagate themselves from brain to brain . . .":* quoted from Dawkins 1986:158.

164 *"A wagon with spoked wheels . . .":* quoted from Dennett 1995:348.

164 *"[a vehicle] houses a collection of replicators . . .":* quoted from Dawkins 1982:114.

164 *Memes are "substrate-neutral":* Dennett 1995.

165 *Proteins are cobbled together:* Lowenstein 1999:70.

167 *Behavioral versus mental memes:* Behavioristic memephiles include Deacon (1999), Gatherer (1998), and Benzon (1996). The mentalistic camp, currently dominant, includes Blackmore (1999), Dawkins (1982), Dennett (1995), Lynch (1996), and a host of others, among its members.

169 *Einstein's Tea Party:* The basic outline for this story comes from a posting by Raymond Recchia to the memetics mail-list on June 9, 2000, <http://aldebaran .cpm.aca.mmu.ac.uk/~majordom/memetics/2000/1487 .html>. It has been twisted to serve my purpose, and its interpretation is my own.

173 *Conclusion that behaviors aren't replicators:* It is also worth noting that considering behaviors to be replicators implies that the replication reaction occurs out in the open air, as with speech. But this contradicts the implication, from Replicator Theory, that *very* specialized conditions are required for the catalysis of copied molecules, especially at the beginning: high concentrations of the necessary building-blocks, physical proximity, and the proper three-dimensional alignments between components. Signals of various kinds, flitting about in the open environment, will be subject to conditions too heterogeneous, the materials too diffuse, for a standardized protocol of replication to evolve effectively. Note too that the replication of artifacts like comp-viruses or Xeroxed paper takes place in highly protected, isolated environments, as expected by Replicator Theory.

175 *Mentalism doomed to science of unobservable entities:* Gatherer 1998; Benzon 1996.

176 *Blood-brain barrier:* Junqueira et al. 1998.

PAGE   **CHAPTER 7: MEMES AS A STATE OF MIND**

178 Head quote: quoted in Popper 1982:89.

178 *The first replicator:* Whether in the form of an autocatalytic set (Kaufmann 1986) or as single molecules like RNA (Joyce 1989, 1992) is a question left open here. I am also adopting a "replicator-first" rather than "metabolism-first" position on the origin of life, but only as a matter of narrative convenience.

182 *Each neuron as mini-computer:* Koch 1998.

184 *"when a pig neuron grows up . . .":* quoted from Deacon 1997:200.

185 *Hox genes:* For example, brain structures involved in higher-level learning emerge reliably in ontogeny under the influence of genetic factors such as the homeobox gene Lhx5, which controls the formation of a cortical structure called the hippocampus. See Zhao et al. 1999.

185 *Last phase of neuron pruning during adolescence:* Bourgeois et al. 1999.

186 *Error threshold:* Eigen 1971.

187 *Bigger brains favored when it's hard to code information in genes:* Bonner 1980.

187 *"the notion that behavior is always a reaction . . .":* quoted from Milner 1999:3.

187 *Memory consolidation during sleep:* Nádasdy et al. 1999.

188 *Temporal variation in brain response to same stimulus:* Dennis Meredith 2000 "Mind over Matter,"
<http://www.dukeresearch.duke.edu/lowres/Mindmatt.htm>

188 *The brain as a "solipsistic" organ:* Freeman 1999.

188 *Self-referentiality as a close relative of self-replication:* Hofstadter 1979. I use the word "self-referential" here to suggest the recursive nature of inter-neural

connectivity. I'm not arguing that the brain is making reference through such connections to a self or identity.

190　*New nodes in the neuronal network during adulthood:* Gould et al. 1999.

190　*Synapses being strengthened or withering away as a result of feedback from the environment:* Hebb 1949; Changeux 1985.

190　*Synaptic plasticity:* See Klintsova and Greenough 1999 for an overview.

191　*CREB can be localized to individual synapses:* Yin 1999.

191　*Decentralized protein production and memory:* Barinaga 2000.

192　*Dependent relationship between short- and long-term memory:* John McCrone, personal communication.

193　*"the Darwinism of synapses . . .":* quoted from Changeux 1985:272.

193　*Abstract meme definitions:* Boyd and Richerson 1985:33; Blackmore 1999:66; Durham 1991:188; Dawkins 1982; Wilkins 1998; Hull 2000; Dennett 1995. Blackmore (1999:66) says, "I shall use the term 'meme' indiscriminately to refer to memetic information in any of its many forms; including ideas, the brain structures that instantiate those ideas, the behaviours these brain structures produce, and their versions in books, recipes, maps and written music. As long as that information can be copied by a process we may broadly call 'imitation,' then it counts as a meme."

193　*Memes as physical network of neurons:* Dawkins 1993; Delius 1991; Plotkin 1993.

193　*"if the brain stores information . . .":* quoted from Dawkins 1982:109.

196　*Neuronal nodes as attractors:* On the other hand, some attractor nodes will probably not be memetic. For example, there are reliable features of the world that could get translated into an attractor (e.g., border detection) but that do not ever replicate themselves through the production of signals. They fire incredibly regularly and always return to their designated state, but do not replicate. Examples might include things learned by each individual independently (thus not requiring social transmission)—things like universal regularities in the perceptual environment that produce single copies of these things in every human brain that grows to maturity. Thus being an attractor ensemble is not enough to become a meme: You have to return to state *and* reproduce at the same time. You also have to have another meme as a proximate cause, so that you become part of a lineage, a causal chain.

197　*Definition of neuromeme:* I should point out that I have a number of important predecessors in this area. The first is Manfred Eigen. In an aside on group selection in the brain, he wrote: "If there is mutual enhancement of linked neurons (forming, for example, neuronal groups), any cyclic closure may define a 'self-reproductive' firing unit. This self-reproductive unit may represent a 'template,' but not in the sense that it leads to multiplication of such circuits (e.g., groups). . . . The phenotype of synaptic linkage in a cycle may be excitation or firing density (or coherence of firing). Competition may be introduced by limitation of synaptic growth and inhibition. Finally, mutation may correspond to modifications (either structural or dynamic) within such cycles" (Eigen 1994:44–45). It is not surprising that Eigen should have expressed ideas so similar to the ones I have deduced from a similar framework, including his Replicator Theory (described earlier).

Calvin (1996) discusses information replication as part of the operation of his "Darwinian" machines within brains but does not use the word meme nor connect his discussion to the memetic literature in any way. Calvin is careful to provide a complete account of the evolution of these information units within brains, with mechanisms for variation, selection, and inheritance in a detailed neurobiological model. However, my schematic approach to replication shares little with Calvin's sophisticated neuroscientific approach.

Liane Gabora (1997) has also put forward an argument that memes must be in the brain. However, Gabora takes her cue from complexity theory. In her view, memes do not function in isolation but as parts of a complex conceptual network, or worldview, which should be the basic level at which cultural evolution is analyzed. For her, the tendency to view memes as discrete, identifiable units (or replicators) is problematic. As parts of the brain, memes are intimately involved in highly distributed, highly structured and interconnected networks. So memes spend most of their time as streams of thought that are continuously renewing, revising, and reassessing the constituent memes. It is these more holistic streams of information that should be our primary focus, in her view.

However, the closest predecessor to neuromemetics, as I develop it here, is the earlier view of Juan Delius. Delius is a prominent neuroethologist who, appropriately enough, works on social learning in songbirds. He is particularly interested in explaining the relatively stable aspects of culture as long-term memory traces. He proposes (Delius 1991) that the simultaneous activation of pre- and post-synaptic neurons (a classical Hebbian mechanism) makes neurons what he calls "hot," or more susceptible to firing. His proposal is that memes are clusters of neurons that become "hot" thanks to information acquired through social learning, particularly imitation or instruction. So for Delius behaviors like the production of signals are how memes express themselves phenotypically, mimicking Dawkins's view in this respect. Memes, then, as a form of memory, are physicochemical changes in the states adopted by clusters of neurons—a definition very close to my own. He notes that memes will be encoded as topologically unique patterns of "hotspots," so a trait in one brain will have a different geometrical configuration than that same trait manifests in another brain. Despite this, he argues they have similar functions in the host. Delius believes that many versions of a meme can be generated by a single individual as a form of mental creativity. He is, in other words, also a good neuronal selectionist. He emphasizes the material nature of memes as physical brain structures, modified by the mental activity stimulated by experiences of social learning. He emphasizes the parasitic nature of memetic replication and notes that memes can be beneficial symbionts or detrimental parasites. He discusses various ways in which memes can aggregate into coevolving complexes. All of these propositions should be familiar to the careful reader of this book.

My view differs from that of Delius primarily in the suggestion that social replication takes place through a mechanism other than social learning. Delius also does not describe memetic coevolution with artifacts (the topic of a later chapter). Otherwise we have come to largely convergent positions. What I provide is an extended argument for why we must think of memes in this way, to contrast with Delius's and Eigen's more programmatic statements.

198 *Genetic information as the novel configuration of matter in the double helix:* Jacob 1973:273.

201 *Multiple replicator copies in brain:* as suggested in Calvin 1996.

201 *No two neurons will have exactly the same effect on the larger network of which each is a part:* John McCrone, personal communication.

203 *Sensory feedback into brain as source of consciousness:* Harth 1994; Humphrey 1999.

204 *Local reconsolidation of memory:* Nader et al. 2000.

204 *False memory:* Loftus 1996.

205 *Fluid topographic maps in brain:* Faggin et al. 1997.

206 *Variation in cortical brain regions:* Mountcastle 1998:83.

208 *"Small world" networks:* Watts and Strogatz 1998.

208 *Local "neighborhoods" preserved by link replacement:* Amaral et al. 2000.

212 *Stem cells wired up in adult cortex:* Shors et al. 2001.

213 *Movement of memory with practice:* Jog et al. 1999.

215 *Similarity of "new" replicators:* Two of these novel replicators—prions and memes—are also found in the brain. Not only that, but prions can kill memes, so there is room for real conflict between these two replicators!

215 *Replicate through signals as intermediaries:* Prions don't share this particular feature with memes and comp-viruses.

217 *The representational capacity of cortex rises exponentially with the number of neurons:* Rolls and Treves 1998:261.

217 Quote: Monod 1970.

218 *"You have a lot of ideas . . .":* quoted from Buchanan 1985:95.

218 *"theorists in biology should realize . . .":* quoted from Crick 1988:142.

219 *Neuronal synchrony and behavioral response:* Singer 1999; Steinmetz et al. 2000.

219 *The neuronal representations voted to be less likely interpretations of events are suppressed:* Desimone 1998.

219 *Decay of synchrony as indicator of change in train of thought:* Rodriguez et al. 1999.

220 *Kinds of selection on brain spikes, and hence the memes that make them:* This formulation of specific selection regimes has another very important quality: It abolishes any complaints about memetics being tautological. It's an old saw

among creationists that evolutionary theory is tautological because it says that only the fit survive but then measures fitness by the number of survivors; survival figures as both the cause and the effect in the theory, making it circular. However, this is a caricature. The mechanism Darwin put forward, natural selection, places independent factors—the things doing the selecting, which can range from parasites to predators to asteroids—between the adaptations that *favor* survival and the actual fate of organisms. So evolutionary biology can define fitness independently of what winds up evolving, which saves the theory from being a tautology and makes it acceptable science.

But meme theory would still seem to be susceptible to the tautology complaint. Isn't it just that those ideas that happen to survive get declared winners after the fact in the case of memes? For the meme concept to escape the same problem, we must define *cultural* fitness independently of what happens to be found in the belief pool. If the first four notes of Beethoven's Fifth Symphony ("da da da daah") are declared a powerful meme only because people can commonly be heard humming it, we have achieved no insight; we have said nothing about why that particular musical phrase succeeds in dominating other potentially hummable tunes in people's minds.

When memeticists say that the popularity of "da da da daah" is explained by its psychological appeal, then memetic fitness is being defined by its success in producing replicas of itself. This is an example of just what the creationists complain about. But with neuromemetics, a meme's fitness is due to its ability to duplicate itself physically, which may be independent of its meaning or utility to those that harbor it. Fitness is not a question of whether a meme makes good use of the psychological structures in its environment. It's a question of how reliably a neuron can return to the same state or survive attacks that would convert it to a non-infectious state. The fitness of neuronally defined memes is a function of their kinetics (or rate of reproduction) against competitors, not just their current popularity.

Another reason the tautology argument doesn't stick to this new version of memetics is that what a meme represents in the mind is separated from its physical instantiation in the brain. Signals are presumably evaluated on the basis of their content, not because of the meme that might get constructed should the signal be admitted into the brain. There is a logical, temporal, and physical separation between the evaluation of signals by brains and the fate of neuromemes. So the tautology complaint just won't work against neuromemetics.

220 *"Mental Darwinism":* This perspective leads, however, to another major complaint against memetics: its lack of emphasis on intentionality. Why get rid of all intentionality in the picture of human mental life and social interaction when it is such a good way of explaining how humans work? It is, after all, a viewpoint that has dominated our thinking about thinking for hundreds, if not thousands, of years. And you can predict quite well what someone will do by assuming they want the same things in life you do. Drop-

ping this rich store of intuitions therefore seems a bit rash. But we don't know that intentionalism is right either. It's just a black box theory, based on folk psychology—thinking that other people's minds work like our own, with many beliefs and motivations likely to be shared between us. Perhaps the language of belief and desire is just an implant put in your head by natural selection because it was a convenient gloss on what really happens in the brain! In fact, assuming people are motivated to achieve a particular goal and to choose the course of action most likely to produce the desired consequence allows you to predict very well what other people will do. Our brains are designed to like such explanations for our actions because that's the language the brain uses to model human interaction for itself. But the interpretation of action in terms of intentions was planted there by a history of selection; that's why it seems natural—it is! We think of ourselves as motivated agents with the free will to decide our fates, and we infer that others are like this too. But that's no reason to think that we actually *are* such agents, just because we've adopted this motivational shorthand as a way to explain ourselves to ourselves in everyday discourse. That's not a scientific explanation, just the first one that comes to hand. That's why it's called a "folk" theory—one that sprang up from intuition before scientists got hold of the problem.

221 *Biophysical models of spikes:* See, for example, Koch 1998.

222 *Hierarchical processing in the brain:* Felleman and Van Essen 1991.

223 *Memes can express a range of representations:* The neuroscientific view should thus go some way to assuaging the qualms Ernst Mayr, the eminent evolutionary biologist of Harvard University, has about the idea of a meme. "It seems to me that this word [meme] is nothing but an unnecessary synonym of the term 'concept.'" See Mayr 1997.

223 *Theory of concepts:* See Margolis and Laurence 1999 for a contemporary overview of thinking about concepts.

223 *Neuromemetics is consistent with neuroscientific views of information representation in brains:* Some neuroscientists object to any attempt to break the brain into bits, which is what neuromemetics seems to do (at least analytically), with its talk about isolated states in brains. They argue that the brain must always be treated as a whole (see, for example, Freeman 1999). After all, the neuronal network is one joined-up totality, with every neuron being connected to every other one through some circuit of synapses. The brain is an organ, of a piece. From the perspective of these holists, everything is connected to everything else, and many current models of brain activity indicate that most stimuli have an effect on the entire cortical tangle. Anything interesting must therefore emerge, as a whole, from intrinsically complex developmental processes involving all of these parts. It's pointless, then, to try and tease this dynamic apart. Trying to define brain "states" or concepts as an isolated aspect of the brain is simply not meaningful. When reacting to a stimulus, the whole hierarchy of brain processes acts together, moving over the

course of a few hundred milliseconds from a raw state of representation to one that is focused and imbued with meaning. So any node's state evolves in concert with how the rest of the brain is reacting to a stimulus.

It's a fairly common objection to claim that some phenomenon is so complex and intertwined that you can't break it up into meaningful units. Many anthropologists say the same thing about culture. They argue that there are no units in culture, much less scraps of it that can replicate themselves. Instead culture is considered a complex amalgam of practices and institutions, not just ideas. These anthropologists believe you can't chop it into bits and put them under the microscope, that no bit has meaning outside its context. Culture is rather a vast web of mutual implications, like a maze with no way out. (Of course, at the same time, they admit that culture is learned, and presumably learned in bits and pieces as you acquire knowledge from others through communicative exchanges.)

All neuromemetics asserts is that there are interesting things happening at the level of circuits within the brain—like the replication of information. The existence of these circuits is well recognized by the same neuroscientists who complain about chopping up the neural real estate. Neuromemetics admits that the brain is a network. But think of this network in terms more familiar to us—as a social network. I used the example in Chapter 5 of two ladies, both of whom are treasurers of the Bingo-Players Association in their New York borough and so structurally equivalent in terms of these two associations. We know that neuronal networks, like social networks, are "small world" networks, which means both kinds of network will share many important properties—maybe even divisibility. This suggests that we can legitimately look at what a particular circuit is doing, without having to recognize that everything in the brain is connected to everything else. So it *is* appropriate to make an analogy between bingo clubs in New York and your brain!

The implication is that we may not have to take the whole brain into account at every moment to make sense of what is happening in there. Everyone in New York City is also connected in various ways—they're all bunched together within the city limits, pay city taxes, and share city services. But not every nose in New York needs to be counted whenever the Bingo-Players meet to discuss their affairs. Similarly, to effectively predict what behavioral response is impending, we may not need to measure the state of every neuron in the brain.

Sure, the brain is complex, and, yes, everything is connected to everything else in such a network, at least in some roundabout way. But in the end, if we're going to identify a *replication* mechanism as we did in genetics, we *have* to stick to the particulars of the case. That's why we can't just opt for "global understanding." We may need to think globally, but we have to analyze locally.

224 *Parameters of "brain code":* Rolls and Treves 1998; Singer 1999; Mountcastle 1998.

226 *In competitive environments, simple replicators evolve, as illustrated by Artificial Life simulations:* When selecting for complexity, mutually catalytic sets of elements emerge, which then organize themselves into higher-order mutualistic cooperatives, creating a self-maintaining ecosystem (Fontana and Buss 1994). The ecological conditions in Core Wars and TIERRA (described in an earlier chapter), on the other hand, favored the evolution of small size and low resource use. See also Lenski and Velicer 2224.

227 *Memes unlike other "independent" replicators:* This statement is somewhat unfair to genes, which nowadays have mostly tethered themselves into interdependent teams along chromosomes, having evolved numerous ways of establishing higher orders of organization through cooperation (Haig 1997). Memes, in contrast, have had this condition thrust upon them from their early history as parasites on gene-produced brains.

228 *"Virulent 'mind viruses'":* fostered by authorities such as Dawkins 1993; Blackmore 1999; Brodie 1996; and Lynch 1996.

228 *Memes as symbionts with humans:* Harms 1996.

229 *Parasitism means dependent, not hostile:* Dogiel 1964.

229 *Memes virulent to the degree they are independent:* Dawkins 1982:110.

230 *Selection on memory for food items during evolution:* Milton 1988.

230 *Selection on memory for social skills during evolution:* Humphrey 1976.

231 *Higher-order mental functions centered in quintessential human cognitive adaptation, the prefrontal cortex:* Cziko 1995.

PAGE   **CHAPTER 8: ESCAPE FROM PLANET BRAIN**

232 Head quote: quoted in Deacon 1999.

232 Head quote: Owings and Morton 1998:11.

233 *"Mind the gap":* I was reminded of this phrase—constantly replicated by either a tape-recorded or electronically synthesized human voice loop in the London underground as a warning to those stepping from the platform into trains—by John Constable.

237 *Social signals as vehicles:* Dawkins 1982:109–10.

239 *Extended phenotype notion:* Dawkins 1982.

246 *"Hot" paper citation study:* Hull 1988.

247 *Reconstructive nature of human communication:* Sperber 2000; Sperber and Wilson 1986.

247 *No mental replicator chains possible through social communication:* Sperber 1996, Chapter 6.

249 *Channeling of inference not enough to save memes from degradation:* David Hull, personal communication.

249 *Memetic lineages maintained in the face of significantly error-prone transmission:* Gil-White (2001).

251 *Genes, facing information overload, left the burden to culture:* Bonner 1988:208.

255 *Memes have "feathered their nest" in the brain over time:* Dennett 1991:207.

255 *Communication is commonly defined as the transmission of information:* See, for example, Bradbury and Vehrencamp 1998:353.

256 *The need to infer message sender's intent during human communication:* Sperber and Wilson 1986.

256 *"Quasi-interactions":* Thompson 1994.

257 *"Mechanical" communication theory:* Shannon and Weaver (1965 [1949]).

259 *"Inferential" communication theory:* Sperber and Wilson 1986, 1997.

259 *Early source of Inferential theory:* Grice (1991 [1975]).

260 *The "relevance" maxim:* Sperber and Wilson 1986.

261 *The need to share inference rules for successful communication:* Sperber and Wilson 1986:15.

261 *Evolutionary communication theory:* See Bradbury and Vehrencamp 1998, Dusenbery 1992, and Hauser 1996 for summaries.

261 *Ritualization for efficiency of communication:* Bradbury and Vehrencamp 1998:535.

262 *Tendency to increasingly sophisticated models of sender-receiver psychology:* Although many in media studies have attempted to extend the mathematical branch of communication theory in this way, it is actually unfair, since the developers of the Mechanical approach explicitly cautioned that this formal theory of communication ignores the source or destination of messages, leaving these as psychological black boxes (Cherry 1966:41; Shannon and Weaver 1965 [1949]). It is therefore not intended as a complete theory of human communication (Cherry 1966:52). In a famous lament, Shannon himself cautioned that "workers in other fields should realize that the basic results of the subject are aimed in a very specific direction, a direction that is not necessarily relevant to such fields as psychology, economics, and other social sciences" (IEEE Transactions on Information Theory, December 1955), quoted in Cherry 1966.

263 *When the biological interests of sender and receiver significantly overlap, signaling becomes more efficient:* Johnstone 1997:164.

263 *Arms race between sender and receiver:* Dawkins and Krebs 1979.

264 *Signaling systems may arise as a means of exploiting a preexisting sensory bias:* Johnstone 1997:174; Dusenbery 1992.

264 *Sender-receiver functions in Evolutionary communication:* Bradbury and Vehrencamp 1998:357.

264 *An overarching perspective on the evolution of communication is lacking:* Johnstone 1997:155.

268 *Imitation as a form of social learning:* For general references, see Heyes and Galef 1996; Whiten and Ham 1992.

269 *Cultural learning:* Tomasello et al. 1993.

269 *"Theory of mind" and imitation:* See, for example, Whiten 1991.

270 *Birdsong learning not true imitation:* Tomasello 1996:324.

271 *Birdsong exhibits all the characteristics of cultural evolution:* A selection of

recent articles from a considerable literature could include Baker 1996; Ficken and Popp 1995; Grant and Grant 1996; and Payne 1996.

271 *Admitting birdsong to the culture "club"*: Blackmore (1999:49–50) attempts to get around this problem by admitting birdsong as an example of memetic spread, despite its not being based on true imitation, but this is an obvious kludge, since no animal *between* birds and humans in evolutionary history gets admitted to the "true imitation club."

271 *Imitation is the only form of social learning appropriate as a foundation of human cultural evolution*: Boyd and Richerson 1985:35; Blackmore 1999:47–52. See also Durham 1991:188.

271 *Imitation as meme's mechanism of transmission*: Dawkins 1976:206; Dawkins 1999. However, as Blackmore notes, Dawkins takes this in a "general" way, since he admits that memes "can propagate themselves from brain to brain, from brain to book, from book to brain, from brain to computer," etc. (Dawkins 1986:158).

271 *Imitation doesn't ensure fidelity in the transmission of information*: Heyes and Plotkin 1989; Heyes 1993.

272 *"Mirror" neurons*: Gallese et al. 1996; Rizzolatti and Arbib 1998.

272 *Same neurons activated when viewing and later recalling the* Mona Lisa: Kreiman et al. 2000.

PAGE   CHAPTER 9: THE TECHNO-TANGO

276 Head quote: Moravec 1998, Chapter 1.

276 *Scary techno-futures*: See, for example, Moravec 1998.

278 *Artifacts move from being mental to physical objects*: Barbieri 2001:220.

278 *Artifacts as a second-order consequence of ideas*: Dawkins (1982) has called such things "extended phenotypes," to acknowledge that they are attached to their creators only in the evolutionary sense that the activity of the body put them there in the first place. (This idea was also discussed in the previous chapter, but with respect to signals rather than artifacts.)

279 *Niche construction theory*: Laland et al. 2000.

280 *"Cultural" niche construction theory*: Laland et al. 2001.

281 *A wagon and its memes*: Dennett (1991:204) uses the same example of a wheeled wagon: "A wagon with spoked wheels carries not only grain or freight from place to place; it carries the brilliant idea of a wagon with spoked wheels from mind to mind."

282 *Memes inside artifacts*: Dan Dennett, for example, continues to argue that an abstract, substrate-neutral definition of memes is to be preferred to any attempt to concretize them. See Dennett (in press).

282 *Meme journeying through signals, minds, and artifacts*: This picture satisfies in large part an old desire of Dennett's: "It is tempting to suppose that some concept of *information* could serve eventually to unify mind, matter, and

meaning in a single theory" (Dennett and Haugeland 1987). Meaning here becomes the importation of memes into minds from others, either directly, through straightforward communication, or indirectly, through artifacts.

285 *Complex machines as good interactors:* It's true that tools can also influence the biological success of the genes responsible for their production. The story of Cro-Magnon groups defeating Neanderthals in prehistoric Europe is the most classic case of a more sophisticated tool kit leading to one population's triumph over another. So, technically speaking, tools qualify as interactors according to the original definition of interactors (cited in Chapter 3). But it remains true that artifacts are evolving into new territories, as will become clear as we go along.

293 *Evolution of symbiotic relationships:* Dogiel 1964:21.

294 *An organism is a special kind of host because it can adapt to a parasite:* Dogiel 1964:433–34.

295 *Nano-scaled robot replicators:* Drexler 1981.

297 *Self-assembling buckyballs and nanotechnology:* Richard E. Smalley, "Self-assembly of fullerene tubes and balls." An address presented before The Robert A. Welch Foundation 39th Conference on Chemical Research: Nanophase Chemistry, held in Houston, TX, October 23–24, 1995, <http://cnst.rice.edu/welch95.html>

302 *"Observable" memetics:* Gatherer 1998.

302 *The "Machiavellian intelligence hypothesis":* Byrne and Whiten 1988.

303 *Problems with the "widely shared" criterion of culture:* Aunger 1999.

304 *Chimpanzee cultural traditions:* Whiten et al. 1999.

304 *Variation in chimpanzee termiting techniques:* McGrew 1992.

305 *Social customs in guppies:* Laland and Williams 1998.

305 *"Ratchet effect" in human cultural evolution:* Tomasello 2000.

305 *A prominent point of view on cumulative culture:* The next several paragraphs summarize the argument of Tomasello 2000.

305 *Intelligence and social life do not produce culture:* Boyd and Richerson 1996:78.

307 *"Culture" has a body:* This point has been argued by a variety of recent theories, including niche construction (Odling-Smee 1988), distributed (Hutchins 1994) and situated cognition (Resnick et al. 1991), and the so-called "Russian school" of Vygotsky acolytes (Cole 1998).

307 *Intelligence within group structures:* Hutchins and Hazelhurst 1991.

308 *Cumulative culture need not depend on imitation:* Blackmore (1999), Boyd and Richerson (1985), and Tomasello (2000) all insist it does.

309 *Dynamics of artifact-based culture:* Laland et al. 2001.

310 *Culture "out there" in the environment:* I'm not alone in breaking with purely mentalist cognitivism. Roy D'Andrade, in his masterful intellectual history of the discipline, argues that cognitive anthropology is now in a fourth "phase" of development in which "culture is seen to be particulate, socially distributed, variably internalized, and variably embodied in external forms [i.e., behaviors and artifacts]" (D'Andrade 1995:248).

PAGE   **CHAPTER 10: RETHINKING REPLICATION**

313   *Viroids as parasites-on-parasites:* Diener 1979, 1996.

313   *Super-parasites need help constructing substrate:* You might think that this principle can be extended to genes as well, that genes can be seen as being parasitic on preexisting molecules produced by general chemical processes. DNA then converts these building blocks into a specialized substrate for replication by tying them together into the double helix. It's true that DNA doesn't make nucleotides from scratch. But independent nucleotides are not replicators, and I define super-parasites as being dependent on other *replicators*. So the buck stops with DNA.

314   *"Routes to replication":* The transmission-through-signals route is not available to prions.

314   *"Natural bias" toward a replicator state:* Introducing the category of naturally biased memes is similar to Atran's (1998) division of memes into "core" and "developing" categories, Plotkin's (2000) distinction between "deep" and "surface" memes, and Tooby and Cosmides's (1992) "evoked" as opposed to "transmitted" culture. The former category in each case is a cultural unit with little dependency on social input, and hence recurrent, while the latter is cross-culturally variable.

314   *Semantic primitives like "over" as evidence of cognitive primitives:* Lakoff 1987.

318   *Physical overlap as distinction between biological and cultural replication:* Griesemer (in press a).

319   *Messenger RNA as a signal:* Mazia 1956 (cited in Sarkar 1996).

PAGE   **CHAPTER 11: THE REVOLUTION OF MEMES**

323   Head quote: Dobzhansky 1962.

323   *"there is no consensus opinion amongst geneticists . . .":* quoted in the Introduction to Fox Keller 2000.

324   *Historical ages and their analogical thinking about brains:* Changeux 1995.

327   *Self composed entirely of memes:* See, for example, Blackmore 1999; Gabora 1997.

BIBLIOGRAPHY

Alberts, Bruce, Dennis Bray, Julian Lewis, Martin Raff, Keith Roberts, and James D. Watson (1994; 3rd ed.) Molecular Biology of the Cell. New York: Garland Publishers.

Alexander, Richard (1979) Darwinism and Human Affairs. Seattle: University of Washington Press.

Amaral, L.A.N., A. Scala, M. Barthélémy, and H.E. Stanley (2000) "Classes of small-world networks," PNAS USA 97:11149–52.

Atran, Scott (1998) "Folk biology and the anthropology of science: Cognitive universals and cultural particulars," Behavioral and Brain Sciences.

Aunger, Robert (1999) "Against Idealism/Contra consensus," Current Anthropology 40:S93–S101.

Aunger, Robert (ed.) (2000) Darwinizing Culture: The Status of Memetics as a Science. Oxford: Oxford University Press.

Bailey, Craig H., Maurizio Giustetto, Yan-You Huang, Robert D. Hawkins, and Eric R. Kandel (2000) "Is heterosynaptic modulation essential for stabilizing Hebbian plasticity and memory?" Nature Reviews Neuroscience 1:11–20.

Baker, David (2000) "A surprising simplicity to protein folding," Nature 405:39–42.

Baker, M.C. (1996) "Depauperate meme pool of vocal signals in an island population of singing honeyeaters," Animal Behaviour 51:853–58.

Barbieri, Marcello (2001) The Organic Codes: The Birth of Semantic Biology. Ancona: Pequod.

Barinaga, Marcia (2000) "Synapses call the shots," Science 290:736.

Barkow, Jerome H., Leda Cosmides, and John Tooby (eds.) (1992) The Adapted Mind: Evolutionary Psychology and the Generation of Culture. Oxford: Oxford University Press.

Bateson, Gregory (1972) Steps to an Ecology of Mind: Collected Essays in Anthropology, Psychiatry, Evolution, and Epistemology. Chicago: University of Chicago Press.

Bennett, Charles H. (1995) "Quantum information and computation," Physics Today 48(10):24–30.

Bennett, Charles H., G. Brassard, C. Crepeau, R. Jozsa, A. Peres, and W. Wootters (1993) "Teleporting an unknown quantum state via dual classical and Einstein–Podolsky–Rosen channels," Physical Review Letters 70:1895–99.

Bennett, Charles H., and Rolf Landauer (1985) "The fundamental physical limits of computation," Scientific American 255:38–46.

Benzon, William (1996) "Culture as an evolutionary arena," Journal of Social and Evolutionary Systems 19:321–62. <http://www.newsavanna.com/wlb/CE/Arena/Arenaoo.shtml>

Best, Michael L., and Richard Pocklington (1999) "Meaning as use: Transmission fidelity and evolution in NetNews," Journal of Theoretical Biology 196:278–84.

Betzig, Laura (1999) "People are animals." In: Human Nature: A Critical Reader, Laura Betzig (ed.). Oxford: Oxford University Press, pp. 1–17.

Betzig, Laura, Monique Borgerhoff-Mulder, and Paul Turke (eds.) (1988) Human Reproductive Behaviour: A Darwinian Perspective. New York: Cambridge University Press.

Blackmore, Susan (2000) "The power of memes," Scientific American 283:52–61.

Blackmore, Susan (1999) The Meme Machine. Oxford: Oxford University Press.

Bonner, John Tylor (1988) The Evolution of Complexity by Means of Natural Selection. Princeton: Princeton University Press.

Bonner, John Tylor (1980) The Evolution of Culture in Animals. Princeton: Princeton University Press.

Bontchev, Vesselin (1994) "Are 'good' computer viruses still a bad idea?" EICAR conference.<http://www.informatik.uni-hamburg.de/pub/virus/texts/viruses/goodvir.html>

Bourgeois, Jean-Pierre, Patricia S. Goldman-Rakic, and Pasko Rakic (1999) "Formation, elimination, and stabilization of synapses in the primate cerebral cortex." In: The New Cognitive Neurosciences, Michael S. Gazzaniga (ed.). Cambridge, MA: MIT Press.

Bouwmeester, D., J.W. Pan, K. Mattle, M. Eibl, H. Weinfurter, and A. Zeilinger (1998) "Experimental quantum teleportation." Philosophical Transactions of the Royal Society of London Series A—Mathematical Physical and Engineering Sciences 356: 1733–37.

Bouwmeester, Dik, Jian-Wei Pan, Klaus Mattle, Manfred Eibl, Harald Weinfurter, and Anton Zeilinger (1997) "Experimental quantum teleportation," Nature 390:575.

Boyd, Robert, and Peter J. Richerson (2000) "Memes: Universal acid or a better mousetrap?" In: Darwinizing Culture: The Status of Memetics as a Science, Robert Aunger (ed.). Oxford: Oxford University Press, pp. 143–62.

Boyd, Robert, and Peter J. Richerson (1996) "Why culture is common, but cultural evolution is rare," Proceedings of the British Academy 88: 77–93.

Boyd, Robert, and Peter J. Richerson (1987) "The evolution of ethnic markers," Cultural Anthropology 2:65–79.

Boyd, Robert, and Peter J. Richerson (1985) Culture and the Evolutionary Process. Chicago: University of Chicago Press.

Bradbury, Jack W., and Sandra L. Vehrencamp (1998) Principles of Animal Communication. Sunderland, MA: Sinauer Associates.

Brodie, Richard (1996) Virus of the Mind: The New Science of the Meme. Seattle, WA: Integral Press.

Buchanan, Bruce (1985) "Steps toward mechanizing discovery." In: Logic of Discovery and Diagnosis in Medicine, Kenneth Schaffner (ed.). Berkeley: University of California Press, pp. 94–114.

Buller, David J. (1999) "Evolutionary psychology." In: A Field Guide to the Philosophy of Mind, Marco Nani and Massimo Marraffa (eds.), Roma Tre, Università degli Studi: Società Italiana di Filosofia Analitica.
<http://www.uniroma3.it/kant/field/ep.htm>

Buss, David M. (1998) Evolutionary Psychology: The New Science of the Mind. Boston: Allyn and Bacon.

Buss, Leo (1987) The Evolution of Individuality. Princeton: Princeton University Press.

Byrne, Richard, and Andrew Whiten, eds. (1988) Machiavellian Intelligence: Social Expertise and the Evolution of Intellect in Monkeys, Apes and Humans. Oxford: Oxford University Press.

Calvin, William H. (1996) The Cerebral Code: Thinking a Thought in the Mosaics of the Mind. Cambridge, MA: MIT Press.

Cavalli-Sforza, Luigi L., and Marcus W. Feldman (1981) Cultural Transmission and Evolution: A Quantitative Approach. Princeton: Princeton University Press.

Changeux, Jean-Pierre (1985) Neuronal Man. Princeton: Princeton University Press.

Cherry, Colin (1966; 2nd ed.) On Human Communication: A Review, a Survey and a Criticism. Cambridge, MA: MIT Press.

Chien, Peter, and Jonathan S. Weissman (2001) "Conformational diversity in a yeast prion dictates its seeding specificity," Nature 410:223–27.

Cohen, Frederick B. (1994; 2nd ed.) A Short Course on Computer Viruses. New York: Wiley.

Cohen, Frederick B. (1986) "Computer Viruses," dissertation for the Department of Electrical Engineering, University of Southern California.

Cohen, Frederick B. (1984) "Computer Viruses: Theory and Experiments."
<http://www.all.net/books/virus/>

Cole, Michael (1998) Cultural Psychology: A Once and Future Discipline. Cambridge, MA: Belknap Press.

Cosmides, Leda, and John Tooby (1987) "From evolution to behavior: Evolutionary psychology as the missing link." In: The Latest on the Best, John Dupré (ed.). Cambridge, MA: MIT Press.

Crandall, B.C. (1996) "Global Algorithm 1.5: The Nanotech Future: A Digital Conversation with B.C. Crandall." CTHEORY.
<http://www.ctheory.com/ga1.5-nanotech.html>

Crick, Francis H.C. (1988) What Mad Pursuit: A Personal View of Scientific Discovery. New York: Basic Books.

Crick, Francis H.C. (1970) "Central dogma of molecular biology," Nature 227:561–63.

Crick, Francis H.C. (1958) "On protein synthesis," Symposium of the Society for Experimental Biology 12:138–63.

Cziko, Gary (1995) Without Miracles. Cambridge, MA: MIT Press.

Daly, Martin, and Margo Wilson (1988) Homicide. New York: Aldine de Gruyter.

D'Andrade, Roy (1995) The Development of Cognitive Anthropology. Cambridge: Cambridge University Press.

Darwin, Charles R. (1909) [1859] The Origin of Species. New York: P. F. Collier and Son.

Dawkins, Richard (1999) "Introduction" to Blackmore 1999.

Dawkins, Richard (1993) "Viruses of the mind." In: Dennett and His Critics: Demystifying Mind, B. Dahlbohm, (ed.). Oxford: Blackwell, pp. 13–27.

Dawkins, Richard (1986) The Blind Watchmaker. New York: Norton.

Dawkins, Richard (1982) The Extended Phenotype. Oxford: Oxford University Press.

Dawkins, Richard (1978) "Replicator selection and the extended phenotype," Zeitschrift für Tierpsychologie 47:61–76.

Dawkins, Richard (1976) The Selfish Gene. Oxford: Oxford University Press.

Dawkins, Richard, and John R. Krebs (1979) "Animal signals: Information or manipulation?" In: Behavioral Ecology, John R. Krebs and N.B. Davies (eds.). London: Blackwell, pp. 282–309.

Deacon, Terrence (1999) "Memes as signs: The trouble with memes (and what to do about it)," The Semiotic Review of Books 10(3):1–3.
<http://www.chass.utoronto.ca/epc/srb/srb/10-3edit.html>

Deacon, Terrence (1997) The Symbolic Species. New York: Penguin.

Delius, Juan (1991) "The nature of culture." In: The Tinbergen Legacy, Marian Stamp Dawkins, Timothy R. Halliday, and Richard Dawkins (eds.). London: Chapman and Hall, pp. 75–99.

Dennett, Daniel C. (in press) "From typo to thinko: When evolution graduated to semantic norms." Ms in preparation for volume from the Fyssen Foundation Conference on Evolution and Culture.

Dennett, Daniel C. (2000) Foreword to Aunger 2000.

Dennett, Daniel C. (1999) "The evolution of culture." Charles Simonyi Lecture, Oxford, delivered February 17, 1999.

Dennett, Daniel C. (1995) Darwin's Dangerous Idea: Evolution and the Meanings of Life. London: Allen Lane.

Dennett, Daniel C. (1991) Consciousness Explained. Boston: Little, Brown.

Dennett, Daniel C., and John Haugeland (1987) "Intentionality." In: The Oxford Companion to the Mind, Richard L. Gregory (ed.). Oxford: Oxford University Press.

Desimone, Robert (1998) "Visual attention mediated by biased competition in extrastriate visual cortex," Philosophical Transactions of the Royal Society of London Series B—Biological Sciences 353:1245–55.

Dewdney, A.K. (1984) "In the game called Core War hostile programs engage in the battle of bits," Scientific American [reprinted in The Armchair Universe: An Exploration of Computer Worlds, New York: W.H. Freeman 1988].

Dhurandhar, N.V., B.A. Israel, J.M. Kolesar, G.F. Mayhew, M.E. Cook, and R.L. Atkinson (2000) "Increased adiposity in animals due to a human virus," International Journal of Obesity 24:989–96.

Dick, Philip K. (1982) Do Androids Dream of Electric Sheep? New York: Gallantine.

Diener, Theodor O. (1996) "Origin and evolution of viroids and viroid like satellite RNAs," Virus Genes 11:119–31.

Diener, Theodor O. (1979) Viroids and Viroid Diseases. New York: Wiley.

Dobzhansky, Theodosius (1962) Mankind Evolving: The Evolution of the Human Species. New Haven: Yale University Press.

Dogiel, V.A. (1964) General Parasitology. London: Oliver and Boyd.

Drexler, K. Eric (1981) "Molecular engineering: An approach to the development of general capabilities for molecular manipulation," PNAS USA. 78:5275–78.

Durham, William H. (1992) "Applications of evolutionary culture theory," Annual Reviews in Anthropology 21:331–55.

Durham, William H. (1991) Coevolution: Genes, Culture and Human Diversity. Stanford: Stanford University Press.

Dusenbery, David B. (1992) Sensory Ecology: How Organisms Acquire and Respond to Information. New York: W.H. Freeman.

Edelman, Gerald M. (1993) Bright Air, Brilliant Fire: On the Matter of Mind. New York: Basic Books.

Eigen, Manfred (1996) "Prionics, or the kinetic basis of prion diseases," Biophysical Chemistry 63:A1–A18.

Eigen, Manfred (1994) "Selection and the origin of information." In: Selectionism and the Brain, Manfred Eigen, Olaf Sporns and G. Tononi, (eds.). San Diego: Academic Press, pp. 35–52.

Eigen, Manfred (1992) Steps Toward Life: A Perspective on Evolution. Oxford: Oxford University Press.

Eigen, Manfred (1971) "Self-organization of matter and the evolution of biological macromolecules," Naturwissenschaften 58:465–523.

Eliot, T.S. (1963 [1934]) "Choruses from 'The Rock.'" In: Collected Poems 1909–1962. New York: Harcourt Brace.

Faggin, B.M., K.T. Nguyen, and M.A.L. Nicolelis (1997) "Immediate and simultaneous sensory reorganization at cortical and subcortical levels of the somatosensory system," PNAS USA 94:9428–33.

Feldman, Marcus W., and Kevin Laland (1996) "Gene-culture coevolutionary theory," Trends in Ecology and Evolution 11:453–57.

Feldman, Marcus W., and Luigi L. Cavalli-Sforza (1979) "Aspects of variance and covariance analysis with cultural inheritance," Population Biology 15:276–307.

Felleman, D.J., and David Van Essen (1991) "Distributed hierarchical processing in the primate cerebral cortex," Cerebral Cortex 1:1–47.

Ficken, M.S., and J.S. Popp (1995) "Long-term persistence of a culturally transmitted vocalization of the Black-Capped Chickadee," Animal Behaviour 50:683–93.

Flinn, Mark V. (1997) "Culture and the evolution of social learning." Evolution and Human Behavior 18:23–67.

Fontana, Walter, and Leo W. Buss (1994) "The arrival of the fittest: Toward a theory of biological organization," Bulletin of Mathematical Biology 56:1–64.

Fox Keller, Evelyn (2000) The Century of the Gene. Cambridge: Harvard University Press.

Freeman, Walter J. (1999) How Brains Make Up Their Minds. London: Weidenfeld and Nicolson.

Furusawa, A., J.L. Sørensen, S.L. Braunstein, C.A. Fuchs, H.J. Kimble, and E.S. Polzik (1998) "Unconditional quantum teleportation," Science 282:706–9.

Gabora, Liane (1997) "The origin and evolution of culture and creativity," Journal of Memetics—Evolutionary Models of Information Transmission 1. <http://www.cpm.mmu.ac.uk/jom-emit/1997/vol1/gabora_l.html>

Gallese, V., L. Fadiga, L. Fogassi, and G. Rizzolatti (1996) "Action recognition in the premotor cortex," Brain 119:593–609.

Garcia, John, and R. Koelling (1966) "Relation of cue to consequences in avoidance learning," Psychonomic Science 4:123–24.

Gatherer, Derek (1998) "Why the 'thought contagion' metaphor is retarding the progress of memetics," Journal of Memetics—Evolutionary Models of Information Transmission 2. <http://www.cpm.mmu.ac.uk/jom-emit/1998/vol2/gatherer_d.html>

Gaulin, Steven J.C., and Donald McBurney (2000) Psychology: An Evolutionary Approach. New York: Prentice Hall.

Gazzaniga, Michael S. (1994) Nature's Mind: The Biological Roots of Thinking, Emotions, Sexuality, Language and Intelligence. New York: Basic Books.

Gil-White, Francisco J. (2001) "L'evolution culturelle a-t-elle des règles?" La rechérche Hors-Série No. 5(Avril):92–97.

Gladwell, Malcolm (2000) The Tipping Point: How Little Things Can Make a Big Difference. New York: Little, Brown.

Godfrey-Smith, Peter (2000) "The replicator in retrospect," Biology and Philosophy 15(3):403–23.

Godin, Seth (2000) Unleashing the Ideavirus. New York: Do You Zoom, Inc.

Gordon, Sara (1993) "Inside the mind of Dark Avenger" Virus News International. <http://mh101.infi.net/~wtnewton/vinfo/darkmind.html>

Gould, Elizabeth, Alison J. Reeves, Michael S.A. Graziano, and Charles G. Gross (1999) "Neurogenesis in the neocortex of adult primates," Science 286:548–52.

Grant, B.R., and P.R. Grant (1996) "Cultural inheritance of song and its role in the evolution of Darwin's finches," Evolution 50:2471–87.

Grice, H.P. (1991 [1975]) Studies in the Way of Words. Cambridge, MA: Harvard University Press.

Griesemer, James R. (in press a) "Development, culture and the units of inheritance," Philosophy of Science (Proceedings) 67:S348–68.

Griesemer, James R. (in press b) "The informational gene and the substantial body: On the generalization of evolutionary theory by abstraction." In: Varieties of Idealization. Pozman Studies in the Philosophy of the Sciences and the Humanities, Nancy Cartwright and Martin Jones (eds.). Amsterdam: Editions Rodopi Publishers.

Guglielmino, C.R., C. Viganotti, Barry Hewlett, and Luigi L. Cavalli-Sforza (1995) "Cultural variation in Africa: Role of mechanisms of transmission and adaptation," PNAS USA 92:7585–89.

Haig, David (1997; 4th ed) "The social gene." In: Behavioural Ecology: An Evolutionary Approach, John R. Krebs and Nicholas B. Davies (eds.). Cambridge: Blackwell, pp. 284–306.

Harms, William (1996) "Cultural evolution and the variable phenotype," Biology and Philosophy 11:357–75.

Harth, Erich (1994) The Creative Loop: How the Brain Makes a Mind. New York: Perseus Books.

Hauser, Marc (1996) The Evolution of Communication. Cambridge, MA: MIT Press.

Hebb, Donald O. (1949) The Organization of Behavior: A Neuropsychological Theory. New York: Wiley.

Henrich, Joseph (in press) "Cultural transmission and the diffusion of innovations: Adoption dynamics indicate that biased cultural transmission is the predominate force in behavioral change and much of sociocultural evolution." American Anthropologist.

Heyes, Celia M. (1993) "Imitation, culture and cognition," Animal Behaviour 46:999–1010.

Heyes, Celia M., and Bennett G. Galef (1996) Social Learning in Animals: The Roots of Culture. San Diego: Academic Press.

Heyes, Celia M., and Henry C. Plotkin (1989) "Replicators and interactors in cultural evolution." In: What the Philosophy of Biology Is, Michael Ruse (ed.). Dordrecht: Reidel, pp. 139–62.

Hill, Andrew F., Susan Joiner, Jackie Linehan, Melanie Desbruslais, Peter L. Lantos, and John Collinge (2000) "Species-barrier-independent prion replication in apparently resistant species," PNAS USA 97, 10248–10253.

Hill, Kim, and A. Magdalena Hurtado (1996) Ache Life History: The Ecology and Demography of a Foraging People. Hawthorne, NY: Aldine de Gruyter.

Hofbauer, J., and Karl Sigmund (1988) The Theory of Evolution and Dynamical Systems. Cambridge: Cambridge University Press.

Hofstadter, Douglas (1979) Gödel, Escher, Bach. New York: Basic Books.

Hopfield, John J. (1982) "Neural networks and physical systems with emergent collective computational abilities," PNAS USA 79:2554–58.

Hull, David L. (2000) "Taking memetics seriously: Memetics will be what we make it." In: Darwinizing Culture: The Status of Memetics as a Science, Robert Aunger (ed.). Oxford: Oxford University Press, pp. 43–68.

Hull, David L. (1988) Science as a Process. Chicago: University of Chicago Press.

Hull, David L. (1982) "The naked meme." In: Learning, Development and Culture, Henry C. Plotkin (ed.). London: Wiley, pp. 273–327.

Hull, David L. (1980) "Individuality and selection," Annual Review of Ecology and Systematics 11:311–32.

Hull, David L., R.E. Langman, and S.S. Glenn (in press) "A general account of selection: Biology, immunology and behavior," Behavioral and Brain Sciences.

Humphrey, Nicholas (1999) A History of the Mind: Evolution and the Birth of Consciousness. New York: Copernicus Books.

Humphrey, Nicholas (1976) "The social function of intellect." In: Growing Points in Ethology, Patrick Bateson and Robert Hinde (eds.). Cambridge: Cambridge University Press.

Hurst, Laurence D., Anne Atlan, and Bengt O. Bengtsson (1996) "Genetic conflicts," Quarterly Review of Biology 71:317–64.

Husband, Scott, and Toru Shimizu (1999) "Evolution of the avian visual system." In: Avian Visual Cognition, R. Cook (ed.).
<www.pigeon.psy.tufts.edu/avc/husband>

Hutchins, Edwin (1994) Cognition in the Wild. Cambridge, MA: MIT Press.

Hutchins, Edwin, and Brian Hazelhurst (1991) "Learning in the cultural process." In: Artificial Life II, Christopher G. Langton, Charles Taylor, J. Doyne Farmer, and Steen Rasmussen (eds.). Reading, MA: Addison-Wesley, pp. 689–706.

Ingold, Tim (2000) "The poverty of selectionism," Anthropology Today 16(3):1–2.

Ingold, Tim (1986) Evolution and Social Life. Cambridge: Cambridge University Press.

Jacob, Francois (1973) The Logic of Life. New York: Random House.

Jeffreys, Mark (2000) "The meme metaphor," Perspectives in Biology and Medicine 43:227–42. <http://www.press.jhu.edu/journals/perspectives_in_biology_and_medicine/v043/43.2jeffreys.html>

Jog, Mandar S., Yasuo Kubota, Christopher I. Connolly, Viveka Hillegaart, and Ann M. Graybiel (1999) "Building neural representations of habits," Science 286:1745–49.

Johnstone, Rufus (1997; 4th ed.) "The evolution of animal signals." In: Behavioural Ecology: An Evolutionary Approach, John R. Krebs and Nicholas B. Davies (eds.). Cambridge: Blackwell, pp. 155–78.

Joyce, Gerald F. (1992) "Directed molecular evolution," Scientific American 267(6):90.

Joyce, Gerald F. (1989) "RNA evolution and the origins of life," Nature 338:217–24.

Junqueira, Luiz Carlos, Jose Carneiro, Luis Carl Junqueira, and Robert O. Kelley (1998; 9th edition) Basic Histology. New York: McGraw-Hill Professional Publishing.

Kaufmann, Stuart A. (1993) The Origins of Order: Self-Organisation and Selection in Evolution. New York: Oxford University Press.

Kaufmann, Stuart A. (1986) "Autocatalytic sets of proteins," Journal of Theoretical Biology 119:1–24.

Kay, Lily E. (2000) Who Wrote the Book of Life? A History of the Genetic Code. Stanford: Stanford University Press.

Kelly, Kevin (1994) Out of Control: The New Biology of Machines. London: Fourth Estate.

Kendal, Jeremy R., and Kevin N. Laland (2000) "Mathematical models for memetics," Journal of Memetics—Evolutionary Models of Information Transmission 4. <http://www.cpm.mmu.ac.uk/jom-emit/2000/vol4/kendal_jr&laland_kn.html>

Klintsova, Anna Y., and William T. Greenough (1999) "Synaptic plasticity in cortical systems," Current Opinion in Neurobiology 9:203–8.

Koch, Christof (1998) Biophysics of Computation: Information Processing in Single Neurons. Oxford: Oxford University Press.

Koch, Richard (2001) The Natural Laws of Business: Applying the Theories of Darwin, Einstein, and Newton to Achieve Business Success. New York: Doubleday.

Kreiman, Gabriel, Christof Koch, and Itzhak Fried (2000) "Imagery neurons in the human brain," Nature 408:357–61.

Lakoff, George (1987) Women, Fire and Dangerous Things. Chicago: University of Chicago Press.

Laland, Kevin N., John Odling-Smee, and Marcus W. Feldman (2001) "Cultural niche construction and human evolution," Journal of Evolutionary Biology 14:22–33.

Laland, Kevin N., John Odling-Smee, and Marcus W. Feldman (2000) "Niche construction, biological evolution and cultural change," Behavioural and Brain Sciences 23:131–75.

Laland, Kevin N., and K. Williams (1998) "Social transmission of maladaptive information in the guppy," Behavioural Ecology 9(5):493–99.

Landauer, Rolf (1991) "Information is physical," Physics Today 44:23–29.

Laufer, Berthold (1918) "Review of R.H. Lowie Culture and Ethnology," American Anthropologist 20:90 [quoted by Leslie White in his "Foreword" to Evolution and Culture, Marshall D. Sahlins and Elman Service (eds.), Ann Arbor, MI: University of Michigan Press (1960)].

Lee, D.H., K. Severin, Y. Yokobayashi, and M. R. Ghadiri (1997) "Emergence of symbiosis in peptide self-replication through a hypercyclic network," Nature 390:591–94.

Lenski, Richard E., and G.J. Velicer (2000) "Games microbes play," Selection 1:89–95.

Levy, Steven (1992) Artificial Life. New York: Vintage.

Lewontin, Richard C. (1970) "The units of selection," Annual Review of Ecology and Systematics 1:1–18.

Li, Liming, and Susan Lindquist (2000) "Creating a protein-based element of inheritance," Science 287:661–64.

Loftus, Elizabeth F. (1996) Eyewitness Testimony. Cambridge, MA: Harvard University Press.

Lowenstein, Werner R. (1999) The Touchstone of Life: Molecular Information, Cell Communication, and the Foundations of Life. New York: Oxford University Press.

Ludwig, Mark (1990) The Little Black Book of Computer Viruses. Show Low, Arizona: American Eagle Publications.

Lumsden, Charles J., and Edward O. Wilson (1981) Genes, Mind and Culture. Cambridge, MA: Harvard University Press.

Lynch, Aaron (1998) "Units, events and dynamics in memetic evolution," Journal of Memetics—Evolutionary Models of Information Transmission 2. <http://www.cpm.mmu.ac.uk/jom-emit/2000/vol2/lynch_a.html>

Lynch, Aaron (1996) Thought Contagion: How Belief Spreads through Society. New York: Basic Books.

Lynch, Aaron (1991) "Thought contagion as abstract evolution," Journal of Ideas 2:3–10.

Mahner, M., and Mario Bunge (1997) Foundations of Biophilosophy. New York: Springer-Verlag.

Margolis, Eric, and Stephen Laurence, eds. (1999) Concepts: Core Readings. Cambridge, MA: MIT Press.

Maynard Smith, John (2000) "The concept of information in biology," Philosophy of Science 67:177–94.

Maynard Smith, John (1992) "Byte-sized evolution," Nature 355:772–73.

Maynard Smith, John, and Eörs Szathmáry (1995) The Major Transitions in Evolution. Oxford: Oxford University Press.

Maynard Smith, John (1987) "How to model evolution." In: The Latest on the Best, John Dupré (ed.). Cambridge, MA: MIT Press.

Mayr, Ernst (1997) "The objects of selection," PNAS USA 94: 2091–94. <http://www.pnas.org/cgi/content/full/94/6/2091>

Mazia, D. (1956) "Nuclear products and nuclear reproduction." In: Enzymes: Units of Biological Structure and Function, O.H. Gaebler (ed.). New York: Academic Press, pp. 261–78.

McGrew, William C. (1992) Chimpanzee Material Culture: Implications for Human Evolution. Cambridge: Cambridge University Press.

Merzenich, M.M., and R.C. deCharms (1996) "Neural Representations, Experience, and Change" In: The Mind-Brain Continuum, Rodolfo Llinás and Patricia Churchland (eds.). Cambridge, MA: MIT Press, pp. 61–82.

Michod, Richard E. (1999) Darwinian Dynamics: Evolutionary Transitions in Fitness and Individuality. Princeton: Princeton University Press.

Milner, Peter M. (1999) The Autonomous Brain: A Neural Theory of Attention and Learning. Hillsdale, NJ: Lawrence Erlbaum Associates.

Milton, Katherine (1988) "Foraging behaviour and the evolution of primate intelligence." In: Machiavellian Intelligence: Social Expertise and the Evolution of Intellect in Monkeys, Apes, and Humans, Richard Byrne and Andrew Whiten (eds.). Oxford: Oxford University Press, pp. 285–306.

Mitchell, Melanie (in press) "Life and evolution in computers." In: Darwinian Evolution Across the Disciplines, M. McPeek et al. (eds.). <http://www.santafe.edu/~mm/paper-abstracts.html>

Monod, Jacques (1970) Chance and Necessity [quoted in Douglas R. Hoftstadter, "On Viral Sentences and Self-Replicating Structures," in his book Metamagical Themas: Questing for the Essence of Mind and Pattern, 1985, New York: Basic Books].

Mountcastle, Vernon B. (1998) Perceptual Neuroscience: The Cerebral Cortex. Cambridge, MA: Harvard University Press.

Moravec, Hans P. (1998) Robot: Mere Machine to Transcendent Mind. Oxford: Oxford University Press.

Nádasdy Z., H. Hirase, A. Czurkó, J. Csicsvari, and G. Buzsáki (1999) "Replay and time compression of recurring spike sequences in the hippocampus," Journal of Neuroscience 19:9497–507.

Nader, Karim, Glenn E. Schafe, and Joseph E. LeDoux (2000) "Fear memories require protein synthesis in the amygdala for reconsolidation after retrieval," Nature 406:722–26.

Nei, Masatoshi (1986) Molecular Evolutionary Genetics. New York: Columbia University Press.

Netting, Robert M. (1986; 2nd ed) Cultural Ecology. Prospect Heights, IL: Waveland Press.

Odling-Smee, F.J. (1988) "Niche constructing phenotypes." In: The Role of Behaviour in Evolution, Henry C. Plotkin (ed.) Cambridge, MA: MIT Press.

Owings, Donald H., and Eugene S. Morton (1998) Animal Vocal Communication: A New Approach. Cambridge: Cambridge University Press.

Partridge, Linda, and N.H. Barton (2000) "Evolving evolvability," Nature 407:457–58.

Payne, Robert B. (1996) "Song traditions in indigo buntings: Origin, improvisation, dispersal, and extinction in cultural evolution." In: Ecology and Evolution of Acoustic Communication in Birds, D.E. Kroodsma and E.H. Miller (eds.). Ithaca: Cornell University Press, pp. 198–220.

Pinker, Steven (1997) How the Mind Works. London: Penguin.

Pinker, Steven (1994) The Language Instinct. London: Penguin.

Pittenger, C., and Eric Kandel (1998) "A genetic switch for long-term memory," Comptes Rendus De L'Academie Des Sciences Serie III (Life Sciences) 321:91–96.

Plotkin, Henry C. (2000) "Culture and psychological mechanisms." In: Darwinizing Culture: The Status of Memetics as a Science, Robert Aunger (ed.). Oxford: Oxford University Press, pp. 69–82.

Plotkin, Henry C. (1993) Darwin Machines and the Nature of Knowledge. London: Penguin.

Pocklington, Richard, and Best, Michael L. (1997) "Cultural evolution and units of selection in replicating text," Journal of Theoretical Biology 188:79–87.

Popper, Karl R. (1982) The Open Universe: An Argument for Indeterminism. Totowa, NJ: Rowman and Littlefield.

Prusiner, Stanley B. (1995) "The prion diseases," Scientific American 272:48–57.

Rasmussen, Steen, Carsten Knudsen, Ramus Feldberg, and Morten Hindsholm (1990) "The Coreworld: Emergence and evolution of cooperative structures in a computational chemistry," Physica D 42:111–34.

Ray, Thomas S. (1992) "An approach to the synthesis of life," In: Artificial Life II. Christopher G. Langton, Charles Taylor, J. Doyne Farmer, and Steen Rasmussen (eds.). Redwood City, CA: Addison-Wesley, pp. 371–408.

Rebek, Julius (1994) "Synthetic self-replicating molecules," Scientific American 271:34–40.

Reeves, Byron, and Clifford Nass (1996) The Media Equation: How People Treat Computers, Television, and New Media Like Real People and Places. New York: Cambridge University Press.

Rendell, Luke, and Hal Whitehead (2001; in press) "Culture in whales and dolphins," Behavioral and Brain Sciences 24(2).

Resnick, Laura B., J.M. Levine, and S.D. Teasley (eds.) (1991) Perspectives on Socially-Shared Cognition. Washington, DC: American Psychological Association.

Richards, Robert J. (1987) Darwin and the Emergence of Evolutionary Theories of Mind and Behavior. Chicago: University of Chicago Press.

Richerson, Peter J., and Robert Boyd (1978) "A dual inheritance model of the human evolutionary process I: Basic postulates and a simple model," Journal of Social and Biological Structures 1:127–54.

Ridley, Rosalind, and Harry F. Baker (1996) Fatal Protein: The Story of CJD, BSE, and Other Prion Diseases. Oxford: Oxford University Press.

Rizzolatti, G., and M. Arbib (1998) "Language within our grasp," Trends in the Neurosciences 21:188–94.

Rodriguez, E., N. George, J.P. Lachaux, J. Martinerie, B. Renault, and F.J. Varela (1999) "Perception's shadow: Long-distance synchronization of human brain activity," Nature 397:430–33.

Rogers, Everett M. (1995; 3rd ed.) Diffusion of Innovations. New York: Free Press.

Rolls, Edmund T., and Alessandro Treves (1998) Neural Networks and Brain Function. Oxford: Oxford University Press.

Rosen, Emanuel (2000) The Anatomy of Buzz: How to Create Word-of-Mouth Marketing. New York: Doubleday.

Saborio, G.P., C. Soto, R.J. Kascsak, E. Levy, R. Kascsak, D.A. Harris, and B. Frangione (1999) "Cell-lysate conversion of prion protein into its protease-resistant isoform suggests the participation of a cellular chaperone," Biochemical and Biophysical Research Communications 258:470–75.

Sahlins, Marshall D. (1960) "Evolution: Specific and general." In: Evolution and Culture, Marshall D. Sahlins and Elman Service (eds.). Ann Arbor: University of Michigan Press, pp. 12–44.

Sarkar, Sahotra (1996) "Biological information: A skeptical look at some central dogmas of molecular biology." In: The Philosophy and History of Molecular Biology: New Perspectives, Sahotra Sarkar (ed.). Dordrecht: Kluwer Academic Publishers, pp. 187–231.

Schuster, Peter, and Karl Sigmund (1983) "Replicator dynamics," Journal of Theoretical Biology 100:533–38.

Schwartz, Evan I. (1999) Digital Darwinism: Seven Breakthrough Business Strategies for Surviving in the Cutthroat Web Economy. New York: Broadway Books.

Service, Robert F. (1999) "Big Blue aims to crack protein riddle," Science 286:2250.

Shannon, Claude E. (1948) "A mathematical theory of information," Bell System Technical Journal 27:379–423, 623–56.

Shannon, Claude, and Warren Weaver (1965 [1949]) The Mathematical Theory of Communication. Chicago: University of Illinois Press.

Shors, Tracey J., George Miesegaes, Anna Beylin, Mingrui Zhao, Tracy Rydel, and Elizabeth Gould (2001) "Neurogenesis in the adult is involved in the formation of trace memories," Nature 410:372–76.

Singer, Wolf (1999) "Time as coding space?" Current Opinion in Neurobiology 9:189–94.

Smith, Q.R. (2000) "Transport of glutamate and other amino acids at the blood-brain barrier," Journal of Nutrition 130(4S Suppl):1016S–22S.

Smolin, Lee (1997) The Life of the Cosmos. London: Weidenfeld and Nicolson.

Spafford, Eugene H. (1992) "Computer viruses: A form of artificial life?" In: Artificial Life II, Christopher G. Langton, Charles Taylor, J. Doyne Farmer, and Steen Rasmussen (eds.). Reading, MA: Addison-Wesley, pp. 727–46.

Sperber, Dan (2000) "An objection to the memetic approach to culture." In: Darwinizing Culture: The Status of Memetics as a Science, Robert Aunger (ed.). Oxford: Oxford University Press, pp. 163–74.

Sperber, Dan (1996) Explaining Culture: A Naturalistic Approach. Oxford: Blackwell.

Sperber, Dan, and Deirdre Wilson (1997) "Remarks on relevance theory and the social sciences," Multilingua 16:145–51.
<http://perso.club-internet.fr/sperber/rel-soc.htm>

Sperber, Dan, and Deirdre Wilson (1986 [2nd ed. 1995]) Relevance: Communication and Cognition. Oxford: Blackwell.

Steinmetz, P.N., A. Roy, P. Fitzgerald, S.S. Hsiao, E. Niebur, and K.O. Johnson (2000) "Attention modulates synchronized neuronal firing in primate somatosensory cortex," Nature 404:187–90.

Sterelny, Kim (2001) "Niche construction, developmental systems and the extended replicator." In: Cycles of Contingency: Developmental Systems and Evolution, Susan Oyama, Paul E. Griffiths, and Russell D. Gray (eds.). Cambridge, MA: MIT Press.

Sterelny, Kim, and Paul E. Griffiths (1999). Sex and Death. Chicago: University of Chicago Press.

Steward, Julian (1955) Theory of Culture Change. Urbana: University of Illinois Press.

Symons, Donald (1987) "If we're all Darwinians, what's the fuss about?" In: Sociobiology and psychology: Ideas, issues and applications, Charles Crawford, Martin Smith, and Dennis Krebs (eds.). Hillsdale, NJ: Erlbaum, pp. 121–46.

Szathmáry, Eörs (1991) "Simple growth laws and selection consequences," Trends in Ecology and Evolution 6:366–70.

Szathmáry, Eörs, and John Maynard Smith (1997) "From replicators to reproducers: The first major transitions leading to life," Journal of Theoretical Biology 187:555–71.

Telling, Glenn C., Piero Parchi, Stephen J. DeArmond, Pietro Cortelli, Pasquale Montagna, Ruth Gabizon, James Mastrianni, Elio Lugaresi, Pierluigi Gambetti, and Stanley B. Prusiner (1996) "Evidence for the conformation of the pathologic isoform of the prion protein enciphering and propagating prion diversity," Science 274:2079–82.

Thompson, John B. (1994) Media and Modernity. London: Polity Press.

Tomasello, Michael (2000) The Cultural Origins of Human Cognition. Cambridge, MA: Harvard University Press.

Tomasello, Michael (1996) "Do apes ape?" In Social Learning in Animals: The Roots of Culture, Celia M. Heyes and Bennett G. Galef (eds.). San Diego: Academic Press, pp. 319–46.

Tomasello, Michael, A.C. Kruger, and H.H. Ratner (1993) "Cultural learning," Behavioral and Brain Sciences 16:495–552.

Tooby, John, and Leda Cosmides (1995) "Mapping the evolved functional organization of mind and brain." In: The Cognitive Neurosciences, Michael Gazzaniga (ed.). Cambridge, MA: MIT Press.

Tooby, John, and Leda Cosmides (1992) "The psychological foundations of culture." In: The Adapted Mind: Evolutionary Psychology and the Generation of Culture. Jerome H. Barkow, Leda Cosmides, and John Tooby (eds.). Oxford: Oxford University Press, pp. 19–136.

True, H.L., and Susan Lindquist (2000) "A yeast prion provides a mechanism for genetic variation and phenotypic diversity," Nature 407:477–83.

Tuma, Rabiya S. (2000) "It's not just DNA anymore: Prion proteins and hereditary information," HMS Beagle 81 (June 23).

Valéry, Paul (1989) The Art of Poetry (Collected Works, Vol. 7). Edited by J. Mathews; translated by Denise Folliot. Princeton: Princeton University Press.

Van Gunsteren, W.F., R. Bürgi, C. Peter, and X. Daura (2001) "The key to solving the protein-folding problem lies in an accurate description of the denatured state," Angewandte Chemie International Edition 40:352.

Vayda, Andrew P., and Bonnie J. McCay (1975) "New directions in ecology and ecological anthropology," Annual Review of Anthropology 4:293–306.

Wasserman, Stanley, and Katherine Faust (1994) Social Network Analysis: Methods and Applications. Cambridge: Cambridge University Press.

Watts, Duncan J., and Steven H. Strogatz (1998) "Collective dynamics of 'small-world' networks," Nature 393:440–42.

Weiner, Norbert (1961; 2nd ed.) Cybernetics: or Control and Communication in the Animal and the Machine. Cambridge, MA: MIT Press.

White, Steve R., Jeffrey O. Kephart, and David M. Chess (1995) "Computer Viruses: A Global Perspective," Proceedings of the 5th Virus Bulletin International Conference, Abingdon, England: Virus Bulletin Ltd.

Whiten, Andrew (ed.) (1991) Natural Theories of Mind: Evolution, Development and Simulation of Everyday Mindreading. London: Blackwell.

Whiten, Andrew, Jane Goodall, William C. McGrew, T. Nishida, Vernon Reynolds, Y. Sugiyama, C.E. Tutin, Richard W. Wrangham, and Christophe Boesch (1999) "Chimpanzee cultures," Nature 399:682–85.

Whiten, Andrew, and R. Ham (1992) "On the nature and evolution of imitation in the animal kingdom: A reappraisal of a century of research." In: Advances in the Study of Behavior, Vol. 21, P.J.B. Slater, J.S. Rosenblatt, C. Beer, and M. Milkinski (eds.). New York: Academic Press, pp. 239–83.

Wilkins, John (1999) "Memes ain't (just) in the head," Journal of Memetics—Evolutionary Models of Information Transmission 3.
    <http://www.cpm.mmu.ac.uk/jom-emit/1999/vol3/ wilkins_j.html>

Wilkins, John (1998) "What's in a meme? Reflections from the perspective of the history and philosophy of evolutionary biology," Journal of Memetics—Evolutionary Models of Information Transmission 2.
    <http://www.cpm.mmu.ac.uk/jom-emit/1998/vol2/wilkins_j.html>

Williams, George C. (1995) "A Package of Information." In: The Third Culture: Beyond the Scientific Revolution, John Brockman (ed.). New York: Simon and Schuster.

Williams, George C. (1992) Natural Selection: Domains, Levels, and Challenges. New York: Oxford University Press.

Wilson, D.S. (1999) "Flying over uncharted territory" [review of Susan Blackmore's The Meme Machine], Science 285:206.

Winterhalder, Bruce, and Eric Alden Smith (eds.) (1981) Hunter-Gatherer Foraging Strategies: Ethnographic and Archaeological Analyses. Chicago: University of Chicago Press.

Wootters, W.K., and W.H. Zurek (1982) "A single quantum cannot be cloned," Nature 299:802.

Wright, Robert (1994) The Moral Animal: Why We Are the Way We Are: The New Science of Evolutionary Psychology. New York: Vintage.

Yin, Jerry Chi-Ping (1999) "Location, location, location: The many addresses of memory formation," PNAS 96:9985–86.

Zhao, Y., H.Z. Sheng, R. Amini, A. Grinberg, E. Lee, S. Huang, M. Taira, and H. Westphal (1999) "Control of hippocampal morphogenesis and neuronal differentiation by the LIM homeobox gene Lhx5," Science 283:1155–58.

Acknowledgments

This book is largely the result of numerous Aha! experiences on waking up in the morning. Most of these early-rising ideas, bubbling into consciousness, were killed off by later ones or simply died of their own memetic defects. But some of these inspirations *have* made it into the final artifact you hold in your hands. Most have even survived the treacherous journey into and back out of the minds of some draft readers before settling here. The memes that remain have therefore successfully wended their way through a complex battery of selective tests and will hopefully replicate far and wide (with my name attached, of course!) now that they have produced signal templates in this medium of the printed page.

I want to thank those people who lent their mental energies to this collective dreck-filtering exercise. My wife, Gillian Bentley, comes first on the list of thankees, not in the traditional last-but-most-important position. This is because she insisted on it as recompense for being the one who primarily had to put up with my ravings on waking up. (She eventually learned to simply ignore me.) She has read, and hated, more versions of this book than anyone else. Most important, she ran the household while I meditated on replication issues, and only occasionally asked me how the book was coming along. She alone can tell me how much I owe her. My kids, Justin and Amy, eventually learned that when daddy was sitting at the computer, it was best not to bother him. Thanks to them for being so clever.

Non-kin who contributed through discussion include Pascal Boyer, John Constable, Terry Deacon, Dan Dennett, James Griesemer, David Hull, Nicholas Humphrey, Mark Lake, Kevin Laland, Neil Manson, India Morrison, John Odling-Smee, Peter Richerson, and Dan Sperber. People who have commented on parts of the book include Susan Blackmore, Terry Deacon, Juan Delius, Barry Keverne, Kevin Laland, Tim Lewens, John McCrone, India Morrison, Henry Plotkin, and John Wilkins. Those who heroically took on the full manuscript were Peter Richerson and John van Wyhe. Many thanks to all for their feedback and forbearance. (I apologize to anyone I have forgotten.)

Richard Dawkins, Dan Dennett, Nicholas Humphrey, and David Sloan Wilson provided important, generous impetus during the early stages of this project. I was also fortunate in having John Brockman and Katinka Matson, agents extraordinaire, as my avatars during virtual negotiations to get these memes widely distributed. At The Free Press my editor, Stephen Morrow, was instrumental in getting this book off the ground and very tactful when pointing out ways to improve it, and he even found ways to curb my more creative impulses. The production supervisor, Loretta Denner, found an able copy editor in Patricia Fogarty and a fine indexer in Martin L. White.

For preprints and reprints, I thank Susan Blackmore, Thomas Diener, Peter Godfrey-Smith, James Griesemer, David Hull, Mark Lake, Kevin Laland, John Odling-Smee, India Morrison, Peter Richerson, Yoshio Sakurai, Sahotra Sarkar, Stephen Shennan, Dan Sperber, John E. Stewart, and Eörs Szathmáry.

Thanks also to those who came to the various public presentations I have made of the ideas in this book and provided such valuable feedback.

For tolerating my haunting of their hallways during the writing of this book, I thank the faculty in the Department of Biological Anthropology, Cambridge, and particularly its head, Nick Mascie-Taylor.

Robert Bailey played no role in this book whatsoever but did ably mentor my earlier incarnation as an empirical scientist studying cultural transmission, which first prompted me to think about many of the issues here. Without that grounding and experience, I doubt the present book would have taken its current form. So I guess Bob *is* somewhat responsible, after all.

Finally, I can't quit without thanking Patrick Bateson, Provost of King's College Cambridge, as well as the general Fellowship, both for their warm collegiality and for an extended period of essentially uninterrupted and unencumbered time for the free exploration of meme-space. Such generosity and dedication to the old-fashioned, gentlemanly pursuit of scientific interests is (unfortunately) truly rare in the contemporary academic world. The College even allowed me to organize a conference that was a crucial forebear to my work on this book. I hope the result is worth the investment. In any case, it is certainly clear (to me at least) that the development of novel meme-complexes, such as appear here, is likely to occur only in such an intellectually liberating environment.

INDEX

translation, 236, 243

transmissible spongiform
    encephalopathies (TSEs)
    bovine spongiform encephalopathy,
        9, 10, 96, 98
    Creutzfeld-Jakob disease, 9, 10,
        96–7, 98
    prions causing, 96–7
    scrapie, 9, 10, 96
transmission of information
    central dogma of molecular biol-
        ogy, 94
    in Chinese whispers, 271
    communication seen as, 255
    in conversion, 314, 315
    cultural selectionism on, 48–58
    debate over significance of, 46–8
    direct and indirect, 304, 329
    errors in, 249–50
    evolutionary psychology on, 41, 42
    evolution as more than, 157
    inheritance as problem of, 147
    mediation of, 134
    memes as not required to explain,
        22, 24–5, 29
    quantum teleportation as example
        of, 141–5, 255, 256
    replication as requiring, 73–4, 91,
        136, 148, 156, 166, 189
    sociobiology on, 33
    as tied to particular substrate, 151
    varieties of, 8
    by word of mouth, 13–14
transporters *(Star Trek),* 142, 145
Turing Test, 153, 154
twins, 183, 225
Tylor, Edward, 30

uncertainty, 150, 258, 292
Universal Darwinism, 72, 81, 82
universals, psychological, 39–40

Valéry, Paul, 62
variation
    acquired, 238–9
    in computer virus evolution, 127
    in cultural selectionism, 50, 51
    evolutionary psychology on, 40
    within populations, 27
    as property of evolution, 25
    in sociobiology's view of culture, 32
Vienna School, 83
viral marketing, 13–17
viroids, 161, 313
viruses
    alien genes introduced by, 157
    cultural traits seen as, 49
    memes compared with, 3, 17–19,
        21, 228, 234
    and prions, 98
    reproduction of, 116, 252
    therapeutic uses of, 114
    transmission of, 234
    See also computer viruses
virusoids, 161
von Neumann, John, 109, 117

wagons, 164, 281–2, 289–90, 296, 354n
Wallace, Alfred Russel, 11
Watson, James, 160, 175, 320–1, 323
Weaver, William, 257, 258, 353n
Weismann, August, 238, 240
Wiener, Norbert, 48, 140
Wilkins, John, 335n
Wilson, E.O., 31–2
Wilson, Margo, 44–5
Windows operating system, 122
WordBasic, 124–5
word-of-mouth, 13–14
World Wide Web, 13, 123–5, 209
Wright, Sewall, 67

yeast, 98, 200